Epidemiology
Biostatistics
AND
Preventive
Medicine Review

KEITH HIATT, Maj, USA, MC, FS
SSN 154-62-3182
AEROSPACE MEDICINE
PC# 30104S

SAUNDERS TEXT AND REVIEW SERIES

EPIDEMIOLOGY
BIOSTATISTICS
AND
PREVENTIVE MEDICINE REVIEW

DAVID L. KATZ, M.D., M.P.H.
Assistant Clinical Professor of Medicine,
Epidemiology, and Public Health
Yale University School of Medicine
New Haven, Connecticut

Associate Director, Preventive Medicine Residency Program
Griffin Hospital
Derby, Connecticut

W.B. SAUNDERS COMPANY
A Division of Harcourt Brace & Company
Philadelphia London Toronto Montreal Sydney Tokyo

W.B. SAUNDERS COMPANY
A Division of Harcourt Brace & Company

The Curtis Center
Independence Square West
Philadelphia, Pennsylvania 19106

Library of Congress Cataloging-in-Publication Data

Katz, David L.

Epidemiology, biostatistics, and preventive medicine review / David L. Katz.

   p.   cm.

ISBN 0-7216-4084-2

1. Epidemiology—Examinations, questions, etc. 2. Medical statistics—Examinations, questions, etc. 3. Medicine, Preventive—Examinations, questions, etc. I. Title.
[DNLM: 1. Epidemiologic Methods—examination questions. 2. Biometry—examination questions. 3. Preventive Medicine—examination questions. WA 18.2 K19e 1997]

RA652.K38   1997   614.4'076—dc20

DNLM/DLC   96-29335

EPIDEMIOLOGY, BIOSTATISTICS, AND PREVENTIVE MEDICINE REVIEW   ISBN 0-7216-4084-2

Copyright © 1997 by W.B. Saunders Company.

All rights reserved. No part of this publication may be reproduced or transmitted in any form or by any means, electronic or mechanical, including photocopy, recording, or any information storage and retrieval system, without permission in writing from the publisher.

Printed in the United States of America.

Last digit is the print number: 9   8   7   6   5   4   3   2   1

PREFACE

You may not be turning these pages with uncontained enthusiasm or breathless anticipation. Biostatistics may not be your favorite hobby, and epidemiology may only rarely give you a good laugh. If you were looking for recreation, I suspect you would be looking elsewhere. You probably *need* to know this material.

I want you to know, therefore, that I had fun writing this book. I know from personal experience that preparing for standardized examinations is tedious at best and at worst, downright oppressive. I can now say that writing a review book for such a purpose is no Sunday picnic, either. I thought if I could have fun writing it, perhaps you could enjoy using it. I did, and I hope you do.

This book, a companion text to *Epidemiology, Biostatistics, and Preventive Medicine* by James Jekel (in collaboration with Joanne Elmore and me), is intended to provide a concise yet comprehensive overview of epidemiology, biostatistics, and preventive medicine. This review, like the book it parallels, is unique in covering epidemiology, biostatistics, *and* preventive medicine. The emphasis here, as in the book by James Jekel, is on the interdependence of these subjects and on the relevance of each to clinical practice. While this review is intended to serve as preparation for the United States Medical Licensing Examination (USMLE), the material should be of value to any student or practitioner in health care wishing to enhance his or her knowledge or understanding of the disciplines covered.

The review chapters provide in outline form the material presented more expansively in the corresponding chapters of the companion book. Each chapter is followed by relevant questions, and a comprehensive practice examination is provided at the end. All questions follow the formats used on the USMLE. The answer to each question is provided, along with a thorough explanation.

While I recognize the necessity of factual memorization in medical education, I believe in the greater long-term value of conceptual understanding. I have drawn from my experiences with students, as well as from my not-forgotten experiences as a student, in an effort to provide the requisite factual material with the greatest degree of conceptual clarity. I will have succeeded in my efforts if this book provides you with a genuine and comfortable understanding of the material, prepares you well for any relevant academic or professional challenges, and occasionally makes you laugh.

ACKNOWLEDGMENTS

I didn't *always* have fun writing this book, and I am grateful to my wife, Catherine, for her patient tolerance of my periodic brooding as I struggled to concoct yet one more "distracter" (a wrong answer reasonable enough to warrant consideration). I am grateful to Marji Toensing, Sharon Maddox, Anne-Marie Shaw, and Martha Gay for insightful editing and constructive critical review of the manuscript at each of its many stages. I thank William R. Schmitt, Editorial Manager at W.B. Saunders Company, for (dare I say it?) the benevolent despotism responsible for actually turning many months of effort into a book.

Above all, I thank my colleague, mentor, and friend, Dr. James Jekel, for providing me with this opportunity. Jim's extraordinary skill as a teacher and his passionate commitment to his students are a constant inspiration to me. I dedicate this book to Jim Jekel and to our shared reverence for teaching.

DAVID L. KATZ, MD, MPH

COMPANION TEXT

The material in this book complements that in *Epidemiology, Biostatistics, and Preventive Medicine* (1996) by James F. Jekel, Joann G. Elmore, and David L. Katz, published by the W.B. Saunders Company. The 21 chapters in this review parallel those of *Epidemiology, Biostatistics, and Preventive Medicine.* The illustrations, explanations, and topics in the companion text are more comprehensive than those in this review. The companion text, however, lacks questions and answers, a comprehensive examination, and an epidemiologic and medical glossary.

CONTENTS

SECTION I

EPIDEMIOLOGY

1 EPIDEMIOLOGIC APPROACHES, CONTRIBUTIONS, AND ISSUES 3

2 EPIDEMIOLOGIC DATA SOURCES AND MEASUREMENTS 9

3 EPIDEMIOLOGIC SURVEILLANCE AND OUTBREAK INVESTIGATION 17

4 THE STUDY OF CAUSATION IN EPIDEMIOLOGIC INVESTIGATION AND RESEARCH 25

5 COMMON RESEARCH DESIGNS USED IN EPIDEMIOLOGY 29

6 ASSESSMENT OF RISK IN EPIDEMIOLOGIC STUDIES 34

SECTION II

BIOSTATISTICS

7 UNDERSTANDING AND REDUCING ERRORS IN CLINICAL MEDICINE 43

8 IMPROVING DECISIONS IN CLINICAL MEDICINE 49

9 DESCRIBING VARIATION IN DATA 59

10 STATISTICAL INFERENCE AND HYPOTHESIS TESTING 69

11 BIVARIATE ANALYSIS 81

| 12 | Sample Size, Randomization, and Probability Theory | 97 |
| 13 | Multivariable Analysis | 106 |

SECTION III

Preventive Medicine and Public Health

14	Introduction to Preventive Medicine	115
15	Methods of Primary Prevention: Health Promotion	121
16	Methods of Primary Prevention: Specific Protection	129
17	Methods of Secondary Prevention	139
18	Methods of Tertiary Prevention	147
19	Special Topics in Prevention	153
20	Public Health Responsibilities and Goals	165
21	Medical Care Policy and Financing	175

SECTION IV

Comprehensive Examination, Epidemiologic and Medical Glossary, and Appendix

Comprehensive Examination 189
    Questions 189
    Answers and Explanations 203

Epidemiologic and Medical Glossary 221

Appendix 231
    Table A. Random Numbers 233
    Table B. Standard Normal-Tail Probabilities (Table of $\geq$ Values) 234
    Table C. Upper Percentage Points for t Distribution 238
    Table D. Upper Percentage Points for Chi-Square Distributions 240

SECTION I

EPIDEMIOLOGY

CHAPTER ONE

EPIDEMIOLOGIC APPROACHES, CONTRIBUTIONS, AND ISSUES

SYNOPSIS

OBJECTIVES
- To define epidemiology and its major subtypes.
- To describe the epidemiologic perspective on etiologic factors of disease.
- To characterize the principal goals and contributions of epidemiology.
- To explain the importance of an ecologic perspective on health and disease.

I. Basic Epidemiologic Concepts
   A. Definitions
      1. **Epidemiology** comes from the Greek language, in which *epi* means upon, *demos* denotes the population, and the combining form *-logy* means the study of. Therefore, epidemiology is the study of factors that determine the occurrence and distribution of disease in a population.
      2. **Classical epidemiology** is population-oriented and explores the community origins of health problems.
      3. **Clinical epidemiology** focuses on health care and explores the factors that affect the diagnosis, treatment, and outcome in patient care.
      4. While **infectious disease epidemiology** is more heavily dependent on laboratory support (e.g., microbiology and serology), **chronic disease epidemiology** is more dependent on complex sampling and statistical methods. The boundary between the two is not always distinct.
   B. The "BEINGS" model—See Table 1-1.
   C. Host, agent, and environment
      1. The natural history of a disease is normally characterized by the interactions among a susceptible **host** (e.g., a human being), a putatively causative **agent** (e.g., a virus), and the aspects of the **environment** that permit exposure of the host and transmission of the agent.
      2. An additional factor contributing to the pattern of certain diseases is a **vector,** which is often an insect and which carries the agent to the host (e.g., the *Anopheles* mosquito transmits malaria to humans).
      3. **Biologic agents** of disease and illness not only include infectious organisms (e.g., bacteria, viruses, fungi) but also include allergens and even, under certain circumstances, vaccines, antibiotics, and foods (such as a high-fat diet).
      4. **Chemical agents** include chemical toxins (e.g., lead) and dusts.
      5. **Physical agents** include kinetic energy, radiation, heat, cold, and noise.
      6. **Social and psychological stressors** may be included among agents that compromise health and contribute to the development of illness.

II. Contributions of Epidemiologists to the Medical Sciences
   A. **Investigating the modes of transmission of a new disease**
      1. Methods of epidemiologic investigation and surveillance often provide the initial hypotheses regarding the cause and natural history of a newly recognized disease.
      2. Specific diseases elucidated by modern epidemiologic methods include, among many others, Lyme disease, legionellosis, AIDS, and Hantavirus pulmonary syndrome.
   B. **Determining preventable causes of disease**—See Table 1-1.
      1. By determining the causes of disease and the modes of transmission, epidemiology identifies the links in the causal chain most amenable to being broken. A single, well-targeted intervention (e.g., vaccine) may interrupt the interaction among causal factors of disease and prevent transmission.
      2. **Herd immunity** results when a vaccine not only prevents the vaccinated person from contracting the

TABLE 1-1. "BEINGS": An Acronym for Remembering the Categories of Preventable Causes of Disease

Biologic factors and **B**ehavioral factors
Environmental factors
Immunologic factors
Nutritional factors
Genetic factors
Services, **S**ocial factors, and **S**piritual factors

---

disease but also prevents him or her from spreading the disease (see Fig. 1-1).
3. **Immunodeficiency** (i.e., a deficiency of the immune system) may be long-term (as in AIDS) or transient (lasting for a short period following some types of infections or administration of certain vaccines).
4. **Heritability** is defined in terms of the proportion of disease cases due to genetic causes.

C. **Determining the natural history of disease**
  1. Epidemiologists make a distinction between the **social and environmental causes** of disease and the **biologic mechanisms** of disease.
  2. Excessive fat intake, smoking, and lack of exercise are social factors that contribute to atherogenesis, but inappropriate cholesterol levels are the biologic mechanism of the disease.
  3. Once the biologic mechanism is initiated, it is predictable across a range of environments.

D. **Studying the biologic spectrum of disease**
  1. Most diseases manifest along a spectrum of severity. The more severe cases of a previously unknown disease are usually the first to come to attention.
  2. The **iceberg phenomenon** denotes that the most severe cases of a disease, though less numerous, are most visible; the often far more abundant, milder cases of an illness may remain undetected, beneath the "surface" (see Fig. 1-2).

E. **Evaluating community health interventions**
  1. **Field trials** are used to evaluate a new vaccine before it is given to the community at large.
  2. **Surveillance** of disease is essential after the introduction of a vaccine to ensure the continued safety and efficacy of the vaccine.

F. **Setting disease control priorities**
  1. **Disease control priorities** are based not only on the current magnitude of a particular health problem but on projections of its effects in the future.
  2. **Cost-effectiveness analysis** (see Chapter 14) may be used to determine how limited resources may be most effectively allocated in attempts to prevent and control disease.

G. **Improving the diagnosis, treatment, and prognosis of clinical disease**
  1. Epidemiologic methods are used to determine the best diagnostic tests, and the accuracy of such tests under specific circumstances.
  2. Similarly, epidemiologic methods may be used to compare alternative treatments and select the most effective one.
  3. By enhancing understanding of the natural history of disease, epidemiologic methods improve prognostication, which is the prediction of outcome.

H. **Improving health services research**
  1. Epidemiologic methods are used to estimate regional needs for health services.
  2. Services are allocated on the basis of effectiveness and cost, both of which are studied by epidemiologists (see Chapter 14).

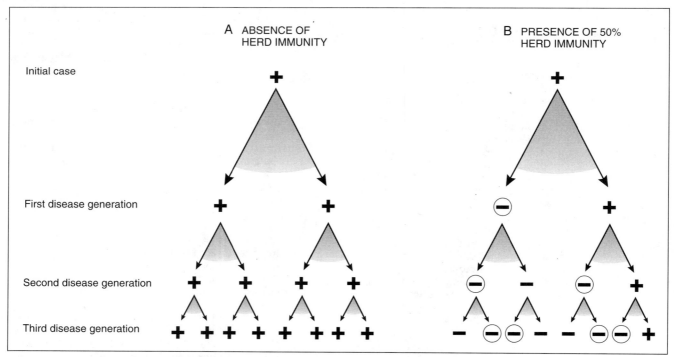

FIGURE 1-1. The effect of herd immunity on the spread of infection. The diagrams illustrate how an infectious disease, such as measles, could spread in a susceptible population if each infected person were exposed to two other persons. In the absence of herd immunity *(diagram A)*, the number of cases doubles each disease generation. In the presence of 50% herd immunity *(diagram B)*, the number of cases remains constant. The plus sign represents an infected person; the minus sign represents an uninfected person; and the circled minus sign represents an immune person who will not pass the infection to others.

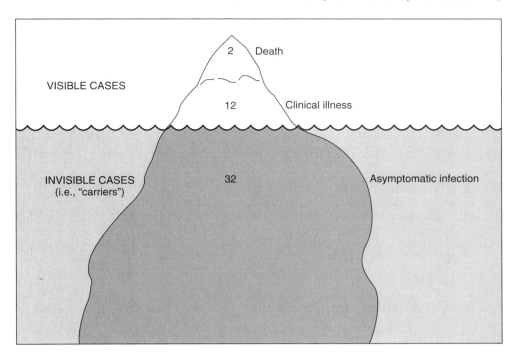

FIGURE 1-2. The iceberg phenomenon, as illustrated by a diphtheria epidemic in Alabama. In epidemics, the number of people with severe forms of the disease (the tip of the iceberg) may be much smaller than the number of people with mild or asymptomatic clinical disease. (Source of data: Jekel, J. F., G. S. Zatlin, and O. F. Gay. *Corynebacterium diphtheriae* survives in a partly immunized group. Public Health Reports 85:310, 1970.)

I. **Providing expert testimony in courts of law**
   1. Epidemiologists may be called upon to testify regarding product hazards and the risks and effects of environmental exposures.
   2. The kinds of lawsuits that may rely on epidemiologic data are those involving claims of damage from general environmental exposures, occupational illnesses, medical liability, and product liability.

III. **Ecologic Issues in Epidemiology**
   A. **Ecologic perspective**
      1. In epidemiology, people are seen not only as individual organisms but also as members of communities, in a **social context.**
      2. **Patterns of disease** affecting individuals characterize and distinguish communities, countries, and even continents.
      3. Epidemiologists are concerned with **human ecology** and, in particular, the impact that health interventions have on disease patterns and on the environment.
   B. **The solution and unintended creation of problems**
      1. An ecologic perspective suggests that changing only one element in an ecosystem is virtually impossible. All interventions, like stones cast into water, send out ripples of secondary effect.
      2. Often, there are adverse consequences of even beneficial interventions (see Table 1–2).

   3. The control of diseases by **immunization** is complex.
      a. The phenomenon of herd immunity is illustrated in Fig. 1–1.
      b. A **"natural booster"** phenomenon (i.e., augmentation of immunity) occurs with periodic exposure to an infectious agent such as *Corynebacterium diphtheriae,* which causes diphtheria; this effect may be lost when immunization prevents exposure. Immunity may wane with time, rendering immunized individuals susceptible should exposure occur.
      c. The great potential benefit of vaccination is demonstrated most vividly by the successful eradication of smallpox.
      d. The injectable **Salk polio vaccine** provided individual immunity to poliomyelitis but did not interrupt transmission (i.e., did not confer herd immunity). The **Sabin oral polio vaccine (OPV)** does provide herd immunity, and has resulted in the apparent eradication of wild virus from the western hemisphere. The OPV, however, which is made from live virus, has produced clinical illness under certain circumstances.
      e. Syphilis is most communicable in its early stages. Early treatment with penicillin prevents disease

TABLE 1-2. Examples of Negative Ecologic Side Effects from the Solution of Earlier Health Problems

Initial Health Problem	Solution	Negative Ecologic Side Effects
Childhood infections	Vaccination.	Decrease in the level of immunity during adulthood, owing to a lack of repeated exposures to infection.
High infant mortality rate	Improved sanitation.	Increase in the population growth rate; appearance of epidemic paralytic poliomyelitis; and appearance of epidemic hepatitis A infection.
Sleeping sickness in cattle	Control of the tsetse fly (the disease vector).	Increase in the area of land subject to overgrazing and drought, owing to an increase in the cattle population.
Malnutrition and the need for larger areas of tillable land	Erection of large river dams (e.g., on the Aswan and Senegal Rivers).	Increase in the rates of some infectious diseases, owing to water system changes that favor the vectors of disease.

progression to late stages but consequently permits reinfection. Reinfection may lead to greater rates of transmission if the highly infectious early stages of the disease occur repeatedly in the same individual.
4. **Improved sanitation** has contributed enormously to the control of infectious disease. It has reduced the infant mortality rate but has had several negative ecologic side effects: an increase in the population growth rate; the appearance of epidemic paralytic poliomyelitis; and the appearance of epidemic hepatitis A infection (see Table 1–2).
5. **Vector control** can alter land use patterns (see Table 1–2). Although control of the tsetse fly reduced the incidence of sleeping sickness in cattle in sub-Saharan Africa, the resulting increase in the cattle population led to overgrazing and drought.
6. **River dam construction** in developing countries increased the amount of available farmland but had the negative effect of changing disease patterns (e.g., increasing the rates of mosquito-borne diseases and schistosomiasis) (see Table 1–2).

C. **Synergism of factors predisposing to disease**
1. By viewing disease patterns ecologically, the interactions between and among diseases may be appreciated. **Synergy** occurs when the impact of two or more diseases on an individual or population is greater than the sum of the separate effects of each individual disease.
2. HIV infection may be more readily spread when other sexually transmitted diseases are present in an individual.
3. There has been increased reactivation of latent tuberculosis and increased transmission of it due to the immunodeficiency associated with HIV disease.
4. Malnutrition and infection often occur in the same populations, and each exacerbates the other.
5. A new strain of viral influenza is produced almost every year as a result of duck and human strains of influenza mingling in pigs and producing genetic variants. This occurs in rural China, where humans, ducks, and pigs live in close proximity. Relatively minor genetic change is termed **antigenic drift**, whereas major change, with the potential to create worldwide influenza epidemics (i.e., pandemics) is termed **antigenic shift**.

QUESTIONS

DIRECTIONS. (Items 1–10): Each of the numbered items or incomplete statements in this section is followed by answers or by completions of the statement. Note that items 1, 2, 4, 7, and 12 are phrased negatively, as indicated by the capitalized word EXCEPT. Select the ONE lettered answer or completion that is BEST in each case. Correct answers and explanations are given at the end of the chapter.

1. All of the following are true statements regarding epidemiology EXCEPT

    (A) it is the study of causal factors in disease
    (B) it is the study of the health of populations
    (C) it is unrelated to the care of individual patients
    (D) it is a science concerned with identifying environmental factors in disease
    (E) it is frequently a first responder in the elucidation of poorly understood health threats

2. All of the following are major types of epidemiology EXCEPT

    (A) classical epidemiology
    (B) infectious epidemiology
    (C) chronic disease epidemiology
    (D) ecologic epidemiology
    (E) clinical epidemiology

3. The "BEINGS" model is a useful paradigm for

    (A) evaluating etiologic factors in disease
    (B) establishing the importance of social contacts
    (C) assessing the impact of herd immunity
    (D) allocating public health resources
    (E) determining the role of social policy on public health

4. All of the following are explicitly included in the "BEINGS" model EXCEPT

    (A) behavioral factors
    (B) iatrogenic factors
    (C) spiritual factors
    (D) genetic factors
    (E) nutritional factors

5. Factors that must interact in order for infectious disease to occur include

    (A) the host and the agent
    (B) the vector and the environment
    (C) the agent and the vector
    (D) genetic factors
    (E) behavioral factors

6. Fatalities associated with the Hantavirus pulmonary syndrome may be an example of

    (A) the first responder effect
    (B) clinical epidemiology
    (C) the iceberg phenomenon
    (D) herd immunity
    (E) iatrogenesis

7. The responsibilities of the epidemiologist may include all of the following EXCEPT

    (A) providing data for genetic counseling
    (B) establishing modes of disease transmission
    (C) rationing health care resources
    (D) analyzing cost-effectiveness
    (E) preventing disease

8. Herd immunity refers to

   (A) immunity naturally acquired in a population
   (B) vaccination of domestic animals to prevent disease transmission to humans
   (C) genetic resistance to species-specific disease
   (D) the prevention of disease transmission to susceptible individuals through acquired immunity in others
   (E) the high levels of antibody present in a population following an epidemic

9. One example of a public health SOLUTION resulting in a new PROBLEM is

   (A) antigenic drift in viral influenza
   (B) desertification in sub-Saharan Africa
   (C) spread of *Legionella pneumophila* through air-conditioning systems
   (D) HIV infection and reactivation of latent tuberculosis
   (E) use of the prostate-specific antigen (PSA) assay to screen for prostate cancer

10. Ethical concerns are raised regarding the assessment of which of the following potentially preventable causes of disease?

    (A) social support networks
    (B) genetic susceptibility
    (C) immunization status
    (D) dietary intake
    (E) smoking history

ANSWERS AND EXPLANATIONS

1. The answer is C: it is unrelated to the care of individual patients. Epidemiology is the study of factors acting upon populations and influencing health. These include the causes of specific diseases, assessments of population health, and the role of environmental factors in health and disease. Often, epidemiology is the first responder when new health threats emerge, and serves to characterize modes of transmission and means of prevention before a complete pathogenetic understanding can be achieved. Clinical epidemiology is a discipline dedicated to improving individual patient care through optimal use of diagnostic tests, therapeutic interventions, and prognostic data. *Epidemiology*, p. 4.

2. The answer is D: ecologic epidemiology. As described in the text, epidemiology has traditionally been divided into chronic disease and infectious disease branches. This distinction may become obsolete as infectious agents are identified among the causes of chronic illness, and infectious illnesses of long duration such as HIV disease are managed as chronic diseases. A more recent distinction between classical and clinical epidemiology emphasizes the emerging importance of epidemiologic principles in improving diagnosis, therapy, and prognostication in the clinical setting. Although an ecologic perspective that permits evaluation of the interactive and synergistic factors involved in disease is important in epidemiology, it does not define a separate type. *Epidemiology*, p. 4.

3. The answer is A: evaluating etiologic factors in disease. "BEINGS" is an acronym that stands for the factors that may play a role in disease development: **b**iologic/**b**ehavioral factors; **e**nvironmental factors; **i**mmunologic factors; **n**utritional factors; **g**enetic factors; **s**ervices, **s**ocial and **s**piritual factors. Often, several of these factors interact to produce illness. The model promotes an ecologic perspective on the etiology of disease. *Epidemiology*, p. 4.

4. The answer is B: iatrogenic factors. Although iatrogenic factors (i.e., causes of illness produced by the physician or health care system) are implicitly included in the "BEINGS" model under the rubric "environment," iatrogenesis is not an explicitly distinguished category of disease etiology. The model explicitly includes **b**iologic/**b**ehavioral factors; **e**nvironmental factors; **i**mmunologic factors; **n**utritional factors; **g**enetic factors; **s**ervices, **s**ocial and **s**piritual factors. *Epidemiology*, p. 4.

5. The answer is A: the host and the agent. In order for disease to occur, a susceptible host must be in an environment in which exposure to the agent can occur. A vector of transmission may or may not be involved. The minimum interaction required for disease transmission to occur involves the agent and the host, and the environment that enables them to come together. Behavioral factors are intrinsic to the host, and genetic factors are included under both host and agent. *Epidemiology*, p. 4.

6. The answer is C: the iceberg phenomenon. Fatalities associated with the Hantavirus pulmonary syndrome may be the most visible manifestation of a disease that is often less severe. Usually, when a new disease is recognized, the initial cases are severe. Cases of a mild illness rarely generate an investigation. This is called the iceberg phenomenon because the majority of cases are initially hidden from view, just as the bulk of an iceberg lies below the water and is not initially seen. *Epidemiology*, p. 9.

7. The answer is C: rationing health care resources. While the responsibilities of the epidemiologist may include recommending cost-effective allocation of health care resources, the rationing of such resources is a political task and falls outside the purview of the epidemiologist. Genetic epidemiologists participate in genetic counseling. Establishing modes of disease transmission and identifying means of preventing disease spread are integral aspects of epidemiology. *Epidemiology*, p. 5.

8. The answer is D: the prevention of disease transmission to susceptible individuals through acquired immunity in others. The concept of herd immunity refers to the prevention of disease transmission to susceptible individuals in a population as a result of acquired immunity among others. If an immunized individual is exposed to an infectious source, transmission will not occur. The contacts of the immunized person, who might *not* be immunized themselves, will be spared exposure, and therefore illness, as a result of this immunity. The characteristics of a particular infection and a particular population determine the level of prevailing immunity required to limit the spread of a disease. For example, the presence of a highly infectious illness such as measles, in a population with multiple exposures such as college students, requires that nearly everyone be immunized to prevent transmission. *Epidemiology*, p. 12.

9. The answer is B: desertification in sub-Saharan Africa. Desertification of sub-Saharan Africa is thought to be largely due to overgrazing of arable land by domestic cattle. The tsetse fly transmitted often fatal illness to domestic cattle, and its control was thought to represent a significant advance. However, the uncontrolled growth of domestic cattle herds has now produced environmental circumstances perhaps more potentially disastrous than before (such as loss of arable land leading to mass malnutrition and starvation). The solutions to public health problems may often harbor unanticipated consequences. Antigenic drift and shift in viral influenza do not result from any particular effort to control the disease, but rather seem to derive from viral properties and environmental conditions in China, where viruses from humans and ducks may mingle in pigs. The spread of *L. pneumophila* through air-conditioning systems did not result from an attempt to solve a public health problem, although it was facilitated by environmental modification. The reactivation of latent tuberculosis resulting from HIV-induced immunocompromise is an example of synergism. While the use of the PSA assay in screening for prostate cancer has proved controversial, the assay is not a public health solution, and no new problem has arisen. *Epidemiology,* pp. 11 and 14.

10. The answer is B: genetic susceptibility. Social support networks, immunization status, dietary intake, and smoking status are all factors that may be modified to prevent disease. The currently available technology permits the identification of genetic susceptibility to diseases such as colon cancer. However, since the ability to modify genetic risk is not nearly as great as the ability to recognize it, ethical concerns have been raised. Little good may come from informing people about risk they cannot readily modify, while harm, such as anxiety or increased difficulty and expense involved in obtaining medical insurance, might result. *Epidemiology,* p. 8.

REFERENCE CITED

Jekel, J. F. Epidemiology, Biostatistics, and Preventive Medicine. Philadelphia, W. B. Saunders Company, 1996.

CHAPTER TWO

EPIDEMIOLOGIC DATA SOURCES AND MEASUREMENTS

SYNOPSIS

OBJECTIVES
- To characterize sources of epidemiologic data.
- To describe measures that are commonly employed in epidemiology.
- To define *risk* as an epidemiologic term, and to present examples of its application.
- To define *rate* as an epidemiologic term, and to present examples of its application.
- To define the rates used to characterize maternal and infant health.

I. Sources of Health Data
   A. Definitions
      1. **Denominator data** define the population at risk.
      2. **Numerator data** define the events or conditions of concern.
   B. International census and health data
      1. Most nations report collected data on births, deaths, and health events to the United Nations, which publishes the *Demographic Yearbook* and the *World Health Statistics Annual*.
      2. Variability in the quality of data reported from different countries is considerable and may limit the utility of international data.
   C. US census and health data
      1. US census
         a. The USA conducts a population census every 10 years, in years ending in zero. Population projections are used for interim years. These are the denominator data for population statistics in the USA.
         b. Numerator data in the USA are collected at the **local, state, and national level** and submitted to the **National Center for Health Statistics (NCHS).**
      2. US health data bases
         a. US Vital Statistics System
            (1) The **US Vital Statistics System** collects national data on births, deaths, causes of death, fetal deaths, marriages, and divorces.
            (2) There is potential for error and inaccuracy in such data. The cause of death on death certificates, for example, is reported as the **underlying cause of death** rather than the **immediate cause of death.** This is subject to interpretation by the reporting physician. Death certificates may be completed in hospitals at night by sleep-deprived resident physicians with little prior knowledge of the patient. The reported cause of death may be inaccurate in 15–20% of death certificates.
         b. **US disease reporting system**
            (1) Data on "reportable" diseases (i.e., those which, by law, must be reported) are collected from local providers of health care nationwide and collated by the **Centers for Disease Control and Prevention (CDC)** in Atlanta, Georgia.
            (2) **Underreporting** occurs due to missed diagnoses, lack of concern for such mild diseases as chickenpox, and reluctance to report diseases such as acquired immunodeficiency syndrome (AIDS), which might have undesirable social consequences for the patient. Despite underreporting, if the proportion of cases reported remains relatively constant, important data are acquired, and changes in the pattern of disease are detectable.
         c. Studies of the **National Center for Health Statistics (NCHS)**
            (1) The NCHS carries out periodic studies and surveys of illness and disability in the USA.
            (2) Examples of ongoing surveys include the National Health Interview Survey (NHIS), the National Health and Nutrition Examination Surveys (NHANES), the National Health Care Survey, and the National Nursing Home Survey.
         d. **Behavioral Risk Factor Surveillance System (BRFSS)**
            (1) The BRFSS coordinates ongoing surveys of behavioral risk factors carried out by state health departments in cooperation with the CDC.
            (2) By 1990, 90% of the population in the USA was accessible to this system.
         e. **Disease registries**
            (1) Disease registries are established, often by state governments, to collect disease-specific data.
            (2) An example of such a registry is the **Connecticut Tumor Registry (CTR),** the oldest population-based cancer registry in the USA.
            (3) Data from such registries are particularly useful for time trend analyses.

II. Epidemiologic Measurements

A. Frequency

1. **Incidence (incident cases)** is the frequency (number) of new occurrences of disease, injury, or death—that is, the number of transitions from well to ill, from uninjured to injured, or from alive to dead—in the study population during the time period being examined.
2. **Prevalence (prevalent cases)** is the number of persons in a defined population who have a specified disease or condition at a point in time, usually the time a survey is done.
3. Difference between point prevalence and period prevalence
   a. **Point prevalence** specifies the number of cases in the study population at one point in time.
   b. **Period prevalence** is the number of persons who had the disease at any time during the specified time interval. Period prevalence is the sum of the point prevalence at the beginning of the interval plus the incidence during the interval.
4. Illustration of morbidity concepts
   a. The concepts of incidence, point prevalence, and period prevalence are illustrated in Fig. 2–1, based on a method devised by Dorn (1957).
   b. On the basis of this illustration, the following calculations can be made:
      (1) There were four incident cases during the year (case nos. 3, 4, 5, and 7).
      (2) The point prevalence at t_1 was four (the prevalent cases were nos. 1, 2, 6, and 8).
      (3) The point prevalence at t_2 was three (case nos. 1, 3, and 5).
      (4) The period prevalence is equal to the point prevalence at t_1 plus the incidence between t_1 and t_2, or, in this case, $4 + 4 = 8$.
      (5) Note that whereas a case could only be an incident case once, a case could be a prevalent case at both time intervals (as with case no. 1 in this figure).
5. The relationship between incidence and prevalence
   a. Prevalence is influenced by the duration of disease, whereas incidence is not.
   b. **Cumulative incidence** refers to the total number of incident cases over a specified period of time. Incident cases resulting in death would be included in the cumulative incidence measure but not in the prevalence.
   c. One formula for prevalence is the following:

   $$P = CI + Im - (D + H + Em)$$

   where P is the prevalence; CI is the cumulative incidence; Im is the number of immigrants with the condition; D is deaths of cases; H is those healed, or cured, of the condition; and Em is emigrants with the condition.
   d. The prevalence can be also approximated with a simpler formula:

   $$\text{Prevalence} = \text{Incidence} \times \text{(Average) Duration}$$

B. Risk

1. Definition of risk
   a. **Risk** is the proportion of persons who are unaffected at the beginning of a study period but who undergo the risk event during the study period.
   b. The **risk event** may be death, disease, or injury.
   c. The persons at risk for the event are called a **cohort,** which is a clearly defined group of persons studied over time.
2. Limitations of the concept of risk
   a. In theory, only the **susceptible population** should be included in the denominator of a risk. For an infectious disease, for example, the denominator should include only exposed individuals lacking protective immunity. However, as the immunity of subjects may not be known, and in population-based studies usually is not, the denominator often includes exposed individuals not truly susceptible to the event (i.e., infection).
   b. The **case fatality ratio** is the proportion of clinically ill persons who die of the condition under study. This is a marker of **virulence.**
   c. The proportion of infected individuals who are clinically ill is often called the **pathogenicity** of the organism.
   d. The proportion of exposed persons who become infected is sometimes called the **infectiousness** of the organism, but this is also influenced by the conditions of exposure.
   e. Risk may fail to convey the **force of mortality,** or its pattern, in a population; therefore, rates are often used instead of risk.

C. Rates

1. Definition of rate
   a. A **rate** is the frequency (number) of events that occur in a defined time period, divided by the average population at risk. The midperiod population is often used as the denominator of a rate.
   b. The formula for a rate is:

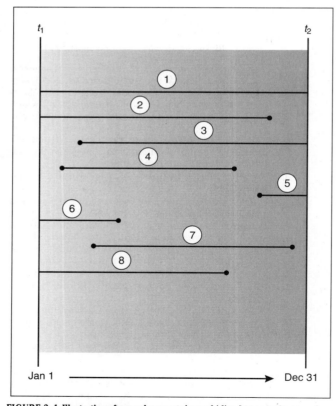

FIGURE 2–1. Illustration of several concepts in morbidity. Lines indicate when eight persons became ill (start of a line) and when they recovered or died (end of a line) between the beginning of a year (t_1) and the end of the same year (t_2). Each person is assigned a case number, which is circled in this figure.

$$\text{Rate} = \frac{\text{Numerator}}{\text{Denominator}} \times \text{Constant multiplier}$$

The **constant multiplier** (e.g., 100, 1000, 10,000, or 100,000) is used to avoid fractions of events, so that rates are easier to discuss and understand.
2. The relationship between risk and rate
   a. Unlike risk, rates can convey differences in the **force of mortality** on a population (see Fig. 2–2).
   b. A rate may be used to approximate risk for the following situations:
      (1) The event in the numerator occurs only once per individual during the study period.
      (2) The proportion of the population affected by the event is small.
      (3) The time interval is relatively short.
3. Criteria for the valid use of the term *rate*
   a. All events counted in the numerator must have happened to persons included in the denominator.
   b. All of the persons counted in the denominator must have been at risk for the events in the numerator (see B. Risk, above, for limitations of this concept).
   c. Before comparisons of rates can be made, all of the following must be true:
      (1) The numerators for all groups being compared must be defined or diagnosed in the same way.
      (2) The constant multipliers being used must be the same.
      (3) The time intervals must be the same.
4. Definitions of specific types of rates
   a. **Incidence rate** is the number of incident cases over a defined study period, divided by the population at risk at the midpoint of that study period.
   b. **Prevalence rate** is actually the *proportion* of persons with a defined disease or condition at the time they are studied; this is not truly a rate, although it is conventionally referred to as one.
   c. **Incidence density** refers to the frequency (density) of new events per **person-time** (e.g., person-months or person-years). A unit of person-time is the observation of one person for that particular interval, of two persons for half the interval, and so on.
D. **Special issues concerning the use of rates**
   1. Rates are usually used to make one of three types of comparisons:
      a. A comparison of an observed rate with a target rate.
      b. A comparison of two different populations at the same time.
      c. A comparison of the same population at two different time periods.
   2. Use of crude rates versus specific rates
      a. **Crude rates** apply to an entire population, without reference to any characteristics of the individuals in it. An example is the crude death rate:

$$\text{Crude death rate} = \frac{\text{Number of deaths (defined place and time period)}}{\text{Midperiod population (same place and time period)}} \times 1000$$

Therefore, the crude death rate in the USA in 1990 was reported as 8.7 per 1000, rather than 0.0087. Crude rates may be misleading because they fail to take such factors as age into account (see Table 2–1).

TABLE 2-1. Crude Death Rate and Life Expectancy for Three Countries

Country (Year)	Crude Death Rate	Life Expectancy at Birth
Colombia (1985–1989)	7.4 per 1000	63.4 years
USA (1988)	8.8 per 1000	71.3 years
Sweden (1989)	10.8 per 1000	74.2 years

Source: United Nations. Demographic Yearbook. New York, United Nations, 1990.

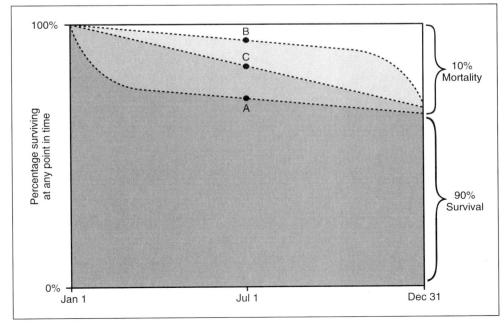

FIGURE 2-2. Illustration of circumstances under which the concept of rate is superior to the concept of risk. Assume that populations A, B, and C are three different populations of the same size; that 10% of each population died in a given year; and that most of the deaths in population A occurred early in the year, most of the deaths in population B occurred late in the year, and the deaths in population C were fairly evenly distributed throughout the year. In all three populations, the risk of death would be the same—that is, 10%—even though the patterns of death differed greatly. However, the rate of death, which is calculated using the midyear population as the denominator, would be the highest in population A, the lowest in population B, and intermediate in population C, reflecting the relative magnitude of the force of mortality in the three populations.

b. **Specific rates** pertain to some homogeneous subgroup of the population such as an age group, gender group, or ethnic group.

c. Specific rates cannot always be used instead of crude rates for the following reasons:
   (1) The frequency of the event in the specific subgroup may not be known.
   (2) The size of the specific subgroups may not be known.
   (3) The numbers at risk in specific subgroups may be too small to provide a stable estimate of the specific rate.

d. An example of a specific rate is the **age-specific death rate (ASDR)**, which is as follows:

$$\text{Age-specific death rate} = \frac{\text{Number of deaths to people in a particular age group (defined place and time period)}}{\text{Midperiod population (same age group, place, and time period)}} \times 1000$$

e. Crude death rates are the sum of ASDRs for each age group, multiplied by the relative size of each age group. In general, any summary weight is the sum of the rates (r_i) times the weights (w_i):

$$\text{Summary rate} = \Sigma w_i r_i$$

3. Standardization of death rates

   a. **Standardized rates** (also called **adjusted rates**) are crude rates that have been modified so as to allow for valid comparisons of rates. This is usually necessary to correct for differing age distributions in different populations.

   b. In **direct standardization**, two populations to be compared are given the same age distribution, which is then applied to the observed ASDRs, providing the number of deaths that *would have* occurred in each of the two populations *if they had been identical* in age distribution (see Box 2-1).

   c. **Indirect standardization** is used if ASDRs are not available in the study population, or if the study population is small and would therefore yield ASDRs that would be statistically unstable. This method applies **standard rates** from the population to the known age distribution of the study group. (**N.B.** Parameters other than age, such as socioeconomic strata, may be standardized.) See Box 2-2.

BOX 2-1. Direct Standardization of Crude Death Rates of Two Populations, Using the Averaged Weights as the Standard Population (Fictitious Data)

Part 1. Calculation of the crude death rates:

	Population A				Population B			
Age Group	Population Size		Age-Specific Death Rate	Expected Number of Deaths	Population Size		Age-Specific Death Rate	Expected Number of Deaths
Young	1,000	×	0.001 =	1	4,000	×	0.002 =	8
Middle-aged	5,000	×	0.010 =	50	5,000	×	0.020 =	100
Older	4,000	×	0.100 =	400	1,000	×	0.200 =	200
Total	10,000			451	10,000			308
Crude death rate		451/10,000 = 4.51%				308/10,000 = 3.08%		

Part 2. Direct standardization of the above crude death rates, with the two populations combined to form the standard weights:

	Population A				Population B			
Age Group	Population Size		Age-Specific Death Rate	Expected Number of Deaths	Population Size		Age-Specific Death Rate	Expected Number of Deaths
Young	5,000	×	0.001 =	5	5,000	×	0.002 =	10
Middle-aged	10,000	×	0.010 =	100	10,000	×	0.020 =	200
Older	5,000	×	0.100 =	500	5,000	×	0.200 =	1,000
Total	20,000			605	20,000			1,210
Standardized death rate		605/20,000 = 3.03%				1,210/20,000 = 6.05%		

BOX 2-2. Indirect Standardization of the Crude Death Rate for Males in a Company, Using the Age-Specific Death Rates for Males in a Standard Population (Fictitious Data)

Part 1. Beginning data:

	Males in the Standard Population			Males in the Company		
Age Group	Proportion of Standard Population	Age-Specific Death Rate	Observed Death Rate	Number of Workers	Age-Specific Death Rate	Observed Number of Deaths
Young	0.40 ×	0.0001 =	0.00004	2,000 ×	? =	?
Middle-aged	0.30 ×	0.0010 =	0.00030	3,000 ×	? =	?
Older	0.30 ×	0.0100 =	0.00300	5,000 ×	? =	?
Total	1.00		0.00334	10,000		48

Observed death rate   0.00334, or 334/100,000   48/10,000, or 480/100,000

Part 2. Calculation of the expected death rate, using indirect standardization of the above rates and applying the age-specific death rates from the standard population to the numbers of workers in the company:

	Males in the Standard Population			Males in the Company		
Age Group	Proportion of Standard Population	Age-Specific Death Rate	Observed Death Rate	Number of Workers	Standard Death Rate	Expected Number of Deaths
Young	0.40 ×	0.0001 =	0.00004	2,000 ×	0.0001 =	0.2
Middle-aged	0.30 ×	0.0010 =	0.00030	3,000 ×	0.0010 =	3.0
Older	0.30 ×	0.0100 =	0.00300	5,000 ×	0.0100 =	50.0
Total	1.00		0.00334	10,000		53.2

Expected death rate   53.2/10,000, or 532/100,000

Part 3. Calculation of the standardized mortality ratio (SMR):

$$\text{SMR} = \frac{\text{Observed death rate for males in the company}}{\text{Expected death rate for males in the company}} \times 100$$

$$= \frac{0.00480}{0.00532} \times 100$$

$$= (0.90)(100) = 90$$

= Males in the company actually had a death rate that was only 90% of the value that would be expected, based on the death rates in the standard population

---

d. The **standardized mortality ratio (SMR)** is the *observed* total events in the study group, divided by the *expected* number of events based on the standard population rates applied to the study group. The constant multiplier for this measure is 100.

4. **Cause-specific rates**
   a. Cause-specific rates provide numerators for comparison that are comparable with regard to diagnosis.
   b. The formula is as follows:

   Cause-specific death rate =

   $$\frac{\text{Number of deaths due to a particular cause (defined place and time period)}}{\text{Midperiod population (same place and time period)}} \times 1000$$

E. **Commonly used rates that reflect maternal and infant health**
   1. Definitions of terms
      a. A **live birth** is the delivery of a product of conception that shows any sign of life after complete removal from the mother. A **sign of life** may be any spontaneous movement, a breath, a cry, or pulsation of the umbilical cord.
      b. An **early fetal death,** or **miscarriage,** is the delivery of a dead fetus during the first 20 weeks of gestation.
      c. An **intermediate fetal death** is delivery of a dead fetus between 20 and 28 weeks' gestation.
      d. After 28 weeks' gestation, delivery of a dead fetus is termed a **late fetal death,** or a **stillbirth.**
      e. An **infant death** is the death of a live-born child before that child's first birthday.
   2. Definitions of specific types of rates
      a. The **crude birth rate** is the number of live births divided by the midperiod population:

      $$\text{Crude birth rate} = \frac{\text{Number of live births (defined place and time period)}}{\text{Midperiod population (same place and time period)}} \times 1000$$

      b. The **crude death rate** and **age-specific death rate** are discussed above under D.2.
      c. The **infant mortality rate (IMR)** is often used as an overall index of the health of a nation:

      $$\text{Infant mortality rate} = \frac{\text{Number of deaths to infants under 1 year of age (defined place and time period)}}{\text{Number of live births (same place and time period)}} \times 1000$$

      d. The **neonatal mortality rate** is the number of deaths to infants under 28 days of age, divided by the number of live births. The **postneonatal mortality rate** is the number of deaths to infants between 28 and 365 days of age, divided by the number of live births minus the number of neonatal deaths. For both of these rates, the conventional constant multiplier is 1000. The neonatal mortality rate is thought to reflect the quality of medical services and prenatal care, while the postneonatal mortality rate reflects the quality of the home environment.
      e. The **perinatal mortality rate** is the number of stillbirths plus the number of deaths to infants under 7 days of age, divided by the number of stillbirths plus the number of live births, multiplied by 1000. The perinatal mortality rate is approximated by the **perinatal mortality ratio,** in which the denominator does not include the stillbirths. The primary use of these measures is to evaluate progress in prenatal and perinatal care.
      f. The **maternal mortality rate (MMR)** is the number of pregnancy-related deaths, divided by the number of live births, multiplied by 100,000. This measure reflects the adequacy of pregnancy-related health care.

QUESTIONS

DIRECTIONS. (Items 1–4): Each of the numbered items or incomplete statements in this section is followed by answers or by completions of the statement. Select the ONE lettered answer or completion that is BEST in each case. Correct answers and explanations are given at the end of the chapter.

Items 1–4

During a given year, 12 cases of disease X are detected in a population of 70,000 college students. Many more of the students have mild symptoms of the disease, such as persistent daydreams about selling coconuts on a Caribbean beach.

1. All of the following are true statements regarding the apparent risk of acquiring disease X EXCEPT that it is

   (A) much less than 1%
   (B) dependent on the accuracy of diagnosis
   (C) possibly influenced by the iceberg phenomenon
   (D) dependent upon reliable reporting
   (E) constant over time

2. Of the detected cases, 7 die. The ratio of 7/12 therefore represents

   (A) the crude death rate
   (B) the pathogenicity
   (C) the standardized mortality ratio
   (D) the case fatality ratio
   (E) 1 − prevalence

3. To report the incidence rate of disease X we would need to know

   (A) nothing more than the information provided
   (B) the midyear population at risk
   (C) the duration of the clinical illness
   (D) the case fatality ratio
   (E) the age distribution of the population

4. To report the prevalence of disease X we would need to know

   (A) the number of current cases at a given time
   (B) the duration of illness
   (C) the cure rate
   (D) losses to follow-up
   (E) the rate at which new cases were developing

DIRECTIONS. (Items 5–14): For each numbered item, select the ONE lettered option that is most closely associated with it. It is generally advisable to begin by reading the list of options. Then, for each item in the set, try to generate the correct answer and locate it in the option list, rather than evaluating each option individually. Each lettered option may be used once, more than once, or not at all.

Items 5–14

(A) age-specific death rate
(B) crude death rate
(C) infant mortality rate
(D) standardized mortality ratio
(E) case fatality rate
(F) not a true rate; actually, a proportion
(G) crude rates
(H) standardized rates
(I) indirect standardization of death rates
(J) specific rates
(K) the two populations to be compared are "given" the same age distribution, which is then applied to the observed ASDRs (age-specific death rates) of each population
(L) cause-specific rates
(M) the number of new cases over a defined study period, divided by the midpoint population at risk
(N) adjusted rates
(O) variable rates
(P) number of total births, divided by the number of women, age 18–42 years, in the midperiod population
(Q) number of live births, divided by the midperiod population

Match each of the epidemiologic measures with the term or statement that best corresponds with it.

5. prevalence rate
6. incidence rate
7. direct standardization of death rates
8. fictitious rates
9. used if ASDRs are not available in the population whose crude death rate is to be adjusted
10. observed total deaths in a population divided by the expected deaths in that population
11. used to create numerators that are homogeneous according to diagnosis
12. useful for studying trends in the causes of death over time
13. crude birth rate
14. often used as overall index of the health status of a nation

ANSWERS AND EXPLANATIONS

1. The answer is E: constant over time. The risk of acquiring disease X is estimated by taking the reported number of cases and dividing it by the population at the beginning of the period of interest. In this case, that is 12/70,000, or 0.017%. However, true risk depends on the accuracy of reporting. Cases of disease X may have gone unreported, undiagnosed, or misdiagnosed. Only the most severe cases may have been reported, resulting in the iceberg phenomenon described in Chapter 1. The risk of acquiring disease X may change over time. The risk can only be defined for a particular period during which data have been obtained. *Epidemiology,* p. 22.

2. The answer is D: the case fatality ratio. The number of cases of a particular condition that result in death, divided by the total number of identified cases of the condition in a specified population, is the case fatality ratio. In this instance, it is 7/12, or, expressed as a percentage, 58.3%. The crude death rate is the number of deaths divided by the midperiod population. Pathogenicity is indicated by the proportion of infected persons with clinical illness. The standardized mortality ratio is the observed deaths in a population subgroup, divided by the expected deaths based on a reference population. The term (1 − prevalence) is not meaningful. *Epidemiology,* p. 23.

3. The answer is B: the midyear population at risk. Incidence rate is the number of new cases in a specified population, during a specified period of time, divided by the midperiod population at risk. We do not know if 70,000 represents the population at the midpoint or beginning of observation, nor do we know if the entire population is at risk for disease X. Sometimes, incidence rates are based on the total population, even though not everyone is at risk, as there is no convenient way to distinguish the susceptible from the immune. An example would be incidence rates of hepatitis B in the USA; only the unimmunized would truly be at risk, but the rate might be reported with the total population in the denominator. *Epidemiology,* p. 23.

4. The answer is A: the number of current cases at a given time. By definition, the prevalence of a condition is the number of cases in a specified population at a particular time. If this information is known, nothing more is required to report the prevalence. However, the prevalence is *influenced* by the duration of illness, the cure or recovery rate, the cause of specific mortality, as well as immigration and emigration. These are factors that influence the number of cases in the study population at any particular time. *Epidemiology,* p. 21.

5. The answer is F: not a true rate; actually, a proportion. As defined in the text, the prevalence rate is actually just the proportion, or percentage, of persons in a specified population with a defined condition at the time of study. *Epidemiology,* p. 25.

6. The answer is M: the number of new cases over a defined study period, divided by the midpoint population at risk. Incidence, or incident cases, is merely the number of new cases. To generate a rate, the number of new cases over a specified period of time is divided by the population at risk at the midpoint of the study period (recall the differences between risks and rates) and then multiplied by a constant, such as 1000 or 10,000, to facilitate expression. *Epidemiology,* p. 25.

7. The answer is K: the two populations to be compared are "given" the same age distribution, which is then applied to the observed ASDRs (age-specific death rates) of each population. In direct standardization, the age-specific death rates are available for the populations

to be compared. The age distribution of a hypothetical "standard" population, often the sum of the populations under comparison, is derived. The ASDRs from each of the populations are applied to the hypothetical age distribution, and these summary rates may be compared because they are free of age bias. As is indicated in the text, crude death rates may be low in developing countries because the population is relatively young. Age standardization is required to demonstrate that in particular age groups, mortality rates tend to be higher in developing, as compared with industrialized, countries. *Epidemiology,* p. 26.

8. The answer is H: standardized rates. Standardized rates are derived from a hypothetical age distribution that does not truly exist in the populations being compared. These rates permit valid comparisons but are, in fact, fictitious. *Epidemiology,* p. 28.

9. The answer is I: indirect standardization of death rates. When the age-specific death rates are not known for the populations to be compared, direct standardization of rates is not feasible. Indirect standardization applies the rates from a reference population (e.g., the USA) to the study populations. The reference rates are applied to the age distribution of the study populations, and the number of deaths expected in each group is calculated by assigning the reference population mortality rates to the study population. The actual death rates in the study groups are then compared with the expected rates. *Epidemiology,* p. 29.

10. The answer is D: standardized mortality ratio. The standardized mortality ratio is often derived from indirect standardization methods. The observed number of deaths in a population subgroup (e.g., age group) is divided by the expected number of deaths in that subgroup based on the reference population. This figure is usually multiplied by 100 to generate a percentage (the percentage of the reference mortality rate experienced by the study group). *Epidemiology,* p. 29.

11. The answer is L: cause-specific rates. For rates to be compared, the populations, or denominators, must be comparable. In order to compare *events* that are similar, the numerators must be derived in the same way. Cause-specific rates provide homogeneous numerator data based on diagnosis. Death rates may be similar in two populations, but the deaths may be due to different conditions, with differing implications for public health management. Cause-specific rates would be useful in attempting to define appropriate allocation of public health resources. *Epidemiology,* p. 29.

12. The answer is L: cause-specific rates. While changes in overall mortality may be observed without identifying the underlying causes, to understand time trends in specific causes of death, cause-specific death rates are required. *Epidemiology,* p. 29.

13. The answer is Q: number of live births, divided by the midperiod population. The crude birth rate, by convention, is the number of live births, divided by the midperiod population. This is an unusual rate, as not everyone in the denominator (i.e., males) can truly be said to be "at risk" for the numerator event. This rate is actually a proportion. *Epidemiology,* p. 31.

14. The answer is C: infant mortality rate. The infant mortality rate is influenced by various aspects of maternal and fetal care, including nutrition, access to prenatal medical care, maternal substance abuse, the home environment, social support networks, and others. The vulnerability of infancy is such that, if infants are able to thrive, the overall health of a nation is thought to be adequate, whereas, if infants are failing, and the resulting infant mortality rate is high, then improvements in health care should be prioritized. *Epidemiology,* p. 31.

REFERENCES CITED

Dorn, H. F. A classification system for morbidity concepts. Public Health Reports 72:1043–1048, 1957.

Jekel, J. F. Epidemiology, Biostatistics, and Preventive Medicine. Philadelphia, W. B. Saunders Company, 1996.

Moriyama, I. M. Inquiring into the diagnostic evidence supporting medical certifications of death. *In* Lilienfeld, A. M. and A. J. Gifford, eds. Chronic Diseases and Public Health. Baltimore, Johns Hopkins University Press, 1966.

CHAPTER THREE

EPIDEMIOLOGIC SURVEILLANCE AND OUTBREAK INVESTIGATION

SYNOPSIS

OBJECTIVES

- To define disease surveillance.
- To discuss the responsibilities and purposes of disease surveillance systems.
- To describe the methods used in disease surveillance.
- To define an epidemic.
- To characterize the steps involved in the investigation of a disease outbreak.
- To discuss the initiation of control measures during and after an outbreak investigation.

I. The Surveillance of Disease
A. Responsibility for surveillance in the USA
1. In the USA, the Centers for Disease Control and Prevention (CDC) is the federal agency responsible for most disease surveillance.
2. The CDC is a first responder in interstate disease outbreaks.
3. State health departments have jurisdiction over intrastate outbreaks and may invite the CDC to assist.

B. Purposes and methods for the surveillance of disease
1. Surveillance is usually considered the foundation of public health disease control efforts.
2. **Passive surveillance** is the reporting of disease by physicians, laboratories, and hospitals on a routine basis.
3. **Active surveillance** occurs when public health officials initiate contact with the above sources to obtain information about diseases of interest.
4. Routine surveillance allows for the establishment of **baseline data,** which facilitates detection of any sudden change in the pattern of disease occurrence.

5. Evaluation of time trends
   a. Long-term secular trends
      (1) **Long-term secular trends** are of greater public health significance than any short-term perturbation in the pattern of disease.
      (2) Changes in the observed pattern may be real, or may be due to changes in the rate of detection, the rate of reporting, or the diagnostic classification, or to a combination of these factors.
      (3) The **semilogarithmic scale** is often used to portray disease time trends. A **logarithmic scale** on the y-axis permits visualization of variation in rates over a wider range without loss of detail. An **arithmetic scale** is used on the x-axis.
   b. Seasonal variation
      (1) **Seasonal variation** is characteristic of many diseases.
      (2) Upper **respiratory infections** are most frequent in the winter and early spring in the northern hemisphere, due to indoor congregation and climatic effects on mucous membranes.
      (3) Diseases spread by **insect or arthropod vectors** are typically seen in the summer and early autumn.
      (4) Disease spread by the **fecal-oral route,** which usually requires contamination of water or food, is most common in the summer.
      (5) The incidence of disease is often displayed during an epidemiologic year. Unlike the **calendar year,** which runs from January of one year to January of the next year, an **epidemiologic year** runs from the month of lowest incidence in one year to the same month in the next year. This places the high incidence months near to the middle of the graph, and avoids splitting any possible peaks in incidence between the two ends of the graph, thereby obscuring the data.
   c. Other types of variation
      (1) Disease may vary by day of the week. Certain accidents are more common on weekends.

(2) Occupational disease is more common on weekdays.
6. Identification and documentation of outbreaks
   a. **Outbreak** is often used to denote a local epidemic, although the terms may be considered interchangeable.
   b. Outbreak is often used to avoid the emotive connotations of the term *epidemic*. (With the recent release of the movie *Outbreak*, the advantages of using the term may have dwindled, however.)
   c. An **epidemic** is defined as the occurrence of any disease at a frequency that is unusual (compared with baseline data) or unexpected. Baseline data are required to establish what is usual or expected. A single case of smallpox would constitute an epidemic, as the disease has been eradicated and no cases are expected.
      (1) The **epidemic threshold** defines the degree of variation from usual patterns for a disease required to qualify as an outbreak. This takes into consideration the variability in the usual pattern of disease.
      (2) The most refined application of this system is in the classification of outbreaks of viral influenza.
7. Evaluation of disease interventions
   a. Surveillance is required not only to monitor the rate of naturally occurring phenomena but also the impact of public health interventions such as vaccines.
   b. Vaccine efficacy can be established only by documenting a decline in the incidence of disease following an immunization campaign. Adverse reactions to a vaccine, or its failure to confer immunity, are also determined by continuing surveillance.
8. Setting of disease control priorities
   a. Disease control priorities are defined by surveillance data.
   b. Policy is determined by disease incidence, the severity of disease, the route of transmission, the potential for effective control, and the rapidity of change in incidence patterns.
9. Study of the natural history of diseases
   a. The public health impact of a familiar disease may change over time.
   b. As an example, tuberculosis, of little concern in recent decades, is again a major public health threat due largely to its association with human immunodeficiency virus (HIV) disease and to the development of multiple drug-resistant strains of the causative mycobacteria.

II. **The Investigation of Epidemics**
   A. **The nature of epidemics**
      1. An **epidemic** is an unusual or unexpected occurrence of disease. The term comes from the Greek words *epi*, meaning "upon," and *demos*, meaning "person" or "population."
      2. **Endemic**, meaning "within the population," refers to a disease that is occurring regularly in a defined population.
      3. Disease outbreaks in animals are **epizootic**; regularly occurring disease in animal populations are **enzootic**.
      4. The **attack rate**, or the proportion of exposed persons that become ill, is the customary measure of disease frequency used to define a disease outbreak and establish its severity.
   B. **Procedures for investigating an epidemic**
      1. The ultimate goal of any outbreak investigation is to control or prevent the spread of disease (see Box 3–1).
      2. Establish the diagnosis
         a. The initial step in outbreak investigation is to **establish the diagnosis.**
         b. The diagnosis may be nonspecific at first, but it must be adequate to determine whether the disease occurrence is in fact unusual, thereby qualifying as an outbreak.
      3. Establish the case definition
         a. A **case definition** characterizes the clinical manifestations of the condition under investigation. It provides the inclusion and exclusion criteria that are used to determine which subjects are, and which are not, cases.
         b. The more restrictive (exclusive) the case definition, the greater the risk of **false negatives** (i.e., individuals wrongly included in the unaffected group).
         c. The more inclusive the definition, the greater the risk of **false positives** (i.e., individuals wrongly included in the affected group).
         d. A case definition may substitute for a diagnosis when the diagnosis is uncertain.
      4. Determine whether an epidemic exists
         a. Once a diagnosis or case definition, or both, is established, the determination must be made as to whether or not the observed cases exceed expectation and truly constitute an epidemic.
         b. When the expected rate of disease is uncertain due to lack of surveillance data, it is more difficult to correctly identify an epidemic.
      5. Characterize the epidemic by time, place, and person
         a. Time
            (1) The **time** of an outbreak is depicted with an **epidemic time curve,** which is a graph with time on the x-axis and the number of new cases on the y-axis.
            (2) Inspection of the epidemic time curve facilitates determination of the following:

BOX 3–1. Disease Surveillance and Outbreak Investigation

Functions of disease surveillance
(1) Establishment of baseline data
(2) Evaluation of time trends
(3) Identification and documentation of outbreaks
(4) Evaluation of disease interventions
(5) Setting of disease control priorities
(6) Study of the natural history of diseases

Procedures for outbreak investigation
(1) Establish the diagnosis
(2) Establish the case definition
(3) Determine whether an epidemic exists
(4) Characterize the epidemic by time, place, and person
(5) Develop hypotheses regarding source, type, and route of spread
(6) Test the hypotheses
(7) Initiate control measures
(8) Initiate specific follow-up surveillance to evaluate the control measures

(a) The **type of exposure** (i.e., common source or propagated).
(b) The **route of spread** (respiratory, fecal-oral, skin-to-skin contact, exchange of blood or body fluids, or via insect or animal vectors).
(c) The timing of the initial exposure.
(d) The incubation period.
(e) The existence of **secondary cases** (which represent person-to-person transmission of disease from primary cases to other persons), in addition to **primary cases** (persons infected initially by a common source) (see Fig. 3–1).
(3) In a **common source exposure**, most cases acquire the condition at the same place and time. Therefore, there is a steep, early peak in the epidemic time curve.
(4) The alternative to common source exposure is usually **person-to-person spread**, also known as a **propagated outbreak**.
(5) The **index case** is the case that introduced the organism into the population experiencing the outbreak.
(6) The **timing of exposure** is determined by defining the earliest and latest cases, and measuring back the extremes of the incubation period, to establish a range. The usual incubation period can be used instead of the extremes. As an alternative, the average incubation period prior to the peak (if any) in an epidemic time curve may be used to establish the probable timing of exposure.
b. Place
(1) The geographic distribution of cases establishes the **place** of an outbreak.
(2) A **spot map** is used to show the residence locations of cases.
(3) **Incidence rates by location** are often used, such as by hospital ward (in a hospital infection outbreak), by work area or classroom (in an occupational or school outbreak), or by block or section of a city (in a community outbreak).
(4) Random clustering of disease in a particular location, by chance alone (**"chance" clusters**), must be distinguished from a true outbreak (**"true" clusters**). This can be difficult, and requires careful consideration of plausibility.
c. Person
(1) The case definition specifies who is a case and defines the **person** aspect of the outbreak.
(2) The characteristics of cases must be defined, including age, sex, ethnicity, immunization status, sources of water and food, occupation, and social contacts, to establish the susceptible population.
6. Develop hypotheses regarding source, type, and route of spread
a. Once the initial data are collected and assessed, hypotheses regarding the source and route of spread, the pattern of spread, and ultimately the causal agent are generated.
b. The **source of infection** is the vehicle (e.g., food, water) or person (the index case) that brought the infection into the community in the first place.
c. The **mode of transmission** of epidemic disease may be respiratory, fecal-oral, vector-borne, skin-to-skin, via the exchange of body fluids, or via fomites (objects in the environment).
d. The **pattern of spread** is the pattern by which infection can be spread from the source to those infected. The primary distinction is between a **common source pattern**, such as occurs when contaminated water is drunk by many people in the same time period, or a **propagated pattern**, in which the infection "propagates" itself by spreading directly from person to person over an extended period of time. A **mixed pattern** also occurs, in which persons who acquired a disease from a common source then spread it to family members or others (the secondary cases) by personal contact. In common source outbreaks, affected persons may have just one short **point source exposure**, or they may have a **continuous common source exposure**.
7. Test the hypotheses
a. **Laboratory studies** are useful to confirm or refute suspected causes of outbreaks. These may include the following:
(1) Cultures from patients, and, if appropriate, from possible vehicles such as food or water.
(2) Stool examinations for ova and parasites.
(3) Serum tests for antibodies to the organism suspected of causing the disease.
(4) Tests for nonmicrobiologic agents, such as toxins or drugs.
b. A **case-control study** may be conducted to test hypotheses.
(1) Cases are compared with controls, which are individuals within the same population who are similar to cases except that they did not acquire the condition under investigation.

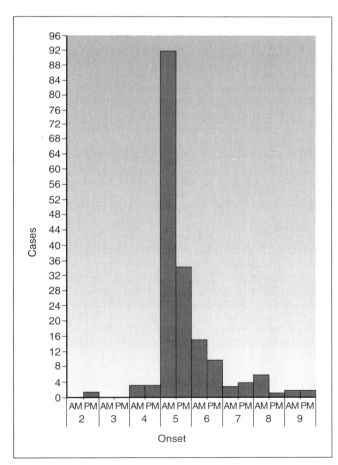

FIGURE 3–1. **Epidemic time curve showing the onset of cases of gastrointestinal disease due to *Shigella boydii* in Fort Bliss, Texas, in November 1976.** Note that the onset is shown in 12-hour periods for dates in November. (Source: Centers for Disease Control. Food and Waterborne Disease Outbreaks: Annual Summary, 1976. Atlanta, Centers for Disease Control, 1977.)

(2) The two groups are compared with regard to exposures that might explain the outbreak.
(3) Any exposures considerably more common among cases than controls might potentially account for the transmission of the disease.
8. Initiate control measures
   a. The implementation of control measures must often be done before the nature of the outbreak is established with certainty.
   b. Initial hypotheses are used as the basis for appropriate control measures.
   c. Control measures include the following:
      (1) **Sanitation,** or modification of the environment.
      (2) **Prophylaxis,** or placing some barrier, such as a vaccine, between susceptible individuals and the suspected agent of disease.
      (3) **Diagnosis and treatment** of active cases, so that they cannot spread the disease to others.
      (4) **Control of any contributing vectors** (e.g., mosquitoes in an outbreak of malaria).
9. Initiate specific follow-up surveillance to evaluate the control measures

QUESTIONS

DIRECTIONS. (Items 1–14): Each of the numbered items or incomplete statements in this section is followed by answers or by completions of the statement. Select the ONE lettered answer or completion that is BEST in each case. Correct answers and explanations are given at the end of the chapter.

1. All of the following are goals of disease surveillance EXCEPT

   (A) to establish baseline incidence and patterns of disease
   (B) to evaluate secular trends in disease
   (C) to identify true epidemics
   (D) to characterize the population at risk
   (E) to prevent adverse health events

2. Arizona, Colorado, and New Mexico report cases of an unexplained respiratory illness with a high case fatality rate. Which one of the following is certainly true regarding this event?

   (A) It is an epidemic.
   (B) It is an example of active surveillance.
   (C) It is not an outbreak, as the condition may be endemic.
   (D) It is appropriately investigated by the CDC.
   (E) It is due to improved surveillance.

3. Recently, cases of "flesh-eating" group A streptococcal disease were reported. To determine whether or not an outbreak was occurring, one would need to know all of the following EXCEPT

   (A) the incubation period
   (B) the usual pattern of this disease
   (C) reporting practices
   (D) diagnostic accuracy
   (E) a case definition

4. An official from the state department of public health visits outpatient clinics and emergency rooms to determine the number of cases of postexposure prophylaxis for rabies. The official's action is an example of

   (A) case finding
   (B) secondary prevention
   (C) active surveillance
   (D) outbreak investigation
   (E) screening

5. An article is published highlighting the long-term consequences of inadequately treated Lyme disease (Shadick et al. 1994). The article is described in the popular press. Patients with joint pains present to their physicians and insist on being tested for Lyme disease. Some of these tests are positive and are reported as cases of Lyme borreliosis. This represents

   (A) recall bias
   (B) active surveillance
   (C) a change in surveillance that will overestimate the likelihood of an outbreak
   (D) a change in surveillance that will have an unknown effect
   (E) a change in reporting that will underestimate prevalence

Items 6–12

A Democratic President invites a dozen or so of his dearest friends, prominent Republican legislators, to a formal luncheon. The salmon mousse is even more popular than highlights from the health care reform movement. Within 24 hours, 11 of the 17 diners experience abdominal pain, vomiting, and diarrhea. The President, who happens not to like salmon because it deadens the taste buds to the subtleties of beef jerky, feels fine. In fact, he goes jogging. Of the 11 guests with symptoms, 4 have fever and 7 do not; 5 have an elevated white blood cell count and 6 do not; 6 ate shrimp bisque and 5 did not; 9 ate salmon mousse and 2 did not; and one goes on to have surgery for acute cholecystitis due to an impacted calculus (stone) in the common bile duct. All of the cases recover within 3 days, with the exception of the senator who underwent surgery; she recovered over a longer period of time. The people at this luncheon had shared no other meals at any time recently.

6. The situation described

   (A) is a disease outbreak
   (B) is not an outbreak, because the usual pattern of disease is not known

(C) is a coincidence until proved otherwise
(D) is due to bacterial infection
(E) should be investigated by the CDC

7. The attack rate is

   (A) 4/11
   (B) 11/17
   (C) 11/13
   (D) 5/11
   (E) 1/17

8. An early priority in investigating this outbreak would be to

   (A) perform stool tests
   (B) submit food samples to the laboratory
   (C) temporarily close the kitchen
   (D) define a case
   (E) perform a case-control study

9. The best case definition for this disease would be

   (A) the onset of abdominal pain and fever following the state luncheon
   (B) vomiting, diarrhea, and an elevated white blood cell count
   (C) abdominal pain, vomiting, and diarrhea within 24 hours of the state luncheon
   (D) acute viral gastroenteritis
   (E) staphylococcal food poisoning

10. Suspecting that the outbreak is a common source exposure due to contaminated food, we attempt to determine which food is responsible. Our initial task is to

    (A) perform a case-control study
    (B) analyze food specimens in the laboratory
    (C) interview the cases to find out what they ate
    (D) examine the food preparers
    (E) close the White House kitchen

11. We are unable to identify a single food that was eaten by every case. We therefore should

    (A) abandon the investigation, since the disease is not very serious
    (B) conclude the investigation but without identifying the source
    (C) conclude that the outbreak was not foodborne
    (D) implicate the food least eaten by those without symptoms
    (E) implicate the food most eaten by those with symptoms

12. We suspect that the salmon mousse is the source of the outbreak. To confirm (or refute) our suspicion, we should

    (A) perform a case-control study in which cases ate salmon mousse and controls did not
    (B) perform a case-control study in which cases became ill and controls did not
    (C) initiate active surveillance
    (D) conduct a prospective study
    (E) identify the causative agent

13. The following curve depicts the epidemic threshold for

    (A) measles
    (B) influenza
    (C) varicella
    (D) pertussis
    (E) rubella

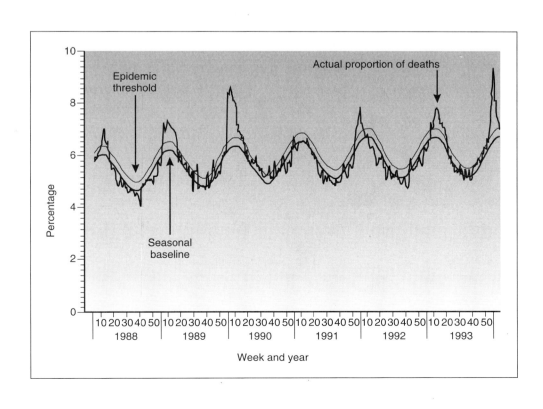

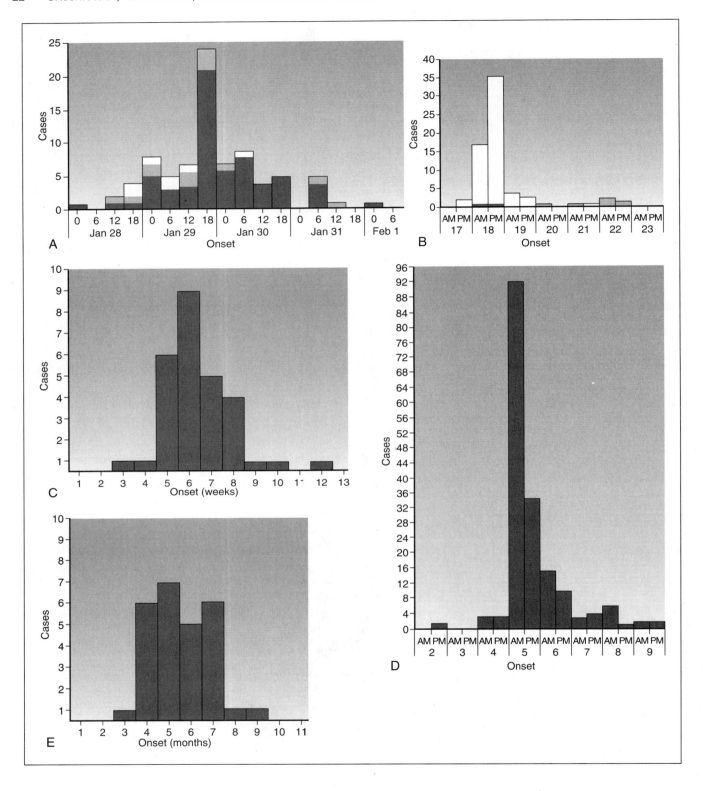

14. Of the epidemic curves shown above, the one that best depicts an outbreak of hepatitis A among members of a tour group to Southeast Asia is

    (A) Figure A

(B) Figure B
(C) Figure C
(D) Figure D
(E) Figure E

ANSWERS AND EXPLANATIONS

1. The answer is D: to characterize the population at risk. Surveillance helps to define the usual patterns and incidence of disease. In order to establish the typical pattern of disease, time trends must be assessed. Epidemics are sudden deviations from the usual pattern of disease, and the ability to note this change is dependent upon the baseline pattern established through surveillance. The ultimate goal of all public health endeavors, directly or indirectly, is to prevent adverse health events, such as the spread of disease. Characterization of the particular population at risk is usually part of an outbreak investigation, after surveillance has detected a change in the pattern of disease. *Epidemiology*, p. 37.
2. The answer is D: It is appropriately investigated by the CDC. The disease alluded to in this question is the Hantavirus pulmonary syndrome. Because fatalities were reported from more than one state, this investigation falls within the jurisdiction of the CDC. From the information provided in the question, one cannot say whether or not this constitutes a true epidemic, as the syndrome is not sufficiently characterized to determine whether a sudden change in the pattern of disease has occurred or not. Unsolicited reporting of cases is passive, not active, surveillance. We do not know whether these cases were reported due to improved surveillance, a true outbreak, or some other reason. Even if the condition were endemic, this would not exclude the possibility of an outbreak. The question provides only enough information to suggest that an investigation by public health epidemiologists is probably indicated. *Epidemiology*, p. 37.
3. The answer is A: the incubation period. To identify an outbreak or epidemic, a sudden change in the pattern of a disease or condition must be detectable. Such a change is only detectable when the usual pattern of disease is known. Of course, with a new disease, the usual pattern would be that no similar cases were known to have occurred. Reporting practices are important, as they can produce a pseudo-outbreak. If a case report generates alarm, reporting of a condition may increase, even if its rate does not change. Diagnostic accuracy and a case definition are required in order that all cases truly represent the condition under study and not a heterogeneous group of diseases with similar clinical manifestations. The incubation period of an infectious disease must be determined to characterize an outbreak but not to establish that one is occurring. *Epidemiology*, p. 40.
4. The answer is C: active surveillance. Whenever a public health official visits health care delivery sites to assess the pattern of a disease or condition, it constitutes active surveillance. Case finding refers to testing a patient group for as yet clinically inapparent disease, such as performing ECGs or chest x-rays in patients being admitted to the hospital without cough or chest pain. Secondary prevention is an effort to prevent clinical consequences of a disease once identified. The administration of postexposure vaccine might qualify as secondary prevention, but the review of such cases by a public health official would not. Finally, we are not provided sufficient information in the question to know whether or not an "outbreak" of rabies exposure is under consideration. *Epidemiology*, p. 37.
5. The answer is D: a change in surveillance that will have an unknown effect. Whenever public attention is focused on a particular health problem, often due to media reports, the percentage of cases reported is likely to rise. However, when the available diagnostic tests are imperfect, as is almost always the case, and certainly is for Lyme disease, some reported cases will be false positives. The number of false positives will rise as more people with vaguely characterized clinical syndromes are tested. In this example, the increased testing may truly be identifying previously undiagnosed cases of Lyme disease, or it may be producing false positives. To distinguish, we would need to know a great deal about the operating characteristics of the tests employed, as well as something about the denominator (i.e., the patients presenting and insisting on testing). Without all of this additional information, we must conclude that surveillance has changed but the effect of the change is, as yet, unknown. *Epidemiology*, p. 37.
6. The answer is A: is a disease outbreak. Although there is no formal surveillance of state luncheons, clearly the situation described represents the unexpected occurrence of disease. The usual pattern of disease in this case derives from common experience rather than surveillance: We do not expect disease to follow lunch. The CDC would not investigate so small and localized an outbreak, unless for some reason it was invited to do so by local public health officials. Perhaps the Republicans would want the FBI to look over the White House kitchen. *Epidemiology*, p. 40.
7. The answer is B: 11/17. The attack rate is the proportion of those exposed who become ill. As the investigation of this outbreak begins, the exposure is likely to be defined as participation in the luncheon. Of the 17 who dine together, 11 become ill. As the outbreak investigation proceeds, the definition of exposure could change if a particular food, such as the salmon mousse, were implicated. Then, the denominator used to calculate the attack rate might change to represent only those exposed to the contaminated food. At present, however, that information is not available to us, and 11/17 represents our best estimate of the attack rate. *Epidemiology*, p. 44.
8. The answer is D: define a case. Establishing a diagnosis or a case definition is the earliest priority in an outbreak investigation. Laboratory tests, control measures, and case-control studies are usually performed later in an investigation, after hypotheses have been generated regarding transmission and cause. *Epidemiology*, p. 44.
9. The answer is C: abdominal pain, vomiting, and diarrhea within 24 hours of the state luncheon. The case definition is required to distinguish between cases and noncases of the disease under investigation. The ideal case definition would permit the inclusion of all cases and the exclusion of all noncases, thereby providing perfect sensitivity and specificity (see Chapter 7). In this instance, the case definition includes the salient clinical features of most affected individuals from the exposed group. We might miss mild forms of the illness or include in our group of cases someone ill for reasons other than the outbreak exposure. The other answers, however, either invoke a specific diagnosis prematurely or include clinical features absent in most of the affected cohort. *Epidemiology*, p. 44.
10. The answer is C: interview the cases to find out what they ate. Even before developing hypotheses about the

source of infection, we would need to characterize this outbreak by time, place, and person. In this fairly simple example, virtually all of that information is contained in our case definition, but that is usually not so. The place and time of exposure were the luncheon, and the cases were the diners described in the example. With enough information to generate the hypothesis that this outbreak is due to contaminated food, we must determine what food to implicate. To do so, we must learn what foods were eaten preferentially by the cases. A simple interview would be the initial step. *Epidemiology,* p. 45.

11. The answer is E: implicate the food most eaten by those with symptoms. The methodology detailed in the text provides the basic means to conduct an outbreak investigation. There is, however, no substitute for thoughtfulness. At the stage of hypothesis generation, it is rare for the exact source of the outbreak to be completely obvious. Often, even if a particular food is suspected in an outbreak such as this one, that food will have been eaten by some people who do not become ill and not eaten by some of the cases. This could occur if the contaminated food were not uniformly contaminated, with some portions free, or relatively free, of the pathogenic agent. Further, the implicated food, contacting other foods on a platter, might contaminate other foods that, in a small proportion of cases, might serve as an alternative source of disease. An outbreak investigation is usually conducted while the outbreak is occurring and before it is clear how severe or protracted it will be. Therefore, if at all possible, the investigation should be conducted to its conclusion. The food least eaten by noncases would be a possible source of the outbreak only if that food were eaten by the majority of cases. (Otherwise, it might simply be a dish that no one liked.) *Epidemiology,* p. 44.

12. The answer is B: perform a case-control study in which cases became ill and controls did not. A case-control study is one in which the cases have the disease under investigation and the controls do not, and both groups are then assessed for prior exposures (see Chapter 5). Cases, as stipulated by our case definition, are luncheon participants with abdominal pain, vomiting, and diarrhea following the luncheon. Controls would be participants at the luncheon who did not become ill. We would then interview both groups about consumption of particular foods, including the salmon mousse. If we found that most cases and few controls had eaten this dish, our hypothesis would be supported. An additional, important means of testing our hypothesis would be to analyze the suspected food in the laboratory, looking for a pathogen or toxin that would explain the clinical syndrome. This step is usually done to confirm a diagnosis and facilitate control measures rather than to establish the source of the outbreak. *Epidemiology,* p. 50.

13. The answer is B: influenza. As discussed in the text, the surveillance for influenza is more elaborate than for any other infectious disease in the USA. This is largely due to the somewhat predictable seasonal variations in the disease, the characteristics of antigenic drift and shift, and the feasibility of producing strain-specific vaccine on a yearly basis. The figure shown demonstrates the variable baseline in influenza incidence and the threshold above which an epidemic is defined. (Source: Centers for Disease Control and Prevention. Update: influenza activity—United States and worldwide, 1993–1994 season, and composition of the 1994–1995 influenza vaccine. Morbidity and Mortality Weekly Report 43:179–183, 1994.) *Epidemiology,* p. 41.

14. The answer is C: Figure C. The epidemic curve graphically reveals the temporal characteristics of a disease outbreak. Included among these are the incubation period (i.e., the time between exposure and the development of disease), the period during which cases develop, and the occurrence of any secondary cases following the initial cluster. Examination of the epidemic curve facilitates the generation of specific hypotheses regarding the cause of an outbreak. The curve for hepatitis A reveals a variable incubation period in the range of 3 weeks, the development of new cases over a period of several weeks, and no indication of secondary spread. Curves A, B, and D characterize outbreaks that develop during a period of hours rather than weeks. Curve E is suggestive of hepatitis B, with an incubation period measured in months. (Source: Figure B, Centers for Disease Control. Shigella Surveillance: Annual Summary, 1971. Atlanta, Centers for Disease Control, 1972; Figure D, Centers for Disease Control. Food and Waterborne Disease Outbreaks: Annual Summary, 1976. Atlanta, Centers for Disease Control, 1977.) *Epidemiology,* p. 45.

REFERENCES CITED

Bennett, W. I., ed. Beyond overeating. New England Journal of Medicine 332:673–674, 1995.

Jekel, J. F. Epidemiology, Biostatistics, and Preventive Medicine. Philadelphia, W. B. Saunders Company, 1996.

Kuczmarski, R. J., et al. Increasing prevalence of overweight among US adults. The National Health and Nutrition Examination Surveys, 1960 to 1991. Journal of the American Medical Association 272:205–211, 1994.

Lindpainter, K. Finding an obesity gene—a tale of mice and men. New England Journal of Medicine 332:679–680, 1995.

Shadick, N. A., et al. The long-term clinical outcomes of Lyme disease: a population-based retrospective cohort study. Annals of Internal Medicine 121:560–567, 1994.

Stamler, J., ed. Epidemic obesity in the United States. Archives of Internal Medicine 153:1040–1044, 1993.

CHAPTER FOUR

THE STUDY OF CAUSATION IN EPIDEMIOLOGIC INVESTIGATION AND RESEARCH

SYNOPSIS

OBJECTIVES
- To define and characterize a causal relationship in epidemiology.
- To describe the steps involved in establishing causality.
- To describe the difficulties involved in correctly identifying causal associations.

I. Types of Causal Relationships
  A. **Causality (a causal relationship)** exists if a factor produces, or contributes to the production of, a specified outcome.
    1. A **sufficient cause** precedes a disease and has the following relationship with it: If the cause is present, the disease will always occur.
    2. A **necessary cause** precedes a disease and has the following relationship with it: If the cause is absent, the disease cannot occur. If the cause is present, the disease may or may not occur.
    3. A **risk factor** is a characteristic which, if present and active, clearly increases the probability of a particular disease in a group of persons who have the factor compared with an otherwise similar group of persons who do not. A risk factor, however, is neither a necessary cause nor a sufficient cause of the disease.
    4. **Direct causality** is present when the factor under consideration exerts its effect without intermediary factors.
    5. **Indirect causality** occurs when one factor influences one or more other factors that are, in turn, directly causal. These intermediary factors are known as **intervening variables.** An example is increased cardiovascular disease (outcome), associated with obesity (causal factor), in part due to the higher blood pressure and higher levels of cholesterol and glucose (intervening variables) that often occur with obesity.
  B. In a **noncausal association,** the relationship between two variables is statistically significant, but no causal relationship exists, either because the temporal relationship is incorrect (the presumed cause comes after, rather than before, the presumed effect) or because another factor is responsible for both the presumed cause and the presumed effect.
II. Steps in the Determination of Cause and Effect
  A. Investigation of the statistical association
    1. The initial step in characterizing a relationship as causal is recognition of an association between or among the putatively causal factors and the effects.
    2. According to **Mill's canons,** association is more likely to be causal if the following are true:
      a. The association is more likely to be causal if it shows **strength**—i.e., the difference is large.
        (1) The association must be strong enough so that it is unlikely to have occurred by chance alone.
        (2) Statistical methods are designed to test the strength of apparent associations.
      b. The association is more likely to be causal if it is **consistent**—i.e., the difference is always observed if the risk factor is present.
      c. The association is more likely to be causal if it is **specific**—i.e., the difference does not appear if the risk factor is not present.

d. The association is more likely to be causal if it is **plausible**—i.e., it makes sense, based on what is known about the natural history of the disease.
e. The association is more likely to be causal if a **dose-response relationship** is revealed—i.e., the risk of disease is greater with stronger exposures to the risk factor.

B. **Investigation of the temporal relationship**
1. To demonstrate causation, the suspected cause must have been present or occurred before the effect (e.g., the disease) developed.
2. An **experimental control group** may be used to demonstrate that cause precedes effect—i.e., that randomization to different exposure states leads to differences in the outcome measure between groups.
3. The onset of "effect" may be difficult to define in chronic diseases such as atherosclerosis.
4. The onset of "cause" may be difficult to define in the case of continuous exposures such as a high-fat diet.

C. **Elimination of all known alternative explanations**
1. In an epidemiologic investigation concerning the causation of disease, even if the presumed causal factor is associated statistically with the disease and occurs before the disease appears, it is necessary to demonstrate that there are no other likely explanations for the association.
2. Proper research methods, such as randomization of subjects in a prospective study (see Chapter 12), can help exclude alternative explanations.
3. Alternative explanations can never be fully ruled out, because advances in science may produce new insights.
    a. **Peptic ulcer disease,** once ascribed, with considerable confidence, to hypersecretion of gastric acid, has been found to be associated with infection by the bacterium *Helicobacter pylori*.
    b. **Cholera** was well explained by the miasma (noxious vapor) theory in the nineteenth century, until the germ theory was advanced to supplant the miasma theory.
    c. There is ongoing debate as to whether or not high levels of iron in blood are causally associated with the development of **cardiovascular disease.** This hypothesis competes with those that emphasize the importance of such factors as dietary fat and the protective benefit of estrogen in women.
    d. Initially, the **eosinophilia-myalgia syndrome** was explained by its association with use of supplements of L-tryptophan, an amino acid. This explanation was replaced when it was found that it was only one brand of L-tryptophan, from a single source, that was associated with the syndrome. A contaminant in this brand of L-tryptophan, rather than the amino acid, was ultimately found to be the cause of eosinophilia-myalgia syndrome.

III. **Common Pitfalls in Causal Research**
A. **Bias**
1. Bias, also known as **differential error,** produces deviations or distortions of data that are consistently in one direction, as opposed to random error.
2. **Measurement bias** results in distorted quantification of exposures or outcome, and is due to improper technique or subjectivity of the measurement scale (e.g., weighing subjects without having them remove their clothing and shoes).
3. **Recall bias** results from differential recall of exposure to causal factors among those who have the disease as compared with those who do not have it. Often, those with an undesirable "effect" (e.g., disease) are more likely to remember a cause than those without such an effect.
4. **Selection bias** occurs when allocation of participants to a study, or a particular study group, is influenced by characteristics of the participants that also influence the probability of the outcome.
    a. **Internal validity** is achieved when the results of a study are true and meaningful for the participants. Internal validity is not compromised by the selection of subjects who are not representative of the population from which they are drawn.
    b. **External validity** is achieved when the results of a study are true and meaningful for the larger population beyond the study participants. External validity is compromised if study participants differ from nonparticipants or otherwise do not represent the larger population from which they are drawn.

B. **Random error**
1. Random error, also known as **nondifferential error,** produces findings that are too high and too low in approximately equal amounts, owing to random factors.
2. Random error is usually less serious than bias, but it reduces statistical power (see Chapters 10 and 12) and increases the likelihood that a true causal relationship will not be demonstrated.

C. **Confounding**
1. Confounding is the confusion of two supposedly causal variables, so that part or all of the purported effect of one variable is actually due to the other.
2. Confounding may result in false identification of a causal relationship, or obscure a true causal relationship.

D. **Synergism** occurs when there is more than one causal factor, and these factors produce a combined influence on the outcome greater than the sum of their individual influences.

E. **Effect modification**
1. Effect modification is also called **interaction.**
2. Effect modification occurs when the strength, or even direction, of the influence of a causal factor on outcome is altered by a third variable, or the **effect modifier.**

IV. **Important Reminders About Risk Factors and Disease**
A. One causal factor may increase the risk for several different diseases (e.g., cigarette smoking is associated with lung cancer, cardiovascular disease, bronchitis, and other diseases).
B. More than one risk factor may contribute to the development of the same disease.

QUESTIONS

DIRECTIONS. (Items 1–10): For each numbered item, select the ONE lettered option that is most closely associated with it and fill in the circle containing the corresponding letter on the answer sheet. It is generally advisable to begin each set by reading the list of options. Then, for each item in the set, try to generate the correct answer and locate it in the option list, rather than evaluating each option individually. Each lettered option may be selected once, more than once, or not at all.

Items 1–10

(A) necessary cause
(B) effect modifier
(C) sufficient cause
(D) intervening variables
(E) recall bias
(F) spectrum bias
(G) measurement bias
(H) biologic plausibility
(I) external validity
(J) synergism
(K) internal validity
(L) confounder
(M) virulence

Match each of the parameters related to the assessment of causality with the appropriate example or related statement.

1. must be associated with both the exposure and the outcome
2. alters the nature of a true relationship between an exposure and an outcome
3. All subjects in a study were weighed while fully dressed.
4. The relative risk for lung cancer in smokers is X; the relative risk for lung cancer in asbestos workers is Y; and the risk in those with both exposures is XY.
5. In a case-control study of lung cancer, coffee consumption was greater among cases than controls.
6. Although there are more TV sets per capita in populations with high rates of cardiovascular disease, the association is not thought to be causal.
7. This is eliminated by a prospective design.
8. An individual is homozygous for the sickle cell gene.
9. Study results are obtained in an unbiased manner.
10. The study population resembles the larger population from which it was drawn.

ANSWERS AND EXPLANATIONS

1. The answer is L: confounder. A confounder is a factor that is associated with both the exposure and the outcome in question. As an example, cigarette smoking is a confounder of the relationship between alcohol and lung cancer. Cigarettes are associated both with the outcome, lung cancer, and with the exposure of interest, alcohol consumption (Bartecchi, MacKenzie, and Schrier, 1994). If one were to assess the relationship between alcohol and lung cancer and not take cigarette smoking into account, alcohol would be found to increase the risk of lung cancer. When the study has controls in place for cigarette smoking (i.e., when varying degrees of alcohol consumption are compared among subjects with comparable cigarette consumption), the association between alcohol and lung cancer disappears. All of the increased risk that appeared to be attributable to alcohol is actually attributable to cigarettes. This is confounding. Note that if cigarettes were causally related to lung cancer but cigarette consumption did not vary with alcohol exposure, there would be no confounding. *Epidemiology,* p. 61.
2. The answer is B: effect modifier. Unlike a confounder, an effect modifier does not obscure the nature of a relationship between two other variables; rather, it changes it. As an example, consider the effect of age on the pharmacologic action of the drug methylphenidate. In children, the drug is used to treat hyperactivity and attention deficit disorder, both of which are conditions in which there is too much activity. In adults, the same drug is used to treat narcolepsy, a condition of extreme daytime somnolence, and, in essence, a paucity of energy and activity. There is a true and measurable effect of methylphenidate on energy and activity levels in both children and adults. This effect is not confounded by age but is altered by it. This is effect modification. Unlike confounders, effect modifiers should not be controlled but rather should be analyzed to enhance understanding of the causal relationship in question. *Epidemiology,* p. 62.
3. The answer is G: measurement bias. Random error will produce some measurements that are too large, some that are too small, and perhaps some that are correct. Such error will contribute to variability within groups, thereby limiting one's ability to detect a significant difference between groups. Random error therefore reduces the power of a study to demonstrate a true difference in outcome (see Chapters 10 and 11). When error is systematic rather than random, statistical power may be preserved, but the validity of the study, the most critical attribute, is threatened. Imagine a study of weight loss in which study and control subjects were weighed fully clothed and with shoes on at enrollment. Then, following the intervention, the control subjects were again weighed fully dressed, but the intervention subjects were weighed after disrobing. This would clearly be a biased and invalid measure of the intervention and resultant weight loss. Bias may threaten internal or external validity, or both. *Epidemiology,* p. 60.
4. The answer is J: synergism. When the combined effect of two (or more) variables on an outcome is greater than the sum of their separate effects, the interaction is called

synergy, or synergism. There is true synergy between cigarette smoking and asbestos exposure on the risk of lung cancer (Harvey and Beattie, 1993). *Epidemiology,* p. 62.

5. The answer is L: confounder. Please refer to the answer provided for question 4, above. Coffee consumption tends to be higher among smokers than nonsmokers. If smoking were not controlled for, an association between coffee consumption and lung cancer might be demonstrated.

6. The answer is H: biologic plausibility. In affluent societies such as the USA, there is a relatively high incidence of cardiovascular disease. This is thought to be largely due to dietary factors and lack of exercise. Other factors, however, distinguish the USA from other countries (for instance, there are more television sets per capita in the USA than in most developing countries). However, before an abundance of an exposure and an abundance of an outcome may be considered related, they must meet the test of biologic plausibility (i.e., the relationship must make sense). There is no plausible way that the presence of a TV set in a home could produce heart disease (unless one considers a resultant state of "couch-potatohood" an intervening variable). While plausibility is required before a causal relationship can be established, an open mind is essential. What is implausible today may become plausible as the science of medicine advances. One such example is the etiology of peptic ulcer disease, long thought to be attributable to hypersecretion of gastric acid, and shown over recent years to result often from infection with the bacterium *Helicobacter pylori* (Fennerty, 1994). *Epidemiology,* p. 57.

7. The answer is E: recall bias. In any retrospective study, typically a case-control study, there is a risk that knowledge of current health status will influence the recall of exposures in the past. For example, the mothers of children born with any sort of congenital anomaly might be more likely to recall toxic exposures during the pregnancy (e.g., to medication) than would mothers of children without such anomalies. In a prospective study, exposure is established at enrollment, before subjects are distinguished on the basis of outcome. Therefore, recall bias is eliminated in a prospective study. *Epidemiology,* p. 60.

8. The answer is C: sufficient cause. A sufficient cause is one that, if present, will invariably cause a particular disease or condition. Such factors are rare; even highly pathogenic microbes fail to cause disease in some exposed individuals. Homozygosity for sickle cell anemia invariably results in the expression of the condition. Therefore, this genotype is a sufficient cause of the disease. *Epidemiology,* p. 55.

9. The answer is K: internal validity. The most important criterion upon which a study may be judged is its internal validity. A study that lacks internal validity is invalid, or unreliable. The results are meaningless due to bias. For example, if in a hypothetical study of orange juice versus orchiectomy (surgical removal of the testes) in the treatment of prostate cancer, men debilitated by illness were assigned to surgery, and men with no overt signs of illness were assigned to orange juice, the better outcome seen with orange juice would be invalid, attributable to the biased design. When bias is eliminated, study results are valid. However, even if results are internally valid, it may not be possible to derive generalizations from them; this is external validity. *Epidemiology,* p. 61.

10. The answer is I: external validity. The external validity of a study is determined by the resemblance the study population bears in terms of relevant characteristics to the larger population from which it was drawn. In a well-designed study of antihypertensive therapy in the prevention of stroke in middle class white males, the validity of the findings for women, the indigent, and other ethnic groups is uncertain. While internal validity defines whether or not a study's results may be trusted, external validity defines the degree to which the results may be considered relevant to individuals other than the study subjects themselves. *Epidemiology,* p. 61.

REFERENCES CITED

Bartecchi, C. E., T. D. MacKenzie, and R. W. Schrier. The human costs of tobacco use. New England Journal of Medicine 330:907–912, 1994.

Fennerty, M. B. *Helicobacter pylori.* Archives of Internal Medicine 154:721–727, 1994.

Harvey, J. C., and E. J. Beattie. Lung cancer. Clinical Symposia 45:2–32, 1993.

Jekel, J. F. Epidemiology, Biostatistics, and Preventive Medicine. Philadelphia, W. B. Saunders Company, 1996.

Pi-Sunyer, F. X. Health implications of obesity. American Journal of Clinical Nutrition 53:1595s–1603s, 1991.

Shadick, N. A., et al. The long-term clinical outcomes of Lyme disease: a population-based retrospective cohort study. Annals of Internal Medicine 121:560–567, 1994.

CHAPTER FIVE
Common Research Designs Used in Epidemiology

SYNOPSIS

OBJECTIVES
- To establish the goals of study design in epidemiology.
- To summarize the study designs employed in epidemiologic research.
- To distinguish between studies designed primarily for hypothesis generation and those designed for hypothesis testing.
- To consider the advantages and disadvantages of different study designs.

I. Functions of Research Design
  A. The basic goal of research design is to permit an unbiased comparison between a group with and a group without a factor under consideration.
  B. A good research design should perform the following functions:
    1. It should permit a comparison of a variable in question (such as disease frequency) between two or more groups at one point in time or, in some cases, between one group before and after receiving an intervention or being exposed to a risk factor.
    2. It should permit quantification of the comparison in absolute terms (as with a risk difference or rate difference) or in relative terms (as with a relative risk or odds ratio) (see Chapter 6).
    3. It should permit the investigators to determine when the risk factor and the disease occurred, in order to determine the temporal sequence of exposure and outcome.
    4. It should minimize biases, confounding, and other problems that would complicate interpretation of the data.

II. Types of Research Design
  A. **Observational designs for generating hypotheses**
    1. Cross-sectional surveys
      a. A **cross-sectional survey** is a survey of a population at a single point in time.
      b. Both **interview surveys** and **mass screening programs** are examples of cross-sectional surveys.
      c. **Advantages** of cross-sectional surveys include the following:
        (1) The studies are fairly quick and easy to perform.
        (2) The studies are useful for determining the prevalence of risk factors and the frequency of prevalent cases of disease for a defined population.
        (3) The studies are useful for assessing the *current* health status of a population.
        (4) The studies are of particular use in infectious disease epidemiology for establishing antibody levels indicative of past exposures and the degree of population immunity.
          (a) **Acute sera** are the first serum samples collected soon after symptoms of an infectious disease occur. **Convalescent sera** are the second serum samples, which are collected 10–28 days later. A significant increase in the serum titer of antibody indicates the specific infection to which exposure recently occurred.
          (b) Similarly, the use of IgG and IgM titers may distinguish between recent and remote exposures. **IgM titers** rise early after infectious exposure and then wane, while **IgG titers** rise later and usually persist.
      d. **Disadvantages** of cross-sectional surveys include the following:
        (1) Temporal relationships remain uncertain because data on exposures and outcomes are obtained simultaneously.
        (2) Indolent and chronic cases of illness are preferentially detected.
          (a) The **Neyman bias,** or **late look bias,** refers to the preferential detection of mild, slowly progressive cases of disease because such cases live longer and can be interviewed, whereas more severe cases of illness result in death and go undetected by the survey.
          (b) **Length bias** results when milder, more indolent cases are detected disproportionately in population screening programs; more aggressive cases have already resulted in death or in symptoms requiring medical intervention.
    2. Cross-sectional ecologic studies
      a. **Cross-sectional ecologic studies** relate the frequency with which some characteristic (e.g., smoking) and some outcome of interest (e.g.,

lung cancer) occur in the same geographically defined population.
  b. **Advantages** of cross-sectional ecologic studies include the following:
     (1) They are quick to perform.
     (2) They are inexpensive to perform.
     (3) Information about individual exposure need not be obtained.
     (4) Information about individual outcome need not be obtained.
  c. **Disadvantages** of cross-sectional ecologic studies include the following:
     (1) The **ecologic fallacy** is the use of ecologic data to draw inferences about causal relationships in individuals. Although the frequency of exposure and outcome is determined in the same *population,* no information regarding the occurrence of exposure and outcome in the same *individual* is provided by these studies.
     (2) Other disadvantages of this design are similar to those listed above under cross-sectional surveys.
 3. Longitudinal ecologic studies
  a. **Longitudinal ecologic studies** use ongoing surveillance or frequent cross-sectional studies to measure trends in disease rates over years in a defined population.
  b. The particular **advantage** of longitudinal ecologic studies is that they may reveal important associations over an extended period of time, such as the relationship between lung cancer and smoking.
  c. **Disadvantages** of this design include the following:
     (1) As with all ecologic studies, no information is provided about exposure and outcome in individuals.
     (2) Other factors that are changing over time can confound associations of interest.
B. **Observational designs for generating or testing hypotheses**
 1. Cohort studies
  a. A cohort is a clearly identified group to be studied.
  b. In a **prospective cohort study,** the investigator assembles the study groups in the present time on the basis of exposure, collects baseline data on them, and continues to collect data over time.
     (1) **Advantages** of prospective cohort studies include the following:
        (a) Data collection may be controlled by the investigator, so that outcome events may be confirmed.
        (b) True estimates of absolute risk may be obtained.
        (c) Many different disease outcomes may be studied.
     (2) **Disadvantages** of prospective cohort studies include the following:
        (a) The costs are high.
        (b) There is a long wait for results.
        (c) Only risk factors (exposures) that have been defined and measured at the beginning of the study can be assessed.
  c. In a **retrospective cohort study,** the investigator goes back into history to define a risk group, and follows the group members up to the present to see what outcomes have occurred.
     (1) **Advantages** of retrospective cohort studies include the following:
        (a) They are less costly than prospective cohort studies.
        (b) Results are available quickly.
     (2) The principal **disadvantage** of retrospective cohort studies relative to prospective studies is that the investigator loses the ability to control the criteria used in the determination of the outcome.
 2. Case-control studies
  a. In a **case-control study,** groups are defined on the basis of outcome, and then assessed for differences in past exposure to possible risk factors.
  b. These studies are essentially designed to compare the "risk of having the risk factor" in the two groups.
  c. **Advantages** of case-control studies include the following:
     (1) They are usually quick to perform.
     (2) They are relatively inexpensive to perform.
     (3) They are useful when the outcome in question is rare.
     (4) Many risk factors (exposures) may be considered.
  d. **Disadvantages** of case-control studies include the following:
     (1) There is a risk of recall bias (see Chapter 4).
     (2) There is difficulty in determining the proper control group.
     (3) The actual risk of the outcome cannot be determined but is estimated by the odds ratio (see Chapters 6 and 7).
C. **Experimental designs for testing hypotheses**
 1. **Randomized controlled clinical trials**
  a. In randomized controlled clinical trials (RCCT), subjects are randomly assigned to one of the following groups: (1) the intervention group, which will receive the experimental treatment, or (2) the control group, which will receive the nonexperimental treatment, consisting either of a placebo (inert substance) or a standard method of treatment.
  b. If neither the subject nor the investigator is made aware of the treatment assignment until the trial is terminated, the trial is a **double-blind trial,** or **double-blind study.**
  c. **Advantages** of RCCTs include the following:
     (1) They are considered the "gold standard" for studying interventions.
     (2) This design minimizes, but does not completely eliminate, bias.
  d. **Disadvantages** of RCCTs include the following:
     (1) They are lengthy and expensive to perform.
     (2) Ethical concerns arise regarding the allocation of subjects to inactive placebo groups.
     (3) There may be contamination of randomized groups by dropouts and therapy changes.
     (4) External validity may be limited if subjects who agree to participate differ in some way from the larger population.
 2. **Randomized controlled field trials**
  a. Randomized controlled field trials are similar to RCCTs but are designed to test a **preventive intervention** such as a vaccine.
  b. Advantages and disadvantages are the same as for RCCTs.

III. Hypothesis Generation and Hypothesis Testing Must Always Be Performed on Different Data Sets

QUESTIONS

DIRECTIONS. (Items 1–10): Each of the numbered items or incomplete statements in this section is followed by answers or by completions of the statement. Select the ONE lettered answer or completion that is BEST in each case. Correct answers and explanations are given at the end of the chapter.

1. The basic goal of epidemiologic research is to

    (A) eliminate all bias
    (B) maximize external validity
    (C) compare two groups that differ by exposure or outcome
    (D) establish causality
    (E) reject the null hypothesis

2. All of the following study designs are useful for hypothesis generation EXCEPT

    (A) randomized controlled field trials
    (B) retrospective cohort studies
    (C) longitudinal ecologic studies
    (D) case-control studies
    (E) cross-sectional surveys

3. As a member of a public health team, you have an interest in controlling measles infection through vaccination. To estimate the level of immunity in your study population and derive data useful for generating vaccination policy, you should

    (A) conduct a case-control study of measles infection
    (B) conduct a randomized trial of measles vaccination
    (C) conduct a retrospective cohort study of measles vaccination
    (D) conduct a cross-sectional survey of vaccination status
    (E) conduct an ecologic study of measles in the population

4. A recently published study (Shadick, et al. 1994) purported to show that a variety of symptoms were more common in subjects with a history of suboptimally treated Lyme disease than in controls. The data were obtained largely by a survey of the participants. Data from such a study might be subject to all of the following EXCEPT

    (A) selection bias
    (B) recall bias
    (C) measurement bias
    (D) late look bias
    (E) random error

5. Cross-sectional surveys are subject to the Neyman bias, or late look bias. This may be explained as

    (A) the tendency to detect only the late stages of a disease, when manifestations are more severe
    (B) the tendency to detect preferentially the more indolent cases of a disease
    (C) the tendency to detect only those cases of a disease that are asymptomatic
    (D) the tendency to detect preferentially fatal illness
    (E) the tendency to find more disease in older cohorts

6. In screening programs, an effect comparable to the late look bias is called

    (A) recall bias, because only severe illness is recalled
    (B) selection bias, because the program selects out severe cases
    (C) selection bias, because the program selects out asymptomatic illness
    (D) spectrum bias, because the cases are clustered at one end of the disease spectrum
    (E) length bias, because cases lasting longer are more apt to be detected

7. In assessing the extent of population exposure to an infectious agent, why is it useful to obtain both IgG and IgM titers?

    (A) It is useful to distinguish remote from recent exposure.
    (B) It is useful to distinguish vaccination from infection.
    (C) It is useful to distinguish cell-mediated from humoral immunity.
    (D) It is useful to establish the degree of immunity.
    (E) Usefulness is limited, as titers decline with time.

8. Which of the following is a measure of the risk of having a risk factor?

    (A) relative risk
    (B) risk ratio
    (C) p value
    (D) odds ratio
    (E) kappa

9. A case-control study may have a particular advantage over a cohort study when the disease in question is

    (A) indolent
    (B) rare
    (C) fatal
    (D) infectious
    (E) virulent

10. In a case-control study of myocardial infarction, which group of subjects would be a poor choice as controls?

    (A) subjects with similar age distribution as cases
    (B) subjects with similar sociodemographic characteristics as cases
    (C) subjects with similar cardiac risk factors as cases
    (D) subjects with no history of myocardial infarction
    (E) subjects admitted to the hospital for noncardiac disease

ANSWERS AND EXPLANATIONS

1. The answer is C: compare two groups that differ by exposure or outcome. In virtually all epidemiologic research, two or more groups of subjects are studied that differ either by exposure to risk factors or by outcome. The basic goal of such research is to compare the frequency of a hypothetically associated outcome or exposure between or among groups. For the study to be valid, bias and confounding must be minimized. *Epidemiology,* p. 66.

2. The answer is A: randomized controlled field trials. Randomized intervention trials represent the "gold standard" for hypothesis *testing.* However, these studies are costly in both time and money. Therefore, they are best reserved for testing hypotheses that already are supported by the results of prior studies of less rigorous design. Often, the randomized controlled trial is the final hurdle before a hypothesis is sufficiently supported to become incorporated into clinical practice. Cross-sectional surveys and longitudinal ecologic studies are among designs appropriate only for hypothesis generation; retrospective cohort studies and case-control studies may be used both to generate and to test hypotheses. Recall that it is not appropriate to generate and test hypotheses with the same set of data. *Epidemiology,* p. 72.

3. The answer is D: conduct a cross-sectional survey of vaccination status. The research question is the foundation for study design. What question is the study to answer? In this example, the question concerns the allocation of public health resources to prevent the spread of measles through vaccination. This question is itself founded on the answers to other questions, such as whether or not there is an effective vaccine to prevent measles. There is an effective measles vaccine when it is administered according to current guidelines (Centers for Disease Control 1991). Your study should answer the following questions: (1) Is there adequate immunity to measles in the population? (2) If immunity is not adequate, how should vaccination policy be directed to optimize protection of the community? A cross-sectional survey of vaccination status (or antibody titers, or both) would be the most expedient, cost-effective means to obtain these answers. Such studies are often useful in setting disease control priorities. *Epidemiology,* p. 66.

4. The answer is D: late look bias. In a sense, bias is important to science because belief, which is very much like bias, is required for a hypothesis to be formulated. In the testing of that hypothesis, however, bias must be diligently avoided. In a retrospective study such as the one described, subjects cannot be randomized as they can be in prospective studies. In almost any retrospective study, there is a risk of selection bias. In the quantification of symptoms, many of which are subjective, measurement bias is possible. Additionally, in any retrospective study in which subjects are surveyed, the category that defines their role in the study may influence recall of relevant events. Random error is possible in any study and is, in fact, never completely avoided. In the study cited in this question, the subjects with a history of Lyme disease might be more likely to recall symptoms suggestive of the disease or its late complications, than would the subjects with no such history, who might dismiss and not remember episodes of minor joint pain. Any form of bias, inadequately addressed, can virtually invalidate the findings of a study. The late look bias is only germane to cross-sectional surveys. It is the tendency of a survey to detect preferentially more indolent, longer-lasting illness rather than aggressive disease, which more rapidly leads to death. *Epidemiology,* p. 67.

5. The answer is B: the tendency to detect preferentially the more indolent cases of a disease. At any given moment, the prevalence of disease is influenced by both the incidence of that disease (the frequency with which new cases arise) and the duration. Diseases of long duration are more apt to accumulate in a population than are diseases that run a short course and result in either recovery or death. Even within the category of a particular illness, such as prostate cancer, the prevalence of the more indolent cases is likely to be higher, due to the accumulation of cases, than the prevalence of aggressive cases that result in a rapid demise. A cross-sectional survey will preferentially detect the indolent, or slowly progressive, cases of disease that tend to accumulate and miss the cases that have recently occurred but already resulted in death. This is the late look bias: one is looking too late to find aggressive cases that have already led to death. *Epidemiology,* p. 66.

6. The answer is E: length bias, because cases lasting longer are more apt to be detected. The intent of any disease screening program is to detect asymptomatic and unrecognized cases. A screening program will usually be prospective and, therefore, not subject to recall bias. Selection bias refers to the selection of subjects for participation, and screening programs may be biased in this way. This would limit external validity but is not a comparable effect to the late look bias. Spectrum bias refers to variability in the sensitivity or specificity, or both, of diagnostic tests with variations in the characteristics of the study population (see, e.g., Lachs, et al. 1992). Length bias, the tendency to detect preferentially long-lasting cases of subclinical illness whenever screening is conducted, is analogous to late look bias. Length bias occurs for the same reasons as late look bias, as discussed in the answer to item 4, above. *Epidemiology,* p. 67.

7. The answer is A: It is useful to distinguish remote from recent exposure. The timing of exposure to an infectious agent may be determined by obtaining acute and convalescent sera, or by measuring both IgG and IgM levels. IgM class antibodies rise early after infection and decline fairly rapidly (within weeks to months), whereas IgG levels rise more slowly and then decline gradually (within years). Antibody levels do not distinguish between wild type infection and vaccination. Antibody levels are a measure of humoral immunity only. The degree of immunity is determined only to a limited degree by the IgG level and does not require that IgM be quantified. The utility of the above method in determining the time of exposure is due to, rather than limited by, the decline in antibody levels over time. *Epidemiology,* p. 67.

8. The answer is D: odds ratio. The odds ratio, derived from a case-control study, indicates the relative frequency of a particular risk factor in those subjects with the outcome in question (cases) and in outcome-free controls. The outcome has already occurred; the risk for developing the outcome therefore cannot be measured directly.

What is measured is exposure, presumably exposure that preceded the outcome and represented a risk factor for it. The odds ratio may be considered the risk of having been exposed in the *past,* given the presence or absence of the outcome *now.* Therefore, the odds ratio, which approximates the risk ratio, or relative risk (the terms are interchangeable), when the disease is rare, may be considered the risk of having the risk factor. *Epidemiology,* p. 69.

9. The answer is B: rare. The groups in a case-control study are assembled on the basis of the outcome. If the outcome is rare, this design is particularly advantageous. Risk factors in a defined group with the rare outcome can be assessed and compared with those for a group without the outcome. If a cohort study is conducted and the outcome is rare, either too few cases will arise to permit meaningful interpretation of the data, or a very large sample size will be required, resulting in great, and often prohibitive, expense. *Epidemiology,* p. 69.

10. The answer is C: subjects with similar cardiac risk factors as cases. The goal of a case-control study is to determine differences in risk factors between groups with and without the outcome. If the two groups were matched on the basis of risk factors for the outcome in question, differences in these factors would be eliminated by design. Matching on the basis of known risk factors to isolate differences in as yet unestablished risk factors is often appropriate. However, if the cases and controls resemble one another too closely, there is a risk of over-matching, in which no differences are detectable and the study becomes useless. This is a particular hazard when known risk factors are correlated with as yet unrecognized risk factors. *Epidemiology,* p. 69.

REFERENCES CITED

Bartecchi C. E., T. D. MacKenzie, and R. W. Schrier. The human costs of tobacco use. New England Journal of Medicine 330:907–912, 1994.

Centers for Disease Control. Update on adult immunization: recommendations of the Immunization Practices Advisory Committee (ACIP). Morbidity and Mortality Weekly Report 40 (No.RR-12):19–22, 1991.

Daily Dietary Fat and Total Food-Energy Intakes—NHANES III, Phase 1, 1988–1991. Morbidity and Mortality Weekly Report 43:116–117, 123–125, 1994.

Jekel, J. F. Epidemiology, Biostatistics, and Preventive Medicine. Philadelphia, W. B. Saunders Company, 1996.

Lachs M. S., et al. Spectrum bias in the evaluation of diagnostic tests: lessons from the rapid dipstick test for urinary tract infection. Annals of Internal Medicine 117:135–140, 1992.

Physicians' Health Study Steering Committee. Final report on the aspirin component of the ongoing Physicians' Health Study. New England Journal of Medicine 321:129–135, 1989.

CHAPTER SIX

Assessment of Risk in Epidemiologic Studies

SYNOPSIS

OBJECTIVES
- To characterize the comparison of risk between groups in epidemiology.
- To distinguish between absolute and relative risk.
- To distinguish risk from odds.
- To describe various measures of the impact of exposures on a population.
- To discuss the application of risk measures to policy formulation and the counseling of patients.

I. Introduction
A. Causal research in epidemiology requires that two fundamental distinctions be made.
  1. The first distinction is between those who do have and those who do not have the risk factor under investigation (the **independent variable**).
  2. The second distinction is between those who do have and those who do not have the disease under investigation (the **dependent variable**).
  3. These distinctions are subject to both random errors and bias.
B. These dichotomies are commonly presented in the form of a **standard 2 × 2 table** (see Table 6–1).

II. Definition of Study Groups
A. In **cohort studies,** the frequency of disease in the exposed subjects is compared with the frequency of disease in the unexposed subjects. Measures of both absolute and relative risk may be derived in these studies.
B. In **case-control studies,** the frequency of the risk factor (exposure) in the cases (diseased) is compared with the frequency of the risk factor (exposure) in the controls (nondiseased). Only measures of relative risk may be derived in these studies.

III. Comparison of Risks in Different Study Groups
A. Absolute differences in risk
  1. A **risk difference (or rate difference)** is the risk (rate) in the exposed group minus the risk (rate) in the unexposed group.
  2. The risk difference is also known as the **attributable risk (AR):**

  $$AR = Risk_{(exposed)} - Risk_{(unexposed)}$$
  $$= [a/(a + b)] - [c/(c + d)]$$

  See Table 6–1 for definitions of $a, b, c,$ and d.
B. Relative differences in risk
  1. The **relative risk (RR),** or **risk ratio (RR),** is the ratio of the risk in the exposed group to the risk in the unexposed group:

  $$RR = Risk_{(exposed)}/Risk_{(unexposed)}$$
  $$= [a/(a + b)]/[c/(c + d)]$$

TABLE 6–1. Standard 2 × 2 Table for Demonstrating the Association between a Risk Factor and a Disease

		DISEASE STATUS		
		Present	Absent	Total
RISK FACTOR STATUS	Present	a	b	$a + b$
	Absent	c	d	$c + d$
	Total	$a + c$	$b + d$	$a + b + c + d$

Interpretation of the cells:

a = subjects with both the risk factor and the disease
b = subjects with the risk factor but not the disease
c = subjects with the disease but not the risk factor
d = subjects with neither the risk factor nor the disease

$a + b$ = all subjects with the risk factor
$c + d$ = all subjects without the risk factor
$a + c$ = all subjects with the disease
$b + d$ = all subjects without the disease

$a + b + c + d$ = all study subjects

BOX 6-1. Equations for Comparing Risks in Different Groups and Measuring the Impact of Risk Factors

(1) Risk difference = Attributable risk (AR)

$$= \text{Risk}_{(exposed)} - \text{Risk}_{(unexposed)}$$

$$= [a/(a + b)] - [c/(c + d)]$$

where a represents subjects with both the risk factor and the disease; b represents subjects with the risk factor but not the disease; c represents subjects with the disease but not the risk factor; and d represents subjects with neither the risk factor nor the disease

(2) Relative risk = Risk ratio (RR)

$$= \text{Risk}_{(exposed)}/\text{Risk}_{(unexposed)}$$

$$= [a/(a + b)]/[c/(c + d)]$$

(3) Odds ratio (OR) $= (a/c)/(b/d)$

$$= ad/bc$$

(4) Attributable risk percent in the exposed

$$= AR\%_{(exposed)}$$

$$= \frac{\text{Risk}_{(exposed)} - \text{Risk}_{(unexposed)}}{\text{Risk}_{(exposed)}} \times 100$$

$$= \frac{RR - 1}{RR} \times 100$$

$$\approx \frac{OR - 1}{OR} \times 100$$

(5) Population attributable risk (PAR) $= \text{Risk}_{(total)} - \text{Risk}_{(unexposed)}$

(6) Population attributable risk percent $= PAR\%$

$$= \frac{\text{Risk}_{(total)} - \text{Risk}_{(unexposed)}}{\text{Risk}_{(total)}} \times 100$$

$$= \frac{(Pe)(RR - 1)}{1 + (Pe)(RR - 1)} \times 100$$

where Pe stands for the effective proportion of the population exposed to the risk factor

---

TABLE 6-2. Measures of Smoking and Lung Cancer Deaths in Adult Males in the USA, 1986

Measure	Amount
*Lung cancer deaths among smokers	191 per 100,000 per year
*Lung cancer deaths among nonsmokers	8.7 per 100,000 per year
†Proportion exposed (Pe) to the risk factor (effective population of smokers, averaged over time)	35%, or 0.35
†Population risk of lung cancer death	72.5 per 100,000 per year
†Relative risk, or RR	22 [191/8.7 = 22]
Attributable risk, or AR	182.3 per 100,000 per year [191 − 8.7 = 182.3]
Attributable risk percent in the exposed, or AR%$_{(exposed)}$	95.4% [182.3/191 × 100 = 95.4]
Population attributable risk, or PAR	63.8 per 100,000 per year [72.5 − 8.7 = 63.8]
†Population attributable risk percent, or PAR%	88% [63.8/72.5 × 100 = 88]

* These rates were calculated from the data marked with a dagger.
† The source of data for measures shown with a dagger is as follows: Centers for Disease Control. Chronic disease reports: deaths from lung cancer—United States, 1986. Morbidity and Mortality Weekly Report 38:501–505, 1989.

A large relative risk that applies to a small segment of the population may actually produce few excess deaths or cases of disease, whereas a small relative risk that applies to a large segment of the population may produce many excess deaths or cases of disease.

2. The **odds ratio (OR)** is the odds of exposure in the diseased group divided by the odds of exposure in the nondiseased group:

$$OR = (a/c)/(b/d)$$
$$= ad/bc$$

When a and c are small relative to b and d (i.e., when the disease is rare, in < 5% of the population), the odds ratio is a good estimate of the risk ratio.

C. Which side is up in the risk ratio and odds ratio?

1. By convention, for both the risk ratio and the odds ratio, the exposed group is usually placed in the numerator.
2. Mathematically, either the exposed group or the unexposed group may be placed in the numerator.

D. Measures of attributable risk

1. There are three commonly employed measures of the impact of exposure: attributable risk percent in the exposed, population attributable risk, and population attributable risk percent. The formulas for these are shown in Box 6–1.
2. The **attributable risk percent in the exposed,** or AR%$_{(exposed)}$, answers this question: among those with the risk factor, what percentage of the total risk for the disease is due to the risk factor? As shown in Box 6–1, there are several methods of calculation. One is based on absolute differences in risk. Another

is based on relative differences in risk (RR). In addition, the odds ratio (OR) from a case-control study may be substituted for the RR in the formula for the relative $AR\%_{(exposed)}$.

3. The **population attributable risk (PAR)** answers this question: in the general population, how much of the total risk for the disease of interest is due to the risk factor (exposure) of interest? The PAR is the risk in the total population minus the risk in the unexposed group.

4. The **population attributable risk percent (PAR%)** answers this question: among the general population, what percentage of the total risk for the disease of interest is due to the risk factor (exposure) of interest? As shown in Box 6–1, the PAR% can be calculated in two ways. If the risk ratio (RR) is used, the proportion of the total population exposed (Pe) must be entered into the formula.

IV. Uses of Risk Assessment Data
A. Application of attributable risk to policy analysis
1. By using the PAR%, the impact of a preventive policy that reduces the percentage of the population exposed or the degree of exposure among those exposed can be derived.
2. Once the costs and expected benefits of an intervention are known, a financial analysis can be conducted using cost-benefit or cost-effectiveness models (see Chapter 14) to determine the optimal (or politically acceptable) allocation of resources.

B. Application of attributable risk to the counseling of patients
1. Attributable risk may be used in the counseling of individual patients to explain how risk will vary with modification of exposure.
2. See Table 6–2 for various ways of expressing the same risk of lung cancer death associated with cigarette smoking.

QUESTIONS

DIRECTIONS. (Items 1–10): Each of the numbered items or incomplete statements in this section is followed by answers or by completions of the statement. Select the ONE lettered answer or completion that is BEST in each case. Correct answers and explanations are given at the end of the chapter.

1. A team of researchers hypothesize that watching "Barney" might lead to epilepsy in childhood. Children with and without epilepsy are compared on the basis of hours spent watching "Barney." Which one of the following statements is a true statement regarding this study?

   (A) Risk factor status is the basis for comparison.
   (B) The risk ratio cannot be calculated directly.
   (C) Absolute and relative measures of risk can be derived.
   (D) The temporal association between exposure and outcome can be established with certainty.
   (E) The use of healthy controls ensures external validity.

2. The researchers in question 1 do not find statistically significant evidence that "Barney" produces epilepsy in childhood. Unwilling, however, to relinquish their line of reasoning, they hypothesize that the parents of children who watch "Barney" are more likely to develop migraine headaches. They assemble two groups of parents: one with children who watch "Barney," and one with children who destroy furniture instead of watching "Barney." All of the following statements regarding this scenario are true EXCEPT

   (A) the relative risk may be calculated directly
   (B) risk factor status is the basis for comparison
   (C) confounders must be controlled to ensure internal validity
   (D) additional risk factors can be assessed as the study progresses
   (E) the temporal association between exposure and outcome should be established

Items 3–7

A study is conducted to determine the effects of drinking Mountain Dew on a teenager's willingness to bungee jump from frightful heights. A total of 500 teenagers is assembled on the basis of bungee-jumping status: 250 are jumpers, and 250 are not. Of the 250 jumpers, 150 report drinking Mountain Dew. Of the nonjumpers, 50 report drinking Mountain Dew. A majority of the nonjumpers report a preference for warm milk.

3. Which of the following is true?

   (A) This is a cohort study.
   (B) The absolute and relative risk of bungee jumping may be determined from this study.
   (C) Jumpers and nonjumpers should be matched for beverage consumption.
   (D) This study can be used to calculate an odds ratio.
   (E) Unanticipated outcomes can be assessed in this study.

4. The absolute difference in the risk of jumping

   (A) is 100
   (B) is 3:1
   (C) is 150
   (D) cannot be calculated, because jumping is the dependent variable
   (E) cannot be calculated, because this is a case-control study

5. The odds ratio calculated from this study will give the odds of

   (A) drinking among nonjumpers to drinking among jumpers
   (B) jumping among drinkers to drinking among jumpers
   (C) nonjumping among drinkers to nondrinking among jumpers
   (D) drinking among jumpers to drinking among nonjumpers
   (E) drinking Mountain Dew to jumping

6. The odds ratio in the above study is
   (A) 2
   (B) 0.2
   (C) 6
   (D) 5
   (E) 0.6

7. The results of this study indicate that
   (A) bungee jumping influences one's choice of beverage
   (B) one's choice of beverage influences one's tendency to bungee jump
   (C) bungee jumping and choice of beverage are associated
   (D) bungee jumping and beverage choice are causally related
   (E) there is no association between warm milk and bungee jumping

Items 8–9

Having obtained your master's degree in public health, not to mention a lifetime supply of Mountain Dew and your own initialized bungee cord, from conducting the study described in items 4–8, above, you decide to pursue a PhD, with further investigation of this provocative subject. This time, you assemble two groups of 250 adolescents based on past history of Mountain Dew consumption and prospectively determine the rate of bungee jumping. You exclude subjects with a prior history of jumping. Over a 5-year period, 135 of the exposed group and 38 of the unexposed group engage in jumping.

8. The relative risk of bungee jumping among the exposed is
   (A) 3.6
   (B) 115
   (C) 2.12
   (D) 6
   (E) 4.8

9. Among drinkers of Mountain Dew, what percentage of the total risk for jumping is due to consumption of Mountain Dew?
   (A) 36%
   (B) 72%
   (C) 2.6%
   (D) 0.36%
   (E) 0%

10. The risk of acquiring infection beta is 312 per 1000 among the unvaccinated and 7.2 per 1000 among the vaccinated. Approximately 80% of the population is exposed to the pathogen every year. As a basis for policy development, you would do all of the following EXCEPT
    (A) consider the clinical significance (severity) of infection
    (B) consider the cost of universal vaccination
    (C) report that vaccination reduces the population risk by 358%
    (D) report that the PAR% is 27
    (E) report a risk of −304.8 per 1000 attributable to vaccination

ANSWERS AND EXPLANATIONS

1. The answer is B: The risk ratio cannot be calculated directly. This is a case-control study. Cases are children with epilepsy, and controls are (or should be) children of similar characteristics but without epilepsy. The use of healthy controls does not ensure that the results will pertain to the general population, and external validity remains uncertain. The difference in the rate of exposure to a putative risk factor ("Barney") is the basis for comparison. In a case-control study, the risk ratio cannot be calculated directly but must be estimated from the odds ratio. The temporal relationship between exposure and outcome is usually not known with certainty in this type of study; both exposure and disease have already occurred at the time of enrollment. Only relative measures of risk may be calculated based on a case-control study. A cohort study is required to obtain absolute risk. *Epidemiology*, p. 75.

2. The answer is D: additional risk factors can be assessed as the study progresses. This is a cohort study. The two groups are assembled on the basis of exposure status (i.e., risk factor status), and then followed for, and compared on the basis of, disease status. Only the risk factors, or exposures, identified at study entry can be assessed as the study progresses. The development of migraine headaches in this case is disease. The temporal association between exposure and outcome should be established to prevent bias. The development of migraine prior to the defined exposure would bias the study, and such applicants should be excluded. In all epidemiologic studies, a thorough attempt must be made to control confounders. The relative risk (of migraine, in this case) may be calculated directly in a cohort study. *Epidemiology*, p. 75.

3. The answer is D: This study can be used to calculate an odds ratio. The study described is a case-control study. Cases are teenagers with a history of bungee jumping, and controls are (or should be) teenagers with generally similar characteristics but no history of bungee jumping. The cases and controls should not be matched for beverage consumption, as this is the exposure of interest. Overmatching is the result when cases and controls are matched on the basis of some characteristic or behavior that is highly correlated with the exposure of interest. Overmatching precludes the detection of a difference in exposure, even if one truly exists. The odds ratio is calculated from a case-control study, and used to estimate relative risk. Absolute risk cannot be determined from a case-control study. In a

case-control study, the outcome defines the group at study entry; while unanticipated risk factors can be assessed, only the outcome variables chosen as the basis for subject selection can be evaluated. *Epidemiology,* p. 77.

4. The answer is E: cannot be calculated, because this is a case-control study. Risk can only be estimated from a case-control study; it cannot be calculated directly. In particular, relative risk can be estimated by the odds ratio; absolute risk cannot be determined from a case-control study, because the groups do not represent the population from which they were drawn. In this study, bungee jumping is the *independent* variable, as the groups were assembled on the basis of their jumping status. The odds to be calculated are the odds of drinking Mountain Dew, given one's history of jumping or not jumping. *Epidemiology,* p. 75.

5. The answer is D: drinking among jumpers to drinking among nonjumpers. The odds ratio in a case-control study is the odds of the exposure in cases relative to controls. The exposure in this study is drinking (specifically, drinking Mountain Dew). Cases are those who bungee jump, while controls are those who do not. The odds of drinking Mountain Dew among cases relative to controls is the outcome of interest. As noted in the text, the odds ratio is essentially a measure of the "risk of having the risk factor." *Epidemiology,* p. 77.

6. The answer is C: 6. As described in the text, the formula for the odds ratio (OR) is $(a/c)/(b/d)$. This is algebraically equivalent to ad/bc, where a is cases with the exposure, b is controls with the exposure, c is cases without the exposure, and d is controls without the exposure. Displaying the data from the above study in a 2 × 2 table is helpful:

		BUNGEE JUMPING	
		Positive	Negative
MOUNTAIN DEW DRINKING	Positive	150 *(a)*	50 *(b)*
	Negative	100 *(c)*	200 *(d)*

OR = (150 × 200)/(50 × 100) = 30,000/5000 = 6. By convention, the odds ratio is expressed with the exposure rate among controls in the numerator. If one were interested in reporting the odds of drinking Mountain Dew in nonjumpers relative to jumpers, the odds ratio could be expressed as 1/6, or 0.17. *Epidemiology,* p. 77.

7. The answer is C: bungee jumping and choice of beverage are associated. The odds of drinking Mountain Dew are 6 times greater among cases than controls in this study. There is, therefore, an apparent association between beverage choice and tendency to jump. There may, in fact, be no true relationship if the findings are due to chance (i.e., not statistically significant), if the study is biased (e.g., if cases are interviewed more thoroughly or differently than controls with regard to drinking history), or if the apparent association is confounded by an unmeasured variable (e.g., alcohol consumption). If the apparent association is true, one still cannot be certain whether bungee jumping predisposes to Mountain Dew consumption, or vice versa, because the temporal sequence of exposure and outcome in a case-control study is uncertain. An association between Mountain Dew and bungee jumping is demonstrated in this study; to reach further conclusions regarding the association would require further study. *Epidemiology,* p. 75.

8. The answer is A: 3.6. The formula for relative risk is

$$RR = [a/(a + b)]/[c/(c + d)]$$

where a, b, c, and d are as defined in the explanation of item 6, above. The 2 × 2 table for this study is as follows:

		BUNGEE JUMPING	
		Positive	Negative
MOUNTAIN DEW DRINKING	Positive	135 *(a)*	115 *(b)*
	Negative	38 *(c)*	212 *(d)*

Note that cells b and d had to be calculated so that the total in each row would be 250. The risk ratio, or relative risk, is

$$RR = [135/(135 + 115)]/[38/(38 + 212)]$$
$$= 0.54/0.152$$
$$= 3.55$$
$$= 3.6$$

Epidemiology, p. 76.

9. The answer is B: 72%. The percentage of total risk due to an exposure among those with the exposure is the attributable risk percent in the exposed, or $AR\%_{(exposed)}$. This can be calculated using the risk ratio, which, in this case, is 3.6. Therefore,

$$AR\%_{(exposed)} = \frac{RR - 1}{RR} \times 100$$
$$= \frac{3.6 - 1}{3.6} \times 100$$
$$= 72.2\%$$

This indicates that among those who drink Mountain Dew, nearly 3/4 of the total risk for bungee jumping is attributable to the beverage. (**N.B.** These data are fictitious. But then again . . .) *Epidemiology,* p. 78.

10. The answer is D: report that the PAR% is 27. All of the choices are correct with the exception of option D. The PAR% in this case is −358%, as provided in option C. With exposure to a preventive measure, such as a vaccine in this case, risk is reduced among the exposed. The risk of the outcome is therefore less among the exposed than among the unexposed, and the attributable risk is negative.

$$PAR\% = \frac{(Pe)(RR - 1)}{1 + (Pe)(RR - 1)} \times 100$$

where Pe is the proportion of the population exposed, and RR is the relative risk. Pe is given as 80% = 0.8. By convention, RR is expressed with the exposed in the numerator. Therefore

$$RR = (7.2/1000)/(312/1000)$$
$$= 0.023$$

$$PAR\% = \frac{(0.8)(0.023 - 1)}{1 + (0.8)(0.023 - 1)} \times 100$$
$$= \frac{-0.7816}{0.284} \times 100$$
$$= \sim -358$$

Therefore, option C (report that vaccination reduces the population risk by 358%) is correct. Option E (report a risk of −304.8 per 1000 attributable to vaccination) is correct, although the attributable risk is of limited utility by itself in policy development, as it fails to consider the level of exposure in the population. The AR is as follows:

$$AR = Risk_{(exposed)} - Risk_{(unexposed)}$$
$$= (7.2/1000) - (312/1000)$$
$$= -304.8/1000$$

This measure is also negative, as it reflects a reduced level of risk resulting from exposure to the vaccine. Option A [consider the clinical significance (severity) of infection] and option B (consider the cost of universal vaccination) are relevant any time a vaccine program is under consideration. The disease, if very mild, may not warrant vaccination. This consideration delayed recent approval of the varicella vaccine (Halloran et al. 1994).

Financial constraints must always be considered. Even an effective vaccine is unlikely to be used if it is prohibitively expensive. *Epidemiology,* p. 79.

REFERENCES CITED

Fletcher A. E. and C. J. Bulpitt. How far should blood pressure be lowered? New England Journal of Medicine 326:251–254, 1992.

Halloran M. E., et al. Theoretical epidemiologic and morbidity effects of routine varicella immunization of preschool children in the United States. American Journal of Epidemiology 140:81–104, 1994.

Jekel, J. F. Epidemiology, Biostatistics, and Preventive Medicine. Philadelphia, W. B. Saunders Company, 1996.

Joint National Committee on Detection, Evaluation, and Treatment of High Blood Pressure. The Fifth Report of the Joint National Committee on Detection, Evaluation, and Treatment of High Blood Pressure (JNC V). Archives of Internal Medicine 153:154–183, 1993.

McGinnis J. M. and W. H. Foege. Actual causes of death in the United States. Journal of the American Medical Association 270:2207–2212, 1993.

SECTION II
BIOSTATISTICS

CHAPTER SEVEN

Understanding and Reducing Errors in Clinical Medicine

SYNOPSIS

OBJECTIVES
- To characterize sources of error in clinical data.
- To describe the basic goals of statistical methods in medicine.
- To define type I and type II errors.
- To define sensitivity and specificity.
- To define the predictive value of a diagnostic test.
- To explain the use of likelihood ratios and related measures.
- To explain the application of receiver operating characteristic curves to biostatistics.
- To discuss the measurement of agreement.

I. Goals of Data Collection and Analysis
  A. **Errors in medicine** do occur and are difficult to eliminate.
  B. **Statistical methods**
    1. Statistical methods are employed to compensate for unavoidable errors in clinical data.
    2. Specifically, statistical methods are used for the following reasons:
      a. They measure and explain overall variation in data, some of which is due to error.
      b. They distinguish between meaningful and random variation in outcome measures.
      c. They facilitate interpretation of clinical data.
        (1) They characterize the accuracy of clinical data.
        (2) They characterize the precision of clinical data.
      d. They measure and compensate for intraobserver and interobserver variability.
  C. **The goals of data collection and analysis** are as follows:
    1. One goal is to promote accuracy and precision.
    2. A second goal is to reduce differential and nondifferential errors.
    3. A third goal is to reduce intraobserver and interobserver variability.

II. Definitions
  A. **Random error**
    1. In random error **(nondifferential error)**, some measurements are too high and some are too low.
    2. If there are enough observations, data with random error can still provide an approximately correct estimate of the mean.
    3. Random error may be characterized by poor accuracy or poor precision, or both.
      a. **Accuracy** is the ability of a measurement to be correct, on the average.
      b. **Precision (reproducibility or reliability)** is the ability of a measurement to give the same result or a very similar result with repeated measurements of the same thing.
    4. Random error can, but usually does not, introduce bias into the results of an analysis.
  B. **Differential error**
    1. Differential error is a nonrandom, systematic, or consistent error in which the values tend to be inaccurate in a particular direction.
    2. An example of differential error is measuring the heights of subjects with their shoes on.
    3. Differential error inevitably results in **bias.**

III. Studying the Accuracy and Usefulness of Screening and Diagnostic Tests
  A. **False-positive and false-negative results**
    1. **Type I error** occurs when data lead one to conclude that something is true when in reality it is not true (e.g., a positive diagnostic test result in a patient in whom a disease is truly absent).
      a. Type I error is also known as **false-positive error.**
      b. Type I error is also known as **alpha error.**
    2. **Type II error** occurs when something is said to be false when in reality it is true (e.g., a negative test

result in a patient who actually has the disease in question).
  a. Type II error is also known as **false-negative error.**
  b. Type II error is also known as **beta error.**
3. Type I and type II error can be influenced by the **stage of disease.**
4. Type I and type II error can be influenced by the **spectrum of disease** in the population under investigation.
5. The magnitude of type I and type II error is largely determined by the **cutoff point** for a test result.
  a. If the cutoff point for the upper limit of normal for a test is set relatively *low*, there will be many false-positive results (a lot of type I error) but few false-negative results (little type II error).
  b. If the cutoff point for the upper limit of normal for a test is set relatively *high*, there will be many false-negative results (a lot of type II error) but few false-positive results (little type I error).
  c. In general, any effort to reduce type I error will increase type II error, and vice versa.
  d. Clinical criteria are used to determine the optimal cutoff point for a particular test under particular circumstances.

B. Sensitivity and specificity
1. **Sensitivity** is the ability of a test to detect a disease when it is present.
  a. Table 7–1 will be used to illustrate sensitivity.
    (1) The first column in Table 7–1 represents all of the diseased subjects, consisting of those with true-positive results (a) and those with false-negative results (c).
    (2) The second column represents all of the nondiseased subjects, consisting of those with false-positive results (b) and those with true-negative results (d).
    (3) When the total in the first column is divided by the total of all of the subjects studied, the result represents the prevalence of the disease in the study population.
  b. The formula for sensitivity is $a/(a + c)$.
  c. The limits of a test's sensitivity are represented by false-negative results, which appear in cell c.
  d. The **false-negative error rate** is $c/(a + c)$.
  e. The sensitivity (cases detected by the test) and the false-negative error rate (cases missed by the test) add up to 1 (100%).
  f. Algebraically, this is shown as follows:

$$[a/(a + c)] + [c/(a + c)] = 1$$

2. **Specificity** is the ability of a test to indicate nondisease when no disease is present.
  a. The formula for specificity is $d/(b + d)$.
  b. The limits of a test's specificity are represented by false-positive results, which appear in cell b.
  c. The **false-positive error rate** is $b/(b + d)$.
  d. The specificity (noncases correctly identified by the test) and the false-positive error rate (noncases incorrectly identified as having the condition) add up to 1 (100%).
  e. Algebraically, this is shown as follows:

$$[d/(b + d)] + [b/(b + d)] = 1$$

TABLE 7–1. Standard 2 × 2 Table Comparing the Test Results and the True Disease Status of the Subjects Tested

		TRUE DISEASE STATUS		
		Diseased	Nondiseased	Total
TEST RESULT	Positive	a	b	a + b
	Negative	c	d	c + d
	Total	a + c	b + d	a + b + c + d

Interpretation of the cells is as follows:

a = subjects with a true-positive test result
b = subjects with a false-positive test result
c = subjects with a false-negative test result
d = subjects with a true-negative test result

$a + b$ = all subjects with a positive test result
$c + d$ = all subjects with a negative test result
$a + c$ = all subjects with the disease
$b + d$ = all subjects without the disease

$a + b + c + d$ = all study subjects

Formulas are as follows:

$a/(a + c)$ = sensitivity

$d/(b + d)$ = specificity

$b/(b + d)$ = false-positive error rate (alpha error rate, type I rate)

$c/(a + c)$ = false-negative error rate (beta error rate, type II rate)

$a/(a + b)$ = positive predictive value

$d/(c + d)$ = negative predictive value

$[a/(a + c)]/[b/(b + d)] = (a/b)/[(a + c)/(b + d)]$ = likelihood ratio positive (LR+)

$[c/(a + c)]/[d/(b + d)] = (c/d)/[(a + c)/(b + d)]$ = likelihood ratio negative (LR−)

$(a + c)/(a + b + c + d)$ = prevalence

3. Sensitivity and specificity are independent of disease **prevalence**, which is the total number of subjects with disease $(a + c)$ divided by the total number of persons studied $(a + b + c + d)$.
  a. The formula for sensitivity includes only subjects with the disease.
  b. The formula for specificity includes only subjects without the disease.

C. Predictive values
1. Sensitivity and specificity do not directly answer two important clinical questions:
  a. If a patient's test result is positive, what is the probability that he or she has the disease being tested?
  b. If the result is negative, what is the probability that the patient does not have the disease?
2. These questions, which are influenced by both sensitivity and specificity, can be answered by following a different direction of analysis: predictive values.
3. The **positive predictive value** indicates what proportion of the subjects who have positive test results have the disease.

a. The formula for positive predictive value is $a/(a + b)$.
b. The denominator in the formula for positive predictive value includes both subjects with and without disease; therefore, *positive predictive value is influenced by prevalence.*
4. The **negative predictive value** indicates what proportion of subjects with negative test results are truly free of the disease.
a. The formula for negative predictive value is $d/(c + d)$.
b. The denominator in the formula for negative predictive value includes both subjects with and without disease; therefore, *negative predictive value is influenced by prevalence.*

D. **Screening**
1. When prevalence is low, even a highly specific test will tend to generate more false-positive results than true-positive results.
2. In community screening, a highly sensitive **screening test** should be used to "rule out" those who do not require additional testing for the disease. The combination of low prevalence and a highly sensitive test results in a very high negative predictive value.
3. Given the preponderance of false-positive results when prevalence is low, a positive result from an initial screening must be verified with additional highly specific testing. A **confirmatory test** is used to "rule in" a diagnosis and should have a high degree of specificity.
4. A **pathognomonic test** is a test that elicits a reaction that is synonymous with having the disease.

E. **Likelihood ratios, odds ratios, and cutoff points**
1. The **likelihood ratio positive (LR+)** is the ratio of the sensitivity of a test to the false-positive error rate of the test.
a. The formula for likelihood ratio positive is as follows:

$$LR+ = [a/(a + c)]/[b/(b + d)]$$

b. The *larger* the LR+, the better the test.
2. The **likelihood ratio negative (LR−)** is the ratio of the false-negative error rate divided by the specificity.
a. The formula for likelihood ratio negative is as follows:

$$LR- = [c/(a + c)]/[d/(b + d)]$$

b. The *smaller* the LR−, the better the test.
3. Likelihood ratios are independent of prevalence.
4. The LR+ is useful in determining the odds of disease after performance of a diagnostic test.
a. The **pretest odds** of disease is the ratio of diseased people to nondiseased people, or $(a + c)/(b + d)$.
b. The **posttest odds** of disease is the pretest odds multiplied by the LR+.

IV. **Receiver Operating Characteristic (ROC) Curves**
A. ROC curves may be used to determine the optimal cutoff point for a test (see Figs. 7–1 and 7–2).
1. The *y*-axis shows the sensitivity of a test.
2. The *x*-axis shows the false-positive error rate.

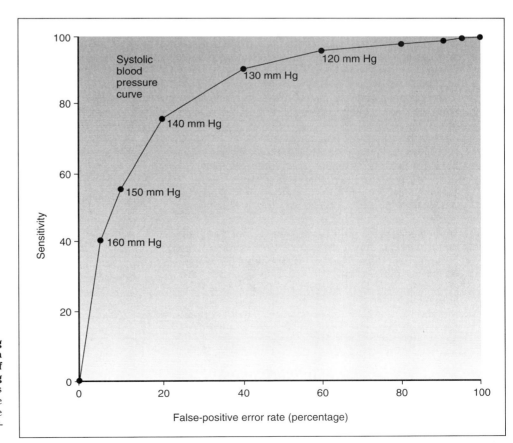

FIGURE 7-1. **Receiver operating characteristic (ROC) curve from a study to determine the best cutoff point for a blood pressure screening program (fictitious data).** Numbers beside the points on the curve are the cutoffs of systolic blood pressure that gave the corresponding sensitivity and false-positive error rate.

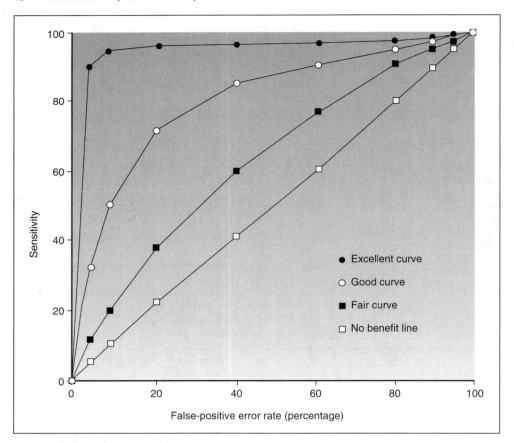

FIGURE 7-2. **Examples of receiver operating characteristic (ROC) curves for four tests.** The uppermost curve is the best of the four.

3. The point closest to the upper left-hand corner of an ROC curve (100% sensitivity and zero false-positive results) represents the best cutoff point for a particular test (see Figs. 7-1 and 7-2).

B. ROC curves may be compared statistically by measuring the area under each curve.

V. Measuring Agreement

A. Observer variability versus agreement

1. An important question both in clinical medicine and in research is the extent to which different observations of the same phenomenon differ.
2. **Intraobserver variability** is a measure of inconsistency in repeated assessments by a single observer.
3. **Interobserver variability** is a measure of disagreement between or among different observers.
4. If there is **intraobserver agreement** or **interobserver agreement,** the data in a study are considered highly reliable and will elicit more confidence.
5. Reliability is not proof of validity; two observers can report the same readings (i.e., show reliability) but be wrong.

B. Overall percent agreement

1. If a test uses a dichotomous variable (i.e., two categories of results, such as positive and negative), the results can be placed in a standard 2 × 2 table (Table 7-2), so that observer agreement can be calculated.
2. In a 2 × 2 table, cells a and d represent agreement, whereas cells b and c represent disagreement.
3. A common way to measure agreement is to calculate the **overall percent agreement,** which is the percentage of the total observations found in cells a and d of the 2 × 2 table.
4. Merely reporting the overall percent agreement is considered inadequate for a number of reasons.
   a. The overall percent agreement does not tell the prevalence of the finding in the subjects studied.
   b. It does not tell how the disagreements occurred: were the positive and negative results evenly distributed between the two observers, or did one observer consistently find more positive results than the other?
   c. Considerable agreement would be expected by chance alone, and the overall percent agreement does not tell the extent to which the agreement improves on chance.

C. Kappa test ratio

1. The kappa test is a measure of the extent to which agreement between two observers improves on chance agreement alone.
2. For any given cell in the 2 × 2 table, the expected number of observations is the row total, multiplied by the column total, with the product divided by the total observations. See Table 7-2 for formulas.
3. The **observed agreement** (A_o) is the sum of the actual number of observations in cells a and d.
4. The **agreement expected by chance** (A_c) is the sum of the expected number of observations in cells a and d.

TABLE 7-2. Standard 2 × 2 Table Comparing the Test Results Reported by Two Observers

		OBSERVER No. 1		
		Positive	Negative	Total
OBSERVER No. 2	Positive	a	b	a + b
	Negative	c	d	c + d
	Total	a + c	b + d	a + b + c + d

Interpretation of the cells is as follows:

a = positive/positive observer agreement
b = negative/positive observer disagreement
c = positive/negative observer disagreement
d = negative/negative observer agreement

Formulas are as follows:

$a + d$ = observed agreement (A_o)

$a + b + c + d$ = maximum possible agreement (N)

$(a + d)/(a + b + c + d)$ = overall percent agreement

$[(a + b)(a + c)]/(a + b + c + d)$ = cell a agreement expected by chance

$[(c + d)(b + d)]/(a + b + c + d)$ = cell d agreement expected by chance

cell a agreement expected by chance + cell d agreement expected by chance = total agreement expected by chance (A_c)

$(A_o - A_c)/(N - A_c)$ = kappa

5. The **maximum possible agreement** is the total number of observations (N).
6. The formula for kappa is as follows:

$$\text{kappa} = (A_o - A_c)/(N - A_c)$$

7. Although the kappa test provides valuable data on observer agreement, two important points should be noted:
   a. While many studies of observer variability involve **dichotomous data** (in which there are two categories of results, such as positive and negative), some studies involve **ordinal data** (in which there are three or more categories of results, such as negative, suspicious, and probable). For ordinal data, a **weighted kappa test** must be used; this test is similar in principle to the unweighted test but is somewhat more complex.
   b. In evaluating the accuracy and usefulness of a laboratory assay, imaging procedure, or any other clinical test, comparing the findings of one observer with those of another observer is not as useful as comparing the findings of an observer with the true status of the disease in the patients being tested. The true disease status, which is used to determine the sensitivity and specificity of tests, is considered to be the "gold standard," and its use is preferable whenever data concerning the true status are available. Even the "gold standard" can contain errors and create the incorrect appearance of considerable error in a test.

QUESTIONS

DIRECTIONS. (Items 1–10): For each numbered item, select the ONE lettered option that is most closely associated with it. To avoid spending too much time on matching sets with large numbers of options, it is generally advisable to begin each set by reading the list of options. Then, for each item in the set, try to generate the correct answer and locate it in the option list, rather than evaluating each option individually. Each lettered option may be selected once, more than once, or not at all. Correct answers and explanations are given at the end of the chapter.

Items 1–10

(A) Accuracy
(B) Precision
(C) Bias
(D) Random error
(E) Alpha error
(F) Beta error
(G) Cutoff point
(H) Sensitivity
(I) Specificity
(J) False-positive error rate
(K) False-negative error rate
(L) Positive predictive value

Match each of the operating characteristics of an epidemiologic study or a diagnostic test with the appropriate description.

1. $c/(a + c)$
2. Ability of a test to detect a disease when it is present
3. Type I error
4. Defines normal and abnormal test results
5. Tendency of a measure to be correct on average
6. $a/(a + c)$
7. Ability to exclude disease when it is absent
8. Nondifferential error
9. The closer to the upper left-hand corner of an ROC curve, the better
10. Differential error

ANSWERS AND EXPLANATIONS

1. The answer is K: false-negative error rate. In a 2 × 2 table, cell c represents false-negative results, or cases in which the condition is actually present but the test result is negative. The sum of cells a and c is all true-positive cases, or the denominator (population) from which false-negative results are derived. The false-negative error rate is the ratio of false-negative results to all cases of the condition, both those detected and those missed by the diagnostic test. *Epidemiology*, p. 90.
2. The answer is H: sensitivity. Sensitivity is defined as the ability of a test to detect true cases of a disease. Numerically, sensitivity is true cases of the disease or condition detected by the test (cell a in a 2 × 2 table), divided by all true cases (cell a plus cell c). *Epidemiology*, p. 89.
3. The answer is E: alpha error. False-positive error is also known as type I error or alpha error. These interchangeable designations refer to situations in which the data indicate that a result (or study hypothesis) is true when it actually is not true. Typically, results attributable to chance or random error account for alpha error, although alpha error may result from bias as well. The value of alpha used in hypothesis testing specifies the level at which statistical significance is defined, and thereby specifies the cutoff for results reported as positive or negative (see Chapter 10). The more stringent the value assigned to alpha (the smaller alpha is) the less likely it is that a false-positive result will occur. Of note, the clinical situation often dictates basic criteria for setting the alpha level. In situations where any new therapy is desperately sought for severely ill patients, a false-positive study of a new drug might be more tolerable than a study of a preventive intervention to be applied to a healthy population. Under most circumstances, the conventional level of alpha employed is 0.05 (see Chapter 10). *Epidemiology*, p. 87.
4. The answer is G: cutoff point. With the exception of tests that are pathognomonic, in which the result is dichotomous and defines the presence or absence of the disease in question, most tests produce results that do not indicate the presence or absence of disease with absolute certainty. The cutoff point indicates the value above or below which a test result is considered abnormal, either leading to a diagnosis or a need for further testing. The cutoff point chosen is influenced by the situational priorities. An initial screening test, for example, should be highly sensitive, and the cutoff point for the upper limit of normal values should be set low. A follow-up test should be highly specific, and the upper limit cutoff point should be set high. *Epidemiology*, p. 89.
5. The answer is A: accuracy. Accuracy is the ability to obtain a test or study result close to the true value. Accuracy is to be distinguished from precision, which is consistency or the ability to obtain reproducible results. An accurate result need not be precise. For example, repeated blood pressure measurements in a group of patients might lead to a correct value of the mean for the group, even if wide variation occurred in individual measurement due to poor technique. A precise result need not be accurate. The same mean weight might be obtained for a group of subjects on consecutive days, but if each measure were obtained with the subjects clothed, all of the results would be erroneously high. Both accuracy and precision are desirable traits in research and diagnostic studies. *Epidemiology*, p. 86.
6. The answer is H: sensitivity. Please refer to the explanation for question 2, above.
7. The answer is I: specificity. Specificity is the capacity of a test to limit positive results only to true cases of the condition in question. Alternatively stated, specificity is the ability of a test to exclude disease when it is absent. A test that is highly specific is a good "rule-in test," as it reliably indicates the presence of the disease when the result is positive. Numerically, specificity is the true-negative test results (cell d in a 2 × 2 table), divided by all negatives (cell b plus cell d). *Epidemiology*, p. 90.
8. The answer is D: random error. Random error is nondifferential because it does not distort data consistently in any one direction. Random error may produce some measurements that are too high and others that are too low. The mean may or may not be distorted by random error. Differential error is bias, as it produces measurements that are consistently too high or too low. *Epidemiology*, p. 87.
9. The answer is G: cutoff point. A cutoff point is used to distinguish normal from abnormal test results. If an upper limit cutoff point is set too high, the normal range will include many subjects with disease. If the upper limit cutoff point is set too low, many normal subjects will have abnormal test results. The optimal cutoff point, where all cases of disease are detected with no false-positive results, is represented by the upper left corner of the ROC curve. Such a point virtually never exists, and is at best approximated. The closer the cutoff point is to the upper left-hand corner of an ROC curve, the greater the sensitivity (ability to detect disease) and the lower the false-positive error rate. *Epidemiology*, p. 89.
10. The answer is C: bias. Bias is differential error, as it distorts data in a particular direction. Weighing subjects after a meal, taking blood pressure readings after caffeine ingestion or exercise, or performing psychometric testing following alcohol ingestion are examples of bias in measurement. As opposed to differential error, random error is equally likely to produce results spuriously high or spuriously low and is therefore nondifferential. *Epidemiology*, p. 86.

REFERENCE CITED
Jekel, J. F. Epidemiology, Biostatistics, and Preventive Medicine. Philadelphia, W. B. Saunders Company, 1996.

CHAPTER EIGHT

IMPROVING DECISIONS IN CLINICAL MEDICINE

SYNOPSIS

OBJECTIVES
- To discuss the use of analytic tools in clinical decision making.
- To discuss Bayes' theorem and its clinical applications.
- To define prior probability and posterior probability and discuss their utility in clinical decision making.
- To describe the technique and clinical applications of decision analysis.
- To discuss the methods, applications, and limitations of meta-analysis.

I. Improving Clinical Decision Making
  A. There is general agreement that effort should be directed at improving clinical decision making.
  B. There is controversy regarding the particular utility of various tools applied to clinical decision making.

II. Bayes' Theorem
  A. Bayes' theorem, originally developed centuries ago by the English clergyman for whom it is named, answers the two important questions that remain unanswered by sensitivity and specificity:
  1. If the test results are positive, what is the probability that the patient has the disease?
  2. If the test results are negative, what is the probability that the patient does not have the disease?
  B. The formula for Bayes' theorem is as follows:

$$p(D+|T+) = \frac{p(T+|D+)p(D+)}{[p(T+|D+)p(D+)] + [p(T+|D-)p(D-)]}$$

  where p denotes probability; D+ means that the patient has the disease in question; D− means that the patient does not have the disease; T+ means that a certain diagnostic test for the disease is positive; T− means that the test is negative; and the vertical line (|) means "conditional upon" what immediately follows.
  1. The **numerator of Bayes' theorem** is the prevalence multiplied by the sensitivity, and describes **cell *a*** (the true-positive results) in a 2 × 2 table (see Box 8–1).
  2. The **denominator of Bayes' theorem** includes the term from the numerator (cell *a*, or the true-positive results), and adds to it the term for the false-positive results (cell *b* from a 2 × 2 table). The false-positive results are described by the false-positive error rate, which is $p(T+|D-)$, multiplied by the prevalence of nondiseased persons, which is $p(D-)$.
  3. Therefore, Bayes' theorem may be rewritten as $a/(a + b)$, which is the formula for the **positive predictive value.**

  C. Bayes' theorem and community screening programs
  1. Bayes' theorem demonstrates that the probability of a positive (or negative) test result being true is dependent on the prevalence of the condition in the population being tested.
  2. In a population with a low prevalence of a particular disease, most positive test results for that disease will be false-positive results.
  3. If a disease is extremely common (prevalent) in a population, many of the negative test results will be false-negative results.
  4. Box 8–1 shows the use of Bayes' theorem in a hypothetical tuberculin screening program.

  D. Bayes' theorem and individual patient care
  1. The prevalence of disease for an individual is a non sequitur because disease is either present or absent.
  2. The probability of disease in a given patient is the clinician's estimate based on social and clinical factors. This estimate is made prior to the performance of laboratory tests and is based on the estimated prevalence of a particular disease among patients with similar signs and symptoms **(prior probability).**
  3. Once testing is initiated to pursue the diagnosis, the results modify the clinical impression of the probability of disease **(posterior probability).**
     a. To obtain the posterior probability, the prior probability is entered into Bayes' theorem as the prevalence term.
     b. To calculate the posterior probability, the sensitivity and specificity of the test in question must be known.
     c. When more than one test is done sequentially to establish a diagnosis, the posterior probability after each test becomes the prior probability for the next test.

> **BOX 8-1. Use of Bayes' Theorem or a 2 × 2 Table to Determine the Positive Predictive Value of a Hypothetical Tuberculin Screening Program**
>
> **Part 1. Beginning data:**
>
> Sensitivity of tuberculin tine test = 96% = 0.96
> False-negative error rate of the test = 4% = 0.04
> Specificity of the test = 94% = 0.94
> False-positive error rate of the test = 6% = 0.06
> Prevalence of tuberculosis in the community = 1% = 0.01
>
> **Part 2. Use of Bayes' theorem:**
>
> $$p(D+|T+) = \frac{p(T+|D+)p(D+)}{[p(T+|D+)p(D+)] + [p(T+|D-)p(D-)]}$$
>
> $$= \frac{(\text{Sensitivity})(\text{Prevalence})}{[(\text{Sensitivity})(\text{Prevalence})] + [(\text{False-positive error rate})(1 - \text{Prevalence})]}$$
>
> $$= \frac{(0.96)(0.01)}{[(0.96)(0.01)] + [(0.06)(0.99)]} = \frac{0.0096}{0.0096 + 0.0594} = \frac{0.0096}{0.0690} = 0.139 = 13.9\%$$
>
> **Part 3. Use of a 2 × 2 table, with numbers based on the assumption that 10,000 persons are in the study:**
>
> | | | TRUE DISEASE STATUS | | | | | |
> |---|---|---|---|---|---|---|---|
> | | | Diseased | | Nondiseased | | Total | |
> | | | Number | (Percentage) | Number | (Percentage) | Number | (Percentage) |
> | TEST RESULT | Positive | 96 | (96) | 594 | (6) | 690 | (7) |
> | | Negative | 4 | (4) | 9,306 | (94) | 9,310 | (93) |
> | | Total | 100 | (100) | 9,900 | (100) | 10,000 | (100) |
>
> Positive predictive value = 96/690 = 0.139 = 13.9%
>
> Source of data: Jekel, J.F., R.A. Greenberg, and B.M. Drake. Influence of prevalence of infection on tuberculin skin testing programs. Public Health Reports 84: 883–886, 1969.

4. It is useful to consider the principles of Bayes' theorem in guiding clinical decisions, even when the formula is not used. This would happen in the following specific situations:
    a. The probability of a disease in a patient is derived from the prevalence of that disease in the population of which the patient is representative.
    b. An estimate of the probability of disease in a patient is made prior to any testing (based on history and physical examination) and is then modified on the basis of test results.
        (1) If the estimated probability of a disease is high enough (high prevalence), then even a negative test result may not dissuade the clinician from treating that disease. This is because most negative results will be false-negative results when the prevalence (prior probability) is very high.
        (2) If the estimated probability of disease is low enough (low prevalence, or low prior probability), even a positive test result may not convince the clinician that the disease is present. The positive result may lead to further testing, however.

5. Box 8–2 shows the use of Bayes' theorem in a clinical setting.

III. Decision Analysis

A. Decision analysis is intended to improve clinical decision making under conditions of uncertainty.
   1. Decision analysis can be used for an individual patient or a general class of patients.
   2. The benefit of decision analysis is to help health care workers understand the following:
      a. The kinds of data that must go into a clinical decision.
      b. The sequence in which decisions have to be made.
      c. The personal values (particularly those of the patients) that must be considered before major decisions are made.

B. Steps in creating a decision tree
   1. There are four logical steps to be taken in constructing a decision tree:
      a. Identify and set limits to the problem.
         (1) Consider possible alternative clinical decisions.

BOX 8-2. Use of Bayes' Theorem or a 2 × 2 Table to Determine the Posterior Probability and the Positive Predictive Value in a Clinical Setting

Part 1. Beginning data:

Sensitivity of the first test = 90% = 0.90
Specificity of the first test = 95% = 0.95
Prior probability of disease = 2% = 0.02

Part 2. Use of Bayes' theorem:

$$p(D+|T+) = \frac{p(T+|D+)p(D+)}{[p(T+|D+)p(D+)] + [p(T+|D-)p(D-)]}$$

$$= \frac{(0.90)(0.02)}{[(0.90)(0.02)] + [(0.05)(0.98)]}$$

$$= \frac{0.018}{0.018 + 0.049} = \frac{0.018}{0.067} = 0.269 = 27\%$$

Part 3. Use of a 2 × 2 table:

		TRUE DISEASE STATUS					
		Diseased		Nondiseased		Total	
		Number	(Percentage)	Number	(Percentage)	Number	(Percentage)
TEST RESULT	Positive	18	(90)	49	(5)	67	(6.7)
	Negative	2	(10)	931	(95)	933	(93.3)
	Total	20	(100)	980	(100)	1,000	(100.0)

Positive predictive value = 18/67 = 0.269 = 27%

---

(2) Establish the sequence in which necessary decisions must be made.
(3) Establish the possible patient outcomes of each decision.

b. Diagram the options.
  (1) **Decision nodes** indicate points at which decisions must be made by clinicians (represented by a square in Fig. 8–1).
  (2) **Chance nodes** indicate points at which clinicians must wait to see the outcomes (represented by a circle in Fig. 8–1).

c. Obtain information concerning each option.
  (1) The **probability of each possible outcome** must either be obtained from studies or estimated.
  (2) The **utility of each final outcome** must be obtained. In decision analysis, the term *utility* is used to mean the value or benefit of a chosen course of action. Utility may be expressed in many ways, including in terms of death or illness rates (in which case smaller rates have greater utility), years of disability-free life, or dollars saved.

d. Compare the utility values and perform a sensitivity analysis.
  (1) Compare the utility values. The decision tree may show that one or another choice is clearly preferable to any other, and that would be strong evidence in favor of that choice. Often, the decision analysis will give two or more outcomes with similar utilities, which means either that better data are needed or that there really are options at that decision node.
  (2) Perform a sensitivity analysis. It is sometimes helpful to perform a sensitivity analysis, which consists of varying the probabilities of occurrence of a particular outcome at various points in the decision tree (one at a time) to see how the overall outcomes and clinical decisions would be affected by these changes. This helps both clinician and patient to see which probabilities and utilities have the largest impact on the outcomes through a reasonable range of values.

2. It may be necessary to **"average out"** some of the data. This is done by multiplying each of the potential risks associated with a particular decision by the probability of that particular outcome, and then summing these products to determine the overall utility of that decision.
3. In complex decision analyses, the objective is to find decisions that are clearly less satisfactory than others and cut off or "prune" the corresponding branches, because they are not rational alternatives.
4. The process of choosing the best branch at each decision node, working back from the right to the left, is called "folding back."

C. Applications of decision trees
  1. Decision analysis can be used to improve clinical decision making in the care of individual patients.

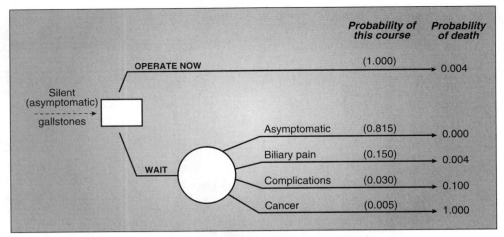

FIGURE 8-1. A decision tree concerning treatment for silent (asymptomatic) gallstones. The decision node, defined as a point where the clinician has to make a decision, is represented by a square. The chance node, defined as a point where the clinician must wait to see the outcome, is represented by a circle. If the clinician decides to operate now, the probability of surgery is 100% (1.000), and the probability of death (the negative utility value) from complications of surgery is 0.04% (0.004, or 1 out of every 250 patients undergoing surgery). If, instead, the clinician decides to wait, there are four possible outcomes, each with a different probability and negative utility value: (1) There is an 81.5% (0.815) probability of remaining asymptomatic, in which case the probability of dying from gallstones would be 0% (0.000). (2) There is a 15% (0.150) probability of developing biliary pain, which would lead to surgery and a 0.4% (0.004) risk of death. (3) There is a 3% (0.030) probability of biliary complications (such as acute cholecystitis or common duct obstruction), with a 10% (0.100) risk of death. (4) There is a 0.5% (0.005) probability of gallbladder cancer, with a 100% (1.000) risk of death. The probabilities of the possible outcomes at the chance node add up to 1 (here, 0.815 + 0.150 + 0.030 + 0.005 = 1.000). (Adapted from Rose, D. N., and J. Wiesel. Letter to the editor. New England Journal of Medicine 308:221–222, 1983. Copyright 1983. Massachusetts Medical Society. All rights reserved.)

2. Decision analysis can also be applied to public health policy decisions (e.g., determining the most cost-effective vaccination strategy).

IV. Meta-Analysis
A. Definition and indications
1. Meta-analysis (meaning "analysis among") is being used increasingly in medicine to try to obtain a qualitative or quantitative synthesis of the research literature on a particular subject.
2. Meta-analysis is indicated when the studies on a particular subject have been limited by sample size. A small sample size reduces power, increasing the likelihood of a false-negative error.

B. Selection of the studies to be analyzed
1. The issue of study selection is perhaps the most troublesome issue for those doing meta-analysis.
2. If studies are limited to published studies, studies in peer-reviewed journals, studies that are randomized controlled trials, or studies that meet other inclusion criteria, the data will be of high quality but also may introduce significant selection bias.
3. If studies are included that have used identical or similar methods, it is easy to combine the outcome data, which facilitates data analysis, but more selection bias may be introduced.

C. Types of meta-analysis
1. **Pooled (quantitative) analysis**
   a. Pooled analysis is used to obtain more accurate estimates of such things as therapeutic efficacy than are available from the smaller data sets of individual studies.
   b. For clarity, such pooled analyses should report both relative risks and risk reductions as well as absolute risks and risk reductions.
2. **Methodologic (qualitative) analysis**
   a. Methodologic analysis is used to determine the directional implications (e.g., favorable or unfavorable) of the research on a particular subject, such as a new medical intervention.
   b. The analysis is nonquantitative.

QUESTIONS

DIRECTIONS. (Items 1–15): Each of the numbered items or incomplete statements in this section is followed by answers or by completions of the statement. Select the ONE lettered answer or completion that is BEST in each case. Correct answers and explanations are given at the end of the chapter.

1. The numerator of Bayes' theorem represents

   (A) prevalence
   (B) false-negative results
   (C) sensitivity
   (D) true-positive results
   (E) incidence

2. Bayes' theorem may be used in the assessment of a screening program to determine

   (A) disease prevalence
   (B) positive predictive value
   (C) test sensitivity
   (D) cost-effectiveness
   (E) false-negative results

3. When applying Bayes' theorem to the care of an individual patient, the prior probability is analogous to

   (A) sensitivity
   (B) the odds ratio
   (C) prevalence
   (D) the likelihood ratio
   (E) specificity

4. The application of Bayes' theorem to patient care will generally result in

   (A) more diagnostic testing
   (B) higher cost
   (C) greater sensitivity
   (D) greater specificty
   (E) improved selection of diagnostic studies

5. A patient presents to your office complaining of reduced auditory acuity on the left side. After taking the history and performing a physical examination, and before further diagnostic testing, you estimate a 74% likelihood that the patient has a large yellow fruit in his left ear. This estimate is

   (A) the prior probability
   (B) the prevalence
   (C) the likelihood ratio
   (D) the posterior probability
   (E) the odds ratio

6. To apply Bayes' theorem to a screening program, the following information must be known:

   (A) prevalence, sensitivity, and specificity
   (B) prior probability and the false-positive error rate
   (C) prevalence, specificity, and posterior probability
   (D) sensitivity, prevalence, and prior probability
   (E) prior probability, the false-positive error rate, and the false-negative error rate

Items 7–8

A 62-year-old woman presents to your office complaining of intermittent left-sided chest pain with exertion for the past several months. She has a family history of heart disease (in male relatives only). She has no other known risk factors for heart disease except for mild hyperlipidemia (total cholesterol of 232 mg/dL, with high-density lipoprotein cholesterol of 38 mg/dL). Her blood pressure is 132/68 mm Hg, and her heart rate is 72 beats per minute. On cardiac examination, you detect a physiologically split second heart sound (S2) and a faint midsystolic click without any appreciable murmur. The point of maximal impulse is nondisplaced, and the remainder of the examination is unremarkable. You are concerned that the pain may represent angina pectoris and decide to initiate a workup. You estimate that the odds of the pain being angina are 1 in 3. You order an electrocardiogram (ECG), which reveals nonspecific abnormalities in the ST segments and T waves across the precordial leads. You decide to order a perfusion stress test, the sensitivity of which is 98% and the specificity 85%. The test is positive, demonstrating reversible ischemia in the distribution of the circumflex artery on perfusion imaging, with compatible ECG changes.

7. The prior probability of angina pectoris is

   (A) 25%
   (B) 66%
   (C) 33%
   (D) 75%
   (E) unknown

8. The posterior probability of angina pectoris following the stress test is

   (A) 67%
   (B) 76%
   (C) 33%
   (D) 10%
   (E) unchanged

Items 9–10

Based on the above, you decide to treat your patient for angina pectoris. You prescribe daily aspirin, a long-acting beta-adrenergic blocking agent (atenolol), and sublingual nitroglycerin tablets to be used if pain recurs. Pain does recur, with increasing frequency and severity, despite treatment. You consult a cardiologist (or maybe you are the cardiologist). Cardiac catheterization is recommended. Assume that the sensitivity of cardiac catheterization for coronary disease is 96% and the specificity 99%. The procedure is negative.

9. At this stage of the workup, the prior probability of angina pectoris is

   (A) 33%
   (B) 67%
   (C) 76%
   (D) 15%
   (E) unknown

10. Based on the results of cardiac catheterization, the probability of coronary disease is

    (A) 76%
    (B) 20%
    (C) 33%
    (D) 11%
    (E) unknown

Items 11–12

A pregnancy test is applied to a population of 10,000 women. Based on prior evaluation, the sensitivity of the test is known to be 98% and the specificity 99%. Assume that 0.1% of the subjects is actually pregnant.

11. The probability of pregnancy given a positive test result is

    (A) 0.1
    (B) 50%
    (C) 89%
    (D) 47%
    (E) 8.9%

12. An ultrasound examination is performed on the women with a positive pregnancy test to confirm the diagnosis. The sensitivity of this examination is 90%, and the specificity is 95%. The probability of pregnancy given a positive ultrasound examination is

    (A) 0.1
    (B) 6.4%
    (C) 20%
    (D) 64%
    (E) 75%

Items 13–15

A 75-year-old man has undergone open reduction and internal fixation of a fracture of the left hip, sustained while bungee jumping from a helicopter. Although the helicopter pilot was unaware that the patient had stowed away, the patient has already initiated litigation against the pilot, the helicopter owner, the helicopter manufacturer, the steel company that produced the metal used to construct the helicopter, the mine used to mine the iron used in making the steel used in making the helicopter, the gasoline company responsible for the fuel used to run the helicopter, the United Arab Emirates for exporting the fuel in the first place, and the Federal Aviation Administration for failing to adequately control the use of airspace. Be that as it may, the patient's surgery was uncomplicated until 14 hours following the operation. At that time, the patient suddenly became dyspneic, tachypneic, and tachycardic. His oxygen saturation has suddenly fallen to 80%, and his ECG reveals a right heart strain pattern. You are quite convinced the patient has suffered a pulmonary embolism. You estimate the likelihood of pulmonary embolism at 90%. You order a ventilation-perfusion scan, which has a sensitivity of 80% and a specificity of 60%. The scan is positive, meaning there is a high probability of pulmonary embolism.

13. The probability of pulmonary embolism is now

    (A) 75%
    (B) 90%
    (C) 95%
    (D) 10%
    (E) 99%

14. Had the ventilation-perfusion scan been negative, given the operating characteristics described above, the posterior probability of pulmonary embolism would be

    (A) 75%
    (B) 90%
    (C) 95%
    (D) 1%
    (E) 99%

15. The appropriate management of the patient described is

    (A) treat for pulmonary embolism only if the ventilation-perfusion scan is positive
    (B) treat for pulmonary embolism regardless of the ventilation-perfusion scan results
    (C) treat for pulmonary embolism without ordering a ventilation-perfusion scan
    (D) obtain additional tests if the ventilation-perfusion scan is negative
    (E) initiate definitive therapy immediately

ANSWERS AND EXPLANATIONS

1. The answer is D: true-positive results. Bayes' theorem is essentially an expression of the positive predictive value. Given a particular estimate of the prior probability of disease (the probability of a given disease before testing), the theorem adjusts the estimate based on the results of testing. The numerator of the formula represents true-positive results; the denominator, as in the formula for the positive predictive value, represents all positive test results, both true-positive results and false-positive results. The proportion of all positive test results that are true-positive results provides an estimate of the likelihood of disease if a test is positive (i.e., the positive predictive value). *Epidemiology*, p. 99.

2. The answer is B: positive predictive value. In a screening program, one needs to know how likely disease is when the results of testing are positive. Bayes' theorem is used to establish the positive predictive value of a screening test, based on an estimate of the underlying population prevalence of the disease under investigation. Disease prevalence must be known or estimated to use Bayes' theorem; the theorem is not used to determine prevalence. The test sensitivity is an intrinsic property of the test when applied to a population with given characteristics; it is unaffected by prevalence and is not determined by use of Bayes' theorem. The cost-effectiveness of a screening program is an important consideration but is distinct from the predictive value of the test. Bayes' theorem is based on the proportion of true-positive results to all positive results (both true-positive and false-positive results). False-negative results are irrelevant to use of the theorem. *Epidemiology*, p. 99.

3. The answer is C: prevalence. Conceptually, Bayes' theorem states that the probability of a disease in an individual is partly dependent on the prevalence of that disease in the population of which the individual is a member. For a child with fever in New Haven, Connecticut, the probability of malaria is low. The probability *(prior probability)* is low in this child because the prevalence of malaria is so low in the population. For a child with fever in a refugee population in Zaire, the probability *(prior probability)* of malaria might be high because malaria is prevalent in the population. The prior probability of a disease in an individual is an estimate based on the prevalence of that disease in a population of similar persons.

    To emphasize this point with one more example, the prior probability of pregnancy in a male patient presenting with abdominal distention and nausea in the morning is 0. It is not 0 because you know the status of the individual with certainty; it is 0 because the prevalence of pregnancy in a population of males is 0. When the prior probability of a condition is either 0 or 100%, no further testing for that condition is indicated. *Epidemiology*, p. 99.

4. The answer is E: improved selection of diagnostic studies. When Bayes' theorem is used appropriately in planning patient care, the result is more careful selection and refined use of diagnostic testing. Even when the theorem is used qualitatively, it suggests that, if the prior probability of disease is high enough, no further testing is required. At other times, when successive tests fail to establish a posterior probability that is sufficiently high

for decisive action to be taken, the theorem will compel you to continue testing. Neither sensitivity nor specificity is enhanced by use of the theorem, because both depend in part on the choice of particular diagnostic modalities. Therefore, using Bayes' theorem in planning patient care should result in *better* use of diagnostic tests—sometimes more use, sometimes less use. The overall cost will vary accordingly.

To fortify this explanation with an example, imagine that a 52-year-old woman presents to your office with a cough. You quickly learn that the cough is productive and associated with pleuritic chest pain, fever, and mild dyspnea (breathlessness) on exertion. She has smoked a pack of cigarettes daily for many years. You examine her and find that her temperature is 100.8°F and her respiratory rate is normal at rest. Auscultation of her lungs reveals rhonchi and wheezes but not rales. In a patient with these findings, you estimate that the prior probability of bronchitis is 95%. By reflex—the type of reflex acquired only in medical school—you might order some blood tests and a chest x-ray. But if you pause and think, you may realize that you have already made up your mind to treat this condition with antibiotics and perhaps corticosteroids and a β-agonist inhaler. When the prior probability is sufficiently high, further diagnostic testing will not, and should not, alter your course of action and is therefore not indicated. Conversely, it may be inappropriate to take immediate action when you have a considerable amount of uncertainty about what to do next for your patient and further diagnostic studies would help reduce the uncertainty. If you apply the concept of Bayes' theorem to your clinical care planning, without even using the actual formula, the standards of your practice are likely to improve. *Epidemiology,* p. 100.

5. The answer is A: the prior probability. The prior probability is an estimate of the probability of a given condition in a given patient at a given point in time, prior to further testing. In actual practice, prior and posterior probabilities flow freely into one another. For example, estimating the probability of sinusitis after interviewing a patient gives you the prior probability of disease before you examine the patient. The revision in your estimate after the examination is the posterior probability, but this revision is also the prior probability before any diagnostic testing, such as sinus x-rays, that you may be considering. *Epidemiology,* p. 101.

6. The answer is A: prevalence, sensitivity, and specificity. Bayes' theorem can be converted algebraically into the formula for the positive predictive value, which is $a/(a + b)$ (see II.B.2 and 3 in the outline, above). This is what we are using the theorem to determine, and, therefore, we do not have to know the predictive value before we can use the theorem. The numerator for Bayes' theorem is the sensitivity of the test being used multiplied by the prevalence, and, therefore, both the prevalence and the sensitivity must be known before we can use the theorem. The denominator contains the numerator term plus a term that is the false-positive error rate, or (1 − specificity), multiplied by (1 − prevalence). Choice B (prior probability and the false-positive error rate) is incorrect because the prior probability pertains to an individual patient, not a screening test; for a screening test, the prevalence must be known or estimated. The only choice that includes everything that must be known to use Bayes' theorem, in terms appropriate for a screening program, is choice A. *Epidemiology,* p. 99.

7. The answer is C: 33%. Prior probability is analogous to prevalence and represents an estimate of the likelihood of disease in a patient. In this case, the prior probability is your estimate of the likelihood of angina pectoris prior to any testing. This was stated as odds of 1 in 3 and, therefore, represents a 33% prior probability. *Epidemiology,* p. 101.

8. The answer is B: 76%. There are two methods for calculating the posterior probability: Bayes' theorem and a 2 × 2 table. Both of these methods are shown in the box below. *Epidemiology,* p. 101.

Part 1. Use of Bayes' theorem:
The sensitivity is provided in the vignette as 98%, or 0.98. The prior probability estimate is substituted for the prevalence, and is 33%, or 0.33. The false-positive error rate is (1 − specificity). The specificity is provided as 85%, and the false-positive error rate is therefore 15%, or 0.15. In this case, (1 − prevalence) is the same as (1 − prior probability) and is therefore (1 − 0.33), or 0.67.

$$p(D+|T+) = \frac{p(T+|D+)p(D+)}{[p(T+|D+)p(D+)] + [p(T+|D-)p(D-)]}$$

$$= \frac{(\text{Sensitivity})(\text{Prevalence})}{[(\text{Sensitivity})(\text{Prevalence})] + [(\text{False-positive error rate})(1 - \text{Prevalence})]}$$

$$= \frac{(0.98)(0.33)}{[(0.98)(0.33)] + [(0.15)(0.67)]}$$

$$= \frac{0.32}{0.32 + 0.10} = \frac{0.32}{0.42} = 0.76 = 76\%$$

Part 2. Use of a 2 × 2 table:
To use this method, an arbitrary "sample size" must be chosen. Assuming that the sample size is 100, the following is true: Cell a is the true-positive results, or sensitivity multiplied by prevalence. The prior probability becomes the prevalence. Therefore, cell a is $(0.98)(33) = 32.3$ (rounded to 32). Cells a plus c must sum to 33, so cell c is 0.7 (rounded to 1). Cell d is the true-negative results, which is specificity multiplied by (1 − prevalence), or $(0.85)(67) = 57$. Cells b plus d must sum to 67, so cell b is 10.

		DISEASE	
		Positive	Negative
TEST RESULT	Positive	32	10
	Negative	1	57

Once the 2 × 2 table is established, the formula for positive predictive value, which is $a/(a + b)$, may be used to calculate the posterior probability, which is as follows:

$$a/(a + b) = 32/(32 + 10) = 32/42 = 0.76 = \mathbf{76\%}$$

9. **The answer is C: 76%.** As discussed in the explanation for question 5, above, prior and posterior probabilities may flow freely into one another. At any given stage in a workup, the best estimate of disease probability is the posterior probability after the last examination or test and also the prior probability before the next test that is to be done. For this reason, the answer to questions 8 and 9 is the same. Following the stress test, the posterior probability of angina is 76%. This is also the probability of angina *prior to* any further diagnostic studies and, therefore, the prior probability at this stage of the workup. *Epidemiology,* p. 101.

10. **The answer is D: 11%.** To calculate the posterior probability after a negative test, the formula for negative predictive value, which is $d/(c + d)$, may be used. To use this formula, a 2 × 2 table must first be set up. Assume a sample size of 100. The prior probability, 76%, becomes the prevalence. Cell *a* is the sensitivity multiplied by the prevalence, or $(0.96)(76) = 73$. Cells *a* plus *c* sum to the prevalence, so cell *c* is 3. Cell *d* is the specificity multiplied by $(1 - \text{prevalence})$, or $(0.99)(24) = 23.8$. Cells *b* plus *d* sum to $(1 - \text{prevalence})$, so cell *b* is 0.2.

		DISEASE	
		Positive	Negative
TEST RESULT	Positive	73	0.2
	Negative	3	23.8

The negative predictive value is the probability that coronary disease is truly *absent* given a negative cardiac catheterization. The formula and calculation are as follows:

$$d/(c + d) = 23.8/(3 + 23.8) = 23.8/26.8 = 89\%$$

The probability that the patient does *not* have coronary disease is now 89%. However, we are interested in knowing how likely it is, given the prior probability of 76% and the negative results of the catheterization, that our patient *does* have coronary disease. This is $(1 - \text{negative predictive value})$, or $(1 - 89\%) = (100\% - 89\%) = \mathbf{11\%}$. *Epidemiology,* p. 101.

11. **The answer is E: 8.9%.** This question may be answered either by use of Bayes' theorem or by construction of a 2 × 2 table and calculation of the positive predictive value. Both of these methods are shown in the box below.

Part 1. Use of Bayes' theorem:
The sensitivity is 0.98. The specificity is 0.99. The prevalence is 0.1%, or 0.001. The false-positive error rate is $(1 - \text{specificity})$, or 0.01. The $(1 - \text{prevalence})$ is 0.999.

$$p(D+|T+) = \frac{p(T+|D+)p(D+)}{[p(T+|D+)p(D+)] + [p(T+|D-)p(D-)]}$$

$$= \frac{(\text{Sensitivity})(\text{Prevalence})}{[(\text{Sensitivity})(\text{Prevalence})] + [(\text{False-positive error rate})(1 - \text{Prevalence})]}$$

$$= \frac{(0.98)(0.001)}{[(0.98)(0.001)] + [(0.01)(0.999)]}$$

$$= \frac{0.00098}{(0.00098 + 0.00999)} = \frac{0.00098}{0.01097}$$

$$= 0.089 = 8.9\%$$

Part 2. Use of a 2 × 2 table:
Cell *a* is the true-positive results, which is sensitivity multiplied by prevalence, or $(0.98)(0.001) = 0.00098$. Cells *c* plus *a* must sum to the prevalence, so cell *c* is 0.00002. Cell *d* is the true-negative results, which is specificity multiplied by $(1 - \text{prevalence})$, or $(0.99)(0.999) = 0.989$. Cells *b* plus *d* must sum to $(1 - \text{prevalence})$, so cell *b* is 0.01.

		PREGNANCY STATUS	
		Positive	Negative
TEST RESULT	Positive	0.00098	0.01
	Negative	0.00002	0.989

The posterior probability is derived by calculating the positive predictive value as follows:

$$a/(a + b) = \frac{0.00098}{(0.00098 + 0.01)} = \frac{0.00098}{0.01098} = 0.089 = 8.9\%$$

Despite the use of a highly accurate pregnancy test, the probability of pregnancy in a woman from this population (perhaps a population of women who are being followed after having tubal ligation), given a positive test result, is only 8.9%. This is due to the extremely low prevalence of pregnancy in the cohort, and demonstrates the fundamental concept on which Bayes' theorem is based: the likelihood of any condition in an individual depends largely on the prevalence of that condition in the population of which the individual is representative.

The false-positive error rate is worth noting. This term, as described in Chapter 7, is calculated as follows:

$$b/(b + d) = \frac{0.01}{(0.01 + 0.989)} = \frac{0.01}{0.999} = 0.01$$

Therefore, 1 out of every 100 tests performed on women who are *not* pregnant will be (false) positive. In this population of 10,000 women, the prevalence of pregnancy is 0.001; there are, therefore, approximately 10 pregnant women in the entire population. The remaining 9990 women are not pregnant and are therefore candidates for a false-positive test result. (Recall that only true-negative cases may yield false-positive results.) Of the 9990 pregnancy tests in nonpregnant women, 1 per 100, or 99.9, will be false-positive tests. Therefore, due to the low prevalence of pregnancy in this group, there will be nearly 10 false-positive tests for every true-positive test.

One other issue to consider is that sensitivity and specificity represent the performance characteristics of a test in a particular population or populations. These values may vary as the test is brought into wider use and applied to populations of differing composition. *Epidemiology*, p. 101.

12. The answer is D: 64%. Either Bayes' theorem or a 2 × 2 table can be used to make the calculation. Both of these methods are shown in the box below.

Part 1. Use of Bayes' theorem:
The sensitivity is 0.90. The specificity is 0.95. The prevalence is the prior probability of pregnancy, or 8.9% (0.089). The false-positive error rate is (1 − specificity), or 0.05. The (1 − prevalence) is 0.911.

$$p(D+|T+) = \frac{p(T+|D+)p(D+)}{[p(T+|D+)p(D+)] + [p(T+|D-)p(D-)]}$$

$$= \frac{(\text{Sensitivity})(\text{Prevalence})}{[(\text{Sensitivity})(\text{Prevalence})] + [(\text{False-positive error rate})(1 - \text{Prevalence})]}$$

$$= \frac{(0.90)(0.089)}{[(0.90)(0.089)] + [(0.05)(0.911)]}$$

$$= \frac{0.0801}{(0.0801 + 0.0456)} = \frac{0.0801}{0.1257} = 0.64 = 64\%$$

Part 2. Use of a 2 × 2 table:
Cell *a* is the true-positive results, which is sensitivity multiplied by prevalence, or (0.90)(0.089) = 0.0801. Cells *c* plus *a* must sum to the prevalence, so cell *c* is 0.0089. Cell *d* is the true-negative results, which is specificity multiplied by (1 − prevalence), or (0.95)(0.911) = 0.865. Cells *b* plus *d* must sum to (1 − prevalence), so cell *b* is 0.046.

PREGNANCY STATUS

		Positive	Negative
TEST RESULT	Positive	0.0801	0.046
	Negative	0.0089	0.865

The posterior probability is the positive predictive value, which is calculated as follows:

$$a/(a+b) = \frac{0.0801}{(0.0801 + 0.046)} = \frac{0.0801}{0.1261} = 0.635 = 64\%$$

Note that although the ultrasound study is a less accurate test than the initial pregnancy test (as described here), the probability of pregnancy is much higher following a positive ultrasound study than a positive pregnancy test. This is because the ultrasound study is applied to a population with a much higher prevalence of pregnancy (i.e., only those women who initially tested positive). *Epidemiology*, p. 101.

13. The answer is C: 95%. The calculation can be made with either Bayes' theorem or a 2 × 2 table. Both of these methods are shown in the box below. *Epidemiology*, p. 101.

Part 1. Use of Bayes' theorem:
The sensitivity is 0.80. The specificity is 0.60. The prevalence (prior probability) is 0.90. The false-positive error rate is (1 − specificity), or 0.40. The (1 − prevalence) is 0.10.

$$p(D+|T+) = \frac{p(T+|D+)p(D+)}{[p(T+|D+)p(D+)] + [p(T+|D-)p(D-)]}$$

$$= \frac{(\text{Sensitivity})(\text{Prevalence})}{[(\text{Sensitivity})(\text{Prevalence})] + [(\text{False-positive error rate})(1 - \text{Prevalence})]}$$

$$= \frac{(0.80)(0.90)}{[(0.80)(0.90)] + [(0.40)(0.10)]}$$

$$= \frac{0.72}{(0.72 + 0.04)} = \frac{0.72}{0.76} = 0.947 = 95\%$$

Part 2. Use of a 2 × 2 table:
Cell *a* is the true-positive results, which is sensitivity multiplied by prevalence, or (0.80)(0.90) = 0.72. Cells *c* plus *a* must sum to the prevalence, so cell *c* is 0.18. Cell *d* is the true-negative results, which is specificity multiplied by (1 − prevalence), or (0.60)(0.10) = 0.06. Cells *b* plus *d* must sum to (1 − prevalence), so cell *b* is 0.04.

PULMONARY EMBOLISM

		Positive	Negative
TEST RESULT	Positive	0.72	0.04
	Negative	0.18	0.06

The posterior probability of pulmonary embolism is the positive predictive value, which is calculated as follows:

$$a/(a+b) = \frac{0.72}{(0.72 + 0.04)} = \frac{0.72}{0.76} = 0.947 = 95\%$$

---

14. The answer is A: 75%. The question asks for the probability of *disease* given a negative test result. The negative predictive value provides the probability of *nondisease* given a negative test. Therefore, calculating the (1 − negative predictive value) will provide the answer. This can be done by using the 2 × 2 table in the explanation for question 13, above.

$$d/(c+d) = \frac{0.06}{(0.18 + 0.06)} = \frac{0.06}{0.24} = 0.25$$

The posterior probability of pulmonary embolism is (1 − negative predictive value), or (1 − 0.25) = 0.75 = 75%. The proverbial take-home message here is that the probability of disease remains high after a negative test if the prior probability of disease was very high. Certainty in medicine is elusive at best and unattainable at worst. You will need to decide when to take (therapeutic) action based on the prior probability of disease, and when additional testing is truly required to inform such a decision. *Epidemiology,* p. 101.

15. The answer is D: obtain additional tests if the ventilation-perfusion scan is negative. The patient is very likely to have pulmonary embolism. The incidence of deep venous thrombosis and pulmonary embolism following internal fixation of a hip fracture is high, and the clinical description is classic for pulmonary embolism (Goldhaber and Morpurgo 1992). However, the use of intravenous heparin, essential in the treatment of pulmonary embolism, is hazardous so soon after major surgery. Therefore, before using this therapy, diagnostic certainty must be quite high. Alternative diagnoses, such as right ventricular infarction, pulmonary atelectasis, or even acute bronchospasm, might produce a similar clinical picture, but each would call for a different treatment. One might treat for pulmonary embolism without first obtaining confirmatory test results in this case, but most clinicians would attempt to confirm the diagnosis with a ventilation-perfusion scan. When ordering a test, however, one must always consider all of the possible outcomes; if only one outcome were possible, the test would be unnecessary. Therefore, a test ordered to confirm a diagnostic impression might, in fact, refute that impression when results defy expectation. In such situations, further diagnostic testing is often obtained to resolve the discrepancy. The pursuit of certainty in clinical practice is often frustrating, with unanticipated test results frequently generating the need for further testing. The best course through this labyrinth, in answering this question and in general, is to anticipate all of the possible results of any tests that are ordered and the action indicated by each. Inaction must be considered among the available courses of action. The challenge is to minimize risk to the patient and maximize benefit, despite the constant burden of uncertainty. *Epidemiology,* p. 101.

REFERENCES CITED

Bisno, A. L. Group A streptococcal infections and acute rheumatic fever. New England Journal of Medicine 325:783–793, 1991.

Goldhaber, S. Z., and M. Morpurgo (editors). Report of the World Health Organization/International Society and Federation of Cardiology Task Force on Pulmonary Embolism: diagnosis, treatment, and prevention of pulmonary embolism. Journal of the American Medical Association 268:1727–1733, 1992.

Jekel, J. F. Epidemiology, Biostatistics, and Preventive Medicine. Philadelphia, W. B. Saunders Company, 1996.

CHAPTER NINE

DESCRIBING VARIATION IN DATA

SYNOPSIS

OBJECTIVES
- To describe the goals of statistics with regard to variation in clinical data.
- To describe the sources of variation in clinical data.
- To describe the types of variables used to characterize variation in clinical data.
- To describe the frequency distributions of continuous variables.
- To characterize and explain the parameters that define a frequency distribution—specifically, the measures of central tendency and the measures of dispersion.
- To describe the idiosyncracies of real frequency distributions that deviate from the theoretical and thereby complicate data analysis.
- To describe the means of visually displaying frequency distributions.
- To discuss the impact of units of measurement on frequency distributions, and the use of unit-free (normalized) data.
- To describe the frequency distributions of dichotomous data and proportions.
- To note briefly the application of frequency distributions to nonparametric data.

I. Introduction
  A. Variation and variables
  1. Variation is evident in almost every characteristic of patients, including their physiologic measurements, diseases, diets, environments, and life-styles.
  2. A measure of a single characteristic is called a **variable.**

  B. Uses of statistics in medicine
  1. Statistics describe the patterns of variation in single variables. This will be discussed in this chapter.
  2. Statistics determine when observed differences in data are likely to be real differences (i.e., significant) (see Chapters 10 and 11).
  3. Statistics determine the patterns and strength of association between variables (see Chapters 11 and 13).

II. Sources of Variation in Medicine
  A. Biologic differences include factors such as differences in genes, nutrition, and environmental exposures.
  B. Variation is seen not only in the **presence or absence of disease** but also in the **stages and manifestations of disease.**
  C. Different conditions of measurement often account for the variations observed in medical data (e.g., time of day; ambient temperature or noise; degree of fatigue, anxiety, or hunger in the patient).
  D. Different methods of measurement can produce different results (e.g., the blood pressure measurement derived from the use of an intra-arterial catheter may differ from that derived from use of an arm cuff, but this would not mean that either measurement was necessarily in error).
  E. Measurement error may also produce variations.
  1. Different laboratory methods or instruments may give different readings from the same sample.
  2. Different x-ray machines may produce films of different quality.
  3. Different observers may report different results.
  F. Statistics can help investigators to interpret data despite **random variation,** but statistics cannot correct for errors in the observation or recording of data.

III. Statistics and Variables
  A. Describing variation
  1. The first step in understanding variation is to describe the variation.
  2. Statistics can be thought of as a set of tools for working with data, just as brushes are tools used by an artist for painting.
  3. A person who works with data must understand the different types of variables that exist in medicine.

  B. Quantitative and qualitative data
  1. A **quantitative characteristic** (e.g., systolic blood pressure or serum sodium level) can be characterized using a rigid, dimensional measurement scale.
  2. A **qualitative characteristic** (e.g., coloration of the skin) must be described in detail but is not truly measurable.

3. Some disease manifestations have both quantitative and qualitative characteristics (e.g., heart murmurs and bowel sounds).
4. Any kind of information that can vary is called a **variable**.

C. Types of variables
1. **Nominal variables** are "naming" or categorical variables that have no measurement scale.
    a. Examples include blood groups (O, A, B, and AB), food groups, occupations, and skin coloration.
    b. If skin coloration is the variable being examined, a different number is assigned to each color before the data is entered into a computer data system.
    c. The number is merely a numerical name given to the color and lacks relative quality, value, or rank.
2. **Dichotomous variables (binary variables)** are variables with only two levels (e.g., normal and abnormal skin color).
    a. Directionality (from better to worse) is implied if one outcome is preferable to the other (e.g., living/dead).
    b. Directionality is not always a factor (e.g., male/female, treatment/placebo).
    c. Usually, it makes no difference to the statistical analysis whether or not the dichotomous variable has an implied direction, but the direction may be important for the conclusions drawn from the data.
3. Dichotomous variables and nominal variables are sometimes called **discrete variables** because the different categories are completely separate from each other.
4. **Ordinal variables (ranked variables)** are medical data that can be characterized in terms of more than two values and have a clearly implied direction from better to worse, but are not measured on a continuous measurement scale.
    a. The direction is generally implied. For example, edema may be reported as "none" or 1+, 2+, 3+, or 4+ pitting edema (swelling).
    b. Ordinal variables contain more information than nominal variables because it is possible to see the relationships between ordinal categories and know whether one is more desirable, equally desirable, or less desirable than another.
    c. Because ordinal variables contain more information than nominal variables, the ordinal variables enable more robust (certain) conclusions to be drawn.
5. **Continuous variables (dimensional variables)** are medically important data that are measured on continuous (dimensional) measurement scales (e.g., height, weight, systolic and diastolic blood pressures, and serum glucose levels).
    a. Even more information is contained in continuous data than in ordinal data because continuous data not only show the position of the different observations relative to each other but also show the extent to which one observation differs from another.
    b. Continuous data usually enable investigators to make more detailed inferences than do ordinal or nominal data.
    c. Relationships between continuous variables are not always linear (in a straight line) (e.g., there is not a linear relationship between birth weight and survival of newborns).
6. **Ratio variables**
    a. If a continuous scale has a true 0 point, the variables derived from it can be called ratio variables (e.g., test scores).
    b. Ratio variables are so named because the ratio of one value to another is meaningful (e.g., a score of 80 is two times higher than a score of 40).
7. **Risks and proportions,** which are two important types of variables in medicine, share some characteristics of a discrete variable and some characteristics of a continuous variable.
    a. Just as it makes no sense to say that a fraction of a death occurred, it makes no sense to say that a fraction of a person suffered an event. However, it does make sense to say that a discrete event (e.g., death) or a discrete characteristic (e.g., presence of a murmur) occurred in a fraction of a population.
    b. Risks and proportions are variables created by the ratio of discrete counts in the numerator to counts in the denominator.
    c. Depending on the circumstances, they may be analyzed as discrete variables or as continuous variables.
8. Counts and units of observation
    a. The unit of observation is the source of data (e.g., the persons, animals, or organisms in a medical study).
    b. Units of observations may be arranged in a **frequency table,** with one characteristic on the x-axis, a different characteristic on the y-axis, and the appropriate counts in the cells of the table.
9. Combining data
    a. A continuous variable may be converted to an ordinal variable by grouping units with similar values together.
    b. Information is lost when continuous data are converted to discrete data, but the creation of ratios and percentages may facilitate comparisons.

IV. Frequency Distributions
  A. Frequency distributions of continuous variables
    1. A frequency distribution is a plot of data displaying the value of each data point on one axis and the frequency with which that value occurs on the other axis (see Table 9–1 and Fig. 9–1).
    2. A frequency distribution is partly described by its **range,** which is the numerical distance between the lowest and highest values.
  B. **Real frequency distributions** are those obtained from actual data.
  C. When **theoretical frequency distributions** are used, they are assumed to describe the underlying populations from which a sample is drawn.
    1. Most measurements of continuous data in medicine and biology tend to approximate the theoretical distribution that is known as the **normal distribution** (also called the **gaussian distribution**).
    2. The normal (gaussian) distribution looks something like a bell seen from the side (see Fig. 9–2).
    3. Real distributions, found by gathering real data, are seldom if ever perfectly smooth and bell-shaped.
    4. In textbooks, the smooth, bell-shaped curves are often used to represent the **expected frequency of observations** (on the y-axis) according to their observed value on a measurement scale (on the x-axis).

TABLE 9-1. Serum Levels of Total Cholesterol Reported in 71 Subjects*

Cholesterol Value (mg/dL)	Number of Observations	Cholesterol Value (mg/dL)	Number of Observations	Cholesterol Value (mg/dL)	Number of Observations
124	1	169	1	217	1
128	1	171	4	220	1
132	1	175	1	221	1
133	1	177	2	222	1
136	1	178	2	226	1
138	1	179	1	227	1
139	1	180	4	228	1
146	1	181	1	241	1
147	1	184	2	264	1
149	1	186	1		
151	1	188	2		
153	2	191	3		
158	3	192	2		
160	1	194	2		
161	1	196	2		
162	1	197	2		
163	2	206	1		
164	3	208	1		
165	1	209	1		
166	1	213	1		

* In this data set, the mean is 179.1 mg/dL, and the standard deviation is 28.2 mg/dL.
Source of data: Unpublished findings in a sample of 71 professional persons in Connecticut.

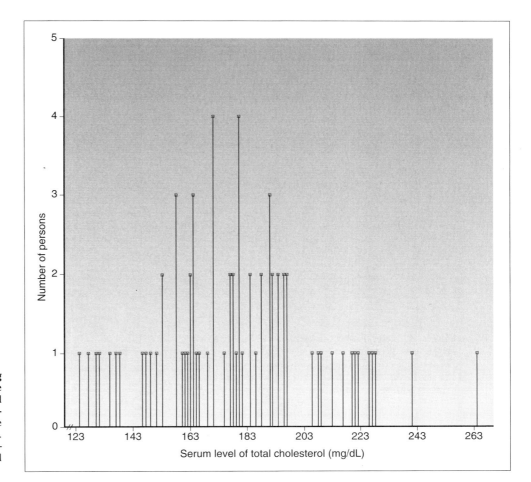

FIGURE 9-1. **Histogram showing the frequency distribution of the serum levels of total cholesterol reported in a sample of 71 subjects.** The data shown here are the same data shown in Table 9-1. (Source of data: Unpublished findings in a sample of 71 professional persons in Connecticut.)

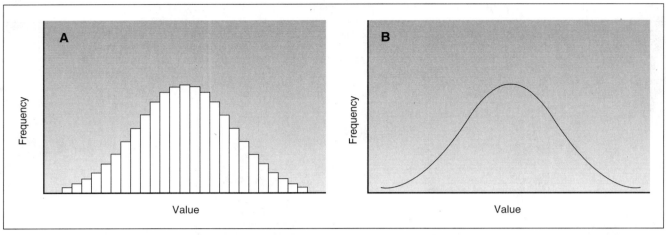

FIGURE 9-2. Illustration of the normal (gaussian) distribution. Diagram A shows a probability distribution of actual data, plotted as a histogram with very narrow ranges, and diagram B shows the way this idea is represented, for simplicity, in textbooks, articles, and tests.

5. Readers should remember, when they see a perfectly smooth, bell-shaped gaussian distribution, that the y-axis is really describing the frequency with which the corresponding values in the x-axis are expected to be found.

V. Parameters of a Frequency Distribution

A. Frequency distributions from continuous data are defined by two types of measures, or parameters: measures of central tendency and measures of dispersion.

B. Measures of central tendency

1. Measures of central tendency locate a scale of observations in space (analogous to a street address for the variable).
2. The first step in examining a distribution is to look for the central tendency of observations. Most types of medical data tend to clump in such a way that the density of observed values is greatest near the center of the distribution.
3. The next step is to examine the distribution in greater detail and look for the mode, the median, and the mean, which are the three measures of central tendency.
4. The **mode** is the most commonly observed value in a distribution (that is, the value with the highest number of observations).
5. The **median** is the middle observation when data have been arranged in order from the lowest to the highest value.
6. The **mean** ($\bar{x}$) is the average value. It is calculated as the sum (Σ) of all of the observed values (x_i) divided by the total number of observations (N_i):

$$\text{Mean} = \bar{x} = \frac{\Sigma(x_i)}{N_i}$$

  a. The sum of the deviations of all observations from the mean equals 0.
  b. The mean of a sample is an *unbiased estimator* of the mean of the population from which it came.
  c. The mean is the *mathematical expectation*.
  d. The sum of the squared deviations of the observations from the mean is *smaller* than the sum of the squared deviations from any other number.
  e. The sum of the squared deviations from the mean is *fixed* for a given set of observations. This property is not unique to the mean, but it is a necessary property of any good measure of central tendency.
  f. The mean is the measure of central tendency used in most statistical analyses.

C. Measures of dispersion

1. Measures of dispersion suggest how widely the observations are spread out around the measures of central tendency (analogous to the property lines for the variable).
2. Measures of dispersion may be based on **percentiles,** which are sometimes called **quantiles.** Percentiles are the percentage of observations below the point indicated when all of the observations are ranked in descending order.
   a. The **overall range** is the distance between the highest value and the lowest value in the data set.
   b. The **interquartile range** is the numerical distance between the 25th and 75th percentiles.
3. Measures of dispersion may be based on the mean.
   a. The **mean deviation** is the average of the absolute value of the deviations of all observations from the mean.

$$\text{Mean deviation} = \frac{\Sigma(|x_i - \bar{x}|)}{N}$$

   b. The **variance** is the sum of the squared deviations from the mean, divided by the number of observations minus 1.

$$\text{Variance} = s^2 = \frac{\Sigma(x_i - \bar{x})^2}{N - 1}$$

   (1) The denominator, $N - 1$, is called the **degrees of freedom.** This represents the number of observations free to vary (and thereby free to contribute to the variance) once the mean is known. The degrees of freedom rather than the total number of observations is required in the denominator in order that the sample variance can serve as an unbiased estimator of the population variance.
   (2) The numerator, $\Sigma(x_i - \bar{x})^2$, is known as the **sum of squares** (abbreviated SS), or the **total sum of squares** (TSS). The TSS represents the total amount of variation in a set of observations (see Box 9-1).
   (3) When the denominator of the equation for variance is expressed as the number of observations minus 1 ($N - 1$), the variance of a random sample is an *unbiased estimator* of the variance of the population from which it was taken.

BOX 9-1. How Do Statisticians Measure Variation?

In statistics, variation is measured as the sum of the squared deviations of the individual observations from an expected value, such as the mean. The mean is the mathematical expectation or expected value of a continuous frequency distribution. The quantity of variation in a given set of observations, therefore, is simply the numerator of the variance, which is the sum of the squares. The sum of the squares (SS) is sometimes called the total sum of the squares (TSS).

For purposes of illustration, assume that the data set consists of the following six numbers: 1, 2, 4, 7, 10, and 12. Assume that x_i denotes the individual observations, $\bar{x}$ is the mean, N is the number of observations, s^2 is the variance, and s is the standard deviation.

Part 1. Tabular representation of the data:

	x_i	$(x_i - \bar{x})$	$(x_i - \bar{x})^2$
	1	−5	25
	2	−4	16
	4	−2	4
	7	+1	1
	10	+4	16
	12	+6	36
Sum, or Σ	36	0	98

Part 2. Graphic representation of the data shown in the third column of the above table—that is, $(x_i - \bar{x})^2$ for each of the six observations:

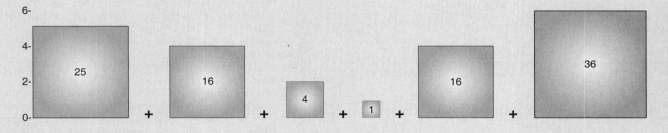

Part 3. Calculation of moments:

$\Sigma(x_i) = 36$ $\Sigma(x_i - \bar{x})^2 = \text{TSS} = 98$

$N = 6$ $s^2 = \text{TSS}/(N - 1) = 98/5 = 19.6$

$\bar{x} = 6$ $s = \sqrt{19.6} = 4.43$

---

(4) The variance of the sum of two independent variables is equal to the *sum* of the variances.

(5) The variance of the difference between two independent variables is also equal to the *sum* of the variances.

c. For simplicity of calculation on a pocket calculator, there is another (but algebraically equivalent) formula used to calculate the **variance.** It is the sum of the squared value of each observation, minus a correction factor (to correct for the fact that it is the absolute values, rather than the deviation from the mean, which are being squared), all divided by $N - 1$:

$$\text{Variance} = s^2 = \frac{\Sigma(x_i^2) - \left[\frac{(\Sigma x_i)^2}{N}\right]}{N - 1}$$

Table 9–2 illustrates the calculation of a variance using this second formula.

d. The **standard deviation** is the square root of the variance.

$$\text{Standard deviation} = s = \sqrt{\frac{\Sigma(x_i - \bar{x})^2}{N - 1}}$$

(1) In a normal distribution, the mean plus or minus one standard deviation includes 68% of the observations.

(2) In a normal distribution, the mean plus or minus two standard deviations includes 95.4% of the observations.

(3) Exactly 95% of the observations in a normal distribution are contained within the range of the mean plus or minus 1.96 standard deviations.

VI. Problems in Analyzing a Frequency Distribution

A. In a normal (gaussian) distribution, the following holds true: Mean = median = mode. This is rarely true for a set of actual observations, however, even if it approximates the normal distribution.

B. **Definitions**

1. **Skewness** is horizontal stretching of a frequency distribution, so that one tail of the plot is longer

TABLE 9-2. Raw Data and Results of Calculations in a Study Concerning Serum Levels of High-Density Lipoprotein (HDL) Cholesterol in 26 Subjects

Parameter	Raw Data or Results of Calculation
Number of observations, or N	26
Initial HDL cholesterol values of the subjects	31, 41, 44, 46, 47, 47, 48, 48, 49, 52, 53, 54, 57, 58, 58, 60, 60, 62, 63, 64, 67, 69, 70, 77, 81, and 90 mg/dL
Highest value	90 mg/dL
Lowest value	31 mg/dL
Mode	47, 48, 58, and 60 mg/dL
Median	$(57 + 58)/2 = 57.5$ mg/dL
Sum of the values, or sum of x_i	1,496 mg/dL
Mean, or $\bar{x}$	$1,496/26 = 57.5$ mg/dL
Range	$90 - 31 = 59$ mg/dL
Interquartile range	$64 - 48 = 16$ mg/dL
Sum of $(x_i - \bar{x})^2$, or TSS	4,298.46 mg/dL squared*
Variance, or s^2	171.94 mg/dL†
Standard deviation, or s	$\sqrt{171.94} = 13.1$ mg/dL

*For a discussion and example of how statisticians measure the total sum of the squares (TSS), see Box 9-1.

†Here, the following formula is used:

$$\text{Variance} = s^2 = \frac{\sum(x_i^2) - \left[\frac{(\sum x_i)^2}{N}\right]}{N - 1} = \frac{90,376 - \frac{2,238,016}{26}}{26 - 1}$$

$$= \frac{90,376 - 86,077.54}{25} = \frac{4,298.46}{25} = 171.94$$

and contains more observations than the other. The plot is skewed to the side (left or right) on which the longer tail resides.
2. **Kurtosis** is vertical stretching of a frequency distribution.
3. **Outliers** are extreme values, widely deviant from the mean.

VII. Methods of Depicting a Frequency Distribution
A. **Histogram**
1. A histogram is a bar graph in which the number of units of observation (e.g., persons) is shown on the y-axis, the measurement values (e.g., cholesterol levels) are shown on the x-axis, and the frequency distribution of data points is portrayed by a series of bars.
2. The area of each bar represents the relative proportion of all observations that fall within the range represented by that bar (see Table 9-1 and Fig. 9-3).

B. **Frequency polygon**
1. A frequency polygon is created by placing a dot along the y-axis for each value along the x-axis, and then connecting the dots with a line (see Table 9-1 and Fig. 9-4).
2. Two advantages of frequency polygons are:
   a. The shape of the distribution is more easily seen in a frequency polygon than in a histogram.
   b. A linear dose-response relationship is more clearly suggested by a frequency polygon than by a histogram.

C. **Line graph**
1. A line graph is recommended when the x-axis represents time and the y-axis represents rates.
2. An **arithmetic line graph** uses an arithmetic scale for both axes (see Chapter 3).
3. A **semilogarithmic line graph** has an x-axis that uses an arithmetic scale and a y-axis that uses a logarithmic scale in order to amplify the lower end of the scale. A semilogarithmic line graph has two advantages:
   a. It reveals the detailed changes in very low rates of disease.
   b. It depicts proportionately similar changes as parallel lines, even if they are very different in absolute magnitude.

D. **Stem and leaf diagram**
1. In a stem and leaf diagram, the **stem** is the vertical column of numbers on the left, and represents the value of the left-hand digit (e.g., 10s or 100s) in the data set.
2. The **leaf** is the set of numbers immediately to the right of the stem and is sometimes separated from the stem by a vertical line. Each number in the leaf represents the next digit in each of the observations.
3. The **#** symbol to the right of the leaf indicates the number of observations in each of the specified ranges (see Table 9-2 and Fig. 9-5).

E. **Quantile**
1. A quantile is a display of data stratified by percentiles.
2. A quantile generally includes the maximum value (the 100% value); the minimum value (the 0% value); the 99%, 95%, 90%, 10%, 5%, and 1% values; the range; the mode; and the interquartile range (from the 25th to the 75th percentile, abbreviated Q3 − Q1) (see Table 9-2 and Fig. 9-5).

F. **Boxplots**
1. Boxplots provide an even briefer way of summarizing the data in a distribution than the stem and leaf diagram does.
2. In the boxplot (see Table 9-2 and Fig. 9-5), the rectangle formed by four plus (+) signs and the horizontal dashes (----) depicts the interquartile range.
3. The two asterisks (*) connected by dashes depict the median.
4. The mean, shown by the smaller plus (+) sign, is very close to the median.
5. Outside of the rectangle, there are two vertical lines, called the "whiskers" of the boxplot. The whiskers show the range where other values might be expected, given the median and interquartile range of the distribution.
6. A quick look at the boxplot will reveal how wide the distribution is, whether or not it is skewed, where the interquartile range falls, how close the median is to the mean, and how many (if any) observations might reasonably be considered outliers.

VIII. Use of Unit-Free (Normalized) Data
A. A frequency distribution is characterized by its **mean** and its **standard deviation;** however, the value of these parameters varies depending on the units of measurement chosen (e.g., a set of weight measurements will produce different values for the mean and the standard

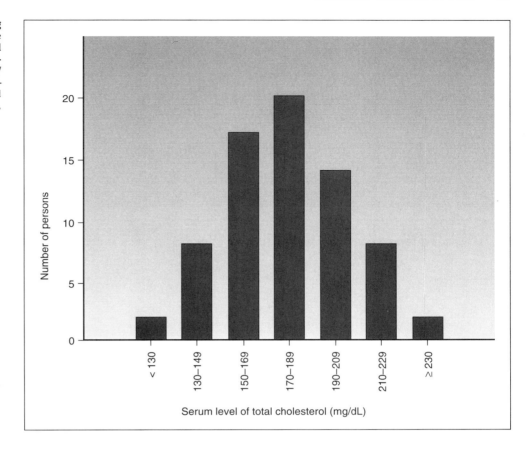

FIGURE 9-3. **Histogram showing the frequency distribution of the serum levels of total cholesterol reported in a sample of 71 subjects, grouped in ranges of 20 mg/dL.** Individual values for the 71 subjects are reported in Table 9-1 and Fig. 9-1. The mean is 179.1 mg/dL, and the median is 178 mg/dL.

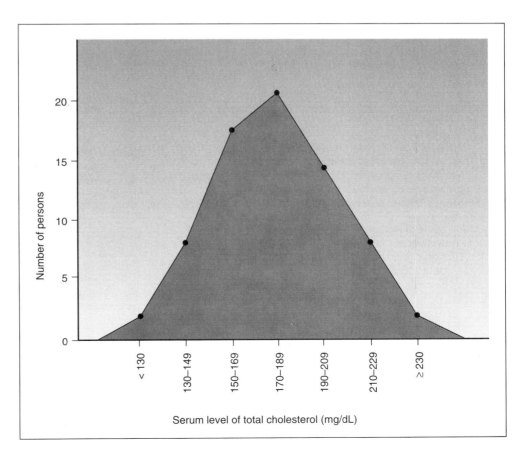

FIGURE 9-4. **Frequency polygon showing the frequency distribution of the serum levels of total cholesterol reported in a sample of 71 subjects, grouped in ranges of 20 mg/dL.** The data in this polygon are the same as the data in the histogram shown in Fig. 9-3. Individual values for the 71 subjects are reported in Table 9-1 and Fig. 9-1.

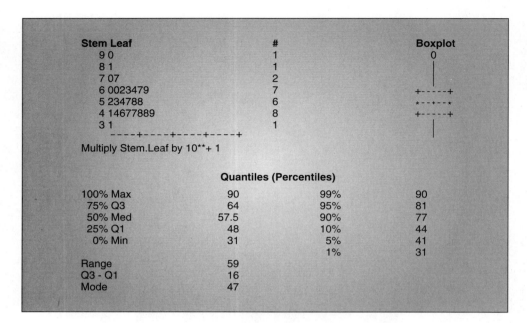

FIGURE 9–5. Stem and leaf diagram, boxplot, and quantiles (percentiles) for the data shown in Table 9–2, as printed out by the Statistical Analysis System (SAS). See the text for a description of how to interpret the data here.

deviation depending on whether the weights are recorded in kilograms or pounds).
- **B.** Unit-free (normalized) data eliminate the distorting effect of measurement units. To normalize data, the following steps are taken:
  1. The mean is set equal to 0 by subtracting the mean from each observation in the data set in whatever units have been chosen.
  2. Each observation is converted to a **z value,** or the number of standard deviations it is above or below the mean. This is done by subtracting the mean ($\bar{x}$) from a given observation (x_i) and dividing the difference by the standard deviation (s):

  $$z_i = \frac{x_i - \bar{x}}{s}$$

  3. A distribution of z values always has a mean of 0 standard deviations and always has a standard deviation of 1 standard deviation.
  4. The z values may be called by various names, including **standard normal deviates.**

IX. Frequency Distributions of Dichotomous Data and Proportions
- **A. Dichotomous data** are characterized by only two possible outcomes.
  1. Dichotomous data can be thought of in terms of flipping a coin.
  2. If the coin is flipped in an unbiased manner, on the average it would be expected to land with the heads side up in half of the flips and with the tails side up in half of the flips, so the probability of heads would equal 0.5 and the probability of tails would equal 0.5.
  3. The sum of all of the probabilities for all of the possible outcomes must equal 1.0.
  4. If a coin is flipped 10 times, the result would very rarely be 10 heads or 10 tails, would somewhat less rarely be a combination of 9 heads and 1 tail, and would most frequently be a combination of 5 heads and 5 tails.
- **B.** The probability of any particular combination of outcomes is obtained from the formula $(a + b)^n$, where a is the probability of the first outcome (i.e., heads), b is the probability of the second outcome (i.e., tails), and n is the number of coin tosses in the trial.
  1. The process of calculating the probability for each of the possible outcomes in a trial is called **expanding the binomial.**
  2. The distribution of probabilities for each combination represents the **binomial distribution.**
  3. When n is large and there is an equal probability of either of the two outcomes, or $a = b = 0.5$, then the binomial distribution approximates the normal (gaussian) distribution.
  4. If the probability of each outcome does not equal 0.5, the binomial distribution becomes skewed, complicating the analysis.
- **C. Binary data** expressed as **proportions** can be analyzed using theory based on the normal distribution.

X. Nonparametric Data
- **A.** Data based on a normal distribution or a binomial distribution are characterized by the mean and the standard deviation, which are known as **parameters.**
- **B.** Nominal (categorical) and ordinal (ranked) data do not necessarily conform to either of the above distributions, and do not permit assumptions to be made regarding the parameters noted. Therefore, these data are **nonparametric data.**
  1. There are **nonparametric tests** for the analysis of nonparametric data (see Chapter 11).
  2. A nonparametric test of particular importance is the **chi-square test** (see Chapter 11).
     a. The data analyzed with a chi-square are counts in frequency tables, which do not follow any particular distribution.
     b. The chi-square does have a particular, consistent distribution, but is nonparametric because the data on which it is performed do not.

QUESTIONS

DIRECTIONS. (Items 1–8): Each of the numbered items or incomplete statements in this section is followed by answers or by completions of the statement. Select the ONE lettered answer or completion that is BEST in each case. Answers and explanations are given at the end of the chapter.

1. A 62-year-old male is rushed to the emergency department by ambulance during an episode of chest pain. The initial evaluation is performed by a triage nurse, who reports to the emergency department attending physician that the patient's pain is probably angina and seems to be severe. This characterization is

   (A) ordinal
   (B) nominal
   (C) parametric
   (D) qualitative
   (E) dichotomous

2. The patient described in item 1, above, reports to the evaluating physician that his pain is an 8 on a scale of 0 (no pain) to 10 (worst pain). After the administration of sublingual nitroglycerin and high-flow oxygen, the patient reports, with obvious relief, that the pain is now a 4 on the same scale. After the administration of morphine sulfate, given as an intravenous push, the pain is gone. This commonly applied scale is

   (A) ordinal
   (B) nominal
   (C) ratio
   (D) dichotomous
   (E) continuous

Items 3–8

A group of 10 volunteers are weighed in a consistent manner before and after they have been on an experimental diet for 6 weeks. This diet consists of apple strudel, tomatilla salsa, and gummy bears. (**N.B.** Don't try this at home!) The weights, in kilograms, are as follows:

Volunteer	Weight Before Diet (kg)	Weight After Diet (kg)
1	81	79
2	79	87
3	92	90
4	112	110
5	76	74
6	126	124
7	80	78
8	75	73
9	68	76
10	78	76

3. The mean weight before the intervention is

   (A) 79
   (B) 86.7
   (C) 76.8
   (D) 91
   (E) 80

4. The median weight after the intervention is

   (A) 71.8
   (B) 81.7
   (C) 86.7
   (D) 89
   (E) 78.5

5. The mode of the weights after the intervention is

   (A) 80
   (B) 78.5
   (C) 86
   (D) 76
   (E) 100

6. The standard deviation of the weights before the intervention is

   (A) 330
   (B) 18.3
   (C) 75.7
   (D) 3026
   (E) 6.1

7. In order to determine whether this diet is effective in promoting weight loss, you intend to perform a statistical test of significance on the differences in weights before and after the intervention. Unfortunately, you don't know how to do this until you get through Chapter 10. What you *do* know now is that, in order to use a parametric test of significance,

   (A) the data in both sets must be normally distributed
   (B) the variances for the two sets of data must be equal
   (C) the means for the two sets of data must be equal
   (D) the distribution of weight in the underlying population must be normal (gaussian)
   (E) the data must not be skewed

8. To determine whether this diet is effective in promoting weight loss, a statistical approach should

   (A) minimize variation due to factors other than the intervention
   (B) produce a binary outcome variable
   (C) compensate for random error
   (D) avoid assumptions about parameters of the frequency distribution
   (E) compensate for selection bias

ANSWERS AND EXPLANATIONS

1. The answer is D: qualitative. A great deal of clinical information is purely descriptive and not specifically quantifiable. Such information, or data, is qualitative. Magnitude is implied by the descriptive modifiers applied to qualitative data, such as "severe" in this case. While we may all readily agree that "severe" is of greater magnitude than "mild," the actual measure of the discrepancy is subjective. Qualitative data need not be free of implied magnitude but are, by definition, devoid of any objective scale of measurement. *Epidemiology,* p. 108.
2. The answer is C: ratio. The scale described uses ratio data. A ratio scale is defined by a continuous variable and a true 0 point. The pain scale has a true 0, indicating the absence of pain. A score of 8 on the scale implies that the pain is twice as severe as pain having a score of 4. Of course, there is a subjective component to all human experience, even when a quantitative system is used in its description. One person's 4 on this scale is likely to be another's 2, or 7. While the exact meaning of "twice as much pain" may be unclear, it is also generally of little importance. This scale provides the practitioner attempting to alleviate pain with the essential information: whether the pain is increasing or decreasing, and by relatively how much. *Epidemiology,* p. 110.
3. The answer is B: 86.7. The mean is the sum of all of the observations in a data set, divided by the number of observations. The sum of the 10 observations (i.e., weights prior to the intervention) is 867. This figure, divided by 10 (the number of observations) yields 86.7. *Epidemiology,* p. 115.
4. The answer is E: 78.5. The median is the middle measure of a data set arranged in ascending or descending order. The data provided here can be rewritten as 73, 74, 76, 76, 78, 79, 87, 90, 110, and 124. In a set of ten, the median is the average of the fifth and sixth data points. In this case, that is 78 (fifth data point) plus 79 (sixth data point) divided by 2, which is 78.5. *Epidemiology,* p. 115.
5. The answer is D: 76. The mode is the most frequently occurring value in a set of data. The value 76 occurs twice in this set, so it is the mode. The mode can occur anywhere in the range represented by a set of data. If more than one value in a data set occurs repeatedly with equal frequency, the set may have more than one mode. *Epidemiology,* p. 115.
6. The answer is B: 18.3. The standard deviation is the square root of the variance. The variance is $\Sigma(x_i - \bar{x})^2$ divided by the degrees of freedom, or $N - 1$. The mean is subtracted from the first observation, and the difference is squared. This is repeated for each observation in the set, and the values obtained are all added together. For this set of data, the result is 3026.1. This figure is divided by $N - 1$, or 9, to yield the variance, which is 336.2. The square root of 336.2 is 18.3. *Epidemiology,* p. 117.
7. The answer is D: the distribution of weight in the underlying population must be normal (gaussian). All so-called parametric tests of significance rely upon assumptions about the parameters that define a frequency distribution—namely, the mean and the standard deviation. To employ parametric methods of statistical analysis, the data being analyzed need not be normally distributed, but the underlying population data from which they are drawn must be. Neither the means nor the variances of two data sets under comparison need to be equal to support the use of parametric methods. In this case, to employ a parametric test of significance, one needs to assume that the distribution of weight in the general population is normal. This assumption is reasonable. *Epidemiology,* p. 111.
8. The answer is A: minimize variation due to factors other than the intervention. There are many sources of variation in clinical data. The goal of statistical analysis in medical research is generally to measure the effect of a particular intervention. The effect of the intervention, therefore, must be isolated from other sources of variation in the data. In this study, for example, subjects might have gained or lost weight for reasons having nothing to do with the particular diet under investigation. A statistical approach should be taken to minimize variation due to factors other than the intervention of interest. For this study, the optimal method of analysis would be to compare each subject to himself or herself, thereby eliminating a great deal of intersubject variability; this will be discussed in Chapter 10. Although it is useful at times to convert a continuous variable into a binary variable, there is no advantage in it here, and considerable information would be lost. Assumptions about the parameters of the underlying frequency distribution—namely, that weight is normally distributed in the general population—are essential to the use of parametric methods, as discussed in the explanation of item 7, above. Random error should be minimized by meticulous study design because once it is introduced, random error cannot be corrected by statistical analysis. An example of random error in this study would be inaccuracies in the measurement of weight. Selection bias (see Chapter 4) is introduced whenever study subjects differ in some important way (such as their willingness to participate in the study) from the larger population they are meant to represent. Statistical methods cannot compensate for selection bias, which tends to reduce the external validity of a study but not compromise its internal validity (see Chapter 10). *Epidemiology,* p. 108.

REFERENCE CITED

Jekel, J. F. Epidemiology, Biostatistics, and Preventive Medicine. Philadelphia, W. B. Saunders Company, 1996.

CHAPTER TEN

Statistical Inference and Hypothesis Testing

SYNOPSIS

OBJECTIVES
- To discuss statistical inference by distinguishing between inductive and deductive reasoning, and between their applications to mathematics and statistics.
- To discuss hypothesis testing and the application of tests of statistical significance.
- To define the alpha level and the *p* value, and to discuss their use in hypothesis testing.
- To distinguish between standard deviation and standard error as measures of variation.
- To discuss the meaning and application of confidence intervals.
- To explain the basic principles underlying tests of statistical significance, and to introduce several important examples of such tests.
- To distinguish between variation within groups and variation between groups, and to describe the role of statistics in explaining variation.
- To distinguish between statistical and clinical significance, and to consider the importance of external validity.

I. Definitions
A. Statistical inference can be defined as the drawing of conclusions from quantitative or qualitative information using the methods of statistics to describe and arrange the data and to test suitable hypotheses.
B. Inductive reasoning (from Latin, meaning "to lead into") seeks to find valid generalizations and general principles from data.
  1. Inductive reasoning proceeds *from the specific* (i.e., from data) *to the general* (i.e., to formulas or conclusions).
  2. In statistics, the data are known but the general formula characterizing the relationship between an independent and a dependent variable is to be determined.
     a. As applied to the formula for a straight line, $y = mx + b$, statistics is used to derive the equation given multiple observations (i.e., values of x and y).
     b. In statistics, the variables x and y are known for all observations but the slope of the line joining them and the y-intercept are unknown and to be determined.
C. Deductive reasoning (from Latin, meaning "to lead out from") forms the basis for mathematics.
  1. Deductive reasoning proceeds *from the general* (i.e., from assumptions, from propositions, and from formulas considered true) *to the specific* (i.e., to specific members belonging to the general category).
  2. In mathematics, the general formula characterizing the relationship between an independent and a dependent variable is known but specific values of the dependent variable are unknown.
     a. As applied to the formula for a straight line, $y = mx + b$, mathematics is used to derive the values of y (or x) given the equation and the value of the other variable.
     b. What is known in statistics is unknown in mathematics, and vice versa.

II. The Process of Testing Hypotheses
A. Asserting the null hypothesis
  1. By convention, the expected outcome is asserted to be *no significant association* (or meaningful difference) between the variables (or outcomes) under consideration: this is the **null hypothesis.**
  2. By convention, the existence of a meaningful difference in outcomes or of a meaningful association between two variables is considered the **alternative hypothesis.**

B. Applying a test of statistical significance

1. Tests of significance are used to determine whether observed differences in outcome or observed associations represent large enough deviations from the null hypothesis to justify *rejecting the null hypothesis,* or whether such deviations are small enough to be readily attributable to random variation.
2. Significance testing is subject to both false-positive error (type I or alpha error) and false-negative error (type II or beta error).
   a. **False-positive error** is the assertion that a hypothesis is true when in reality it is false.
   b. **False-negative error** is the failure to assert that a hypothesis is true when in reality it is true.
   c. Greater care is generally taken to minimize false-positive error in medicine, although both types of error are important. The clinical setting determines the relative importance of each.
   d. The **alpha level** is established as a defense against false-positive error.
      (1) The alpha level represents the maximum probability of false-positive error that the investigator is willing to accept.
      (2) By convention, alpha is set at 0.05, which implies a 5% risk of rejecting the null hypothesis (i.e., asserting the alternative hypothesis) when in fact the alternative hypothesis is false.
   e. The ***p* value** obtained from a statistical test indicates the probability that the observed difference in outcome (or observed association) could have been obtained by chance.
      (1) If the value of *p* obtained is smaller than the predetermined value of alpha, the null hypothesis is rejected, implying that a true difference (or association) does exist.
      (2) If the value of *p* is greater than alpha, the null hypothesis is not rejected, and any difference between the expected and actual outcome is attributed to random variation (chance).

C. Standard deviation and standard error

1. The **standard deviation** (see Chapter 9) is a measure of dispersion, or variation, among the observations within a single data set.
2. The **standard error** is the standard deviation of a *population of sample means* rather than of individual observations. The standard error refers to the variability of means rather than the variability of individual observations.
   a. The formula for calculation of the standard error is as follows:

   $$\text{Standard error} = SE = \frac{SD}{\sqrt{N}}$$

   b. The standard error is predicated on the assumption that the mean for a single set of observations is representative of many such means from many such data sets. The standard deviation of the distribution of these means (standard error) is an unbiased estimate of the standard deviation of the underlying population.
   c. Calculation of the standard error facilitates performance of tests of statistical significance.
   d. Calculation of the standard error provides a measure of error, or uncertainty, in a single mean and permits estimation of the confidence interval (see below) surrounding the estimate.

D. Confidence intervals

1. Whereas the mean ±1.96 standard deviations estimates the range within which 95% of individual observations would be expected to fall, the mean ±1.96 standard errors estimates the range within which 95% of the means of repeated samples would be expected to fall.
2. The mean ±1.96 standard errors is the **95% confidence interval,** the range of values within which the investigator can be 95% certain the true population mean falls (see Table 10–1 and Box 10–1).
3. Confidence intervals may serve as a test of statistical significance for a risk ratio or an odds ratio. If the 95% confidence interval about a risk ratio includes the value of 1 (e.g., 0.92 to 2.70), the risk is not different from 1 with 95% confidence. If the entire 95% confidence interval falls on either side of 1 (e.g., 0.22 to 0.78, or 1.65 to 4.32), the exposure under investigation significantly modifies the outcome risk.

III. Tests of Statistical Significance

A. Tests of statistical significance allow for comparison of two parameters, such as means and proportions, in order to determine whether the difference between them is statistically significant.

B. Critical ratios

1. Critical ratios are a class of tests of statistical significance that depend on dividing some parameter (such as a difference between means) by the standard error (SE) of that parameter.
2. The general formula for tests of significance is as follows:

$$\text{Critical ratio} = \frac{\text{Parameter}}{\text{SE of that parameter}}$$

3. For any critical ratio, the larger the ratio is, the more likely it is that the difference between means

TABLE 10–1. Systolic and Diastolic Blood Pressure Values of 26 Young, Healthy, Adult Subjects

	Blood Pressure		
Subject	Systolic (mm Hg)	Diastolic (mm Hg)	Sex
1	108	62	Female
2	134	74	Male
3	100	64	Female
4	108	68	Female
5	112	72	Male
6	112	64	Female
7	112	68	Female
8	122	70	Male
9	116	70	Male
10	116	70	Male
11	120	72	Male
12	108	70	Female
13	108	70	Female
14	96	64	Female
15	114	74	Male
16	108	68	Male
17	128	86	Male
18	114	68	Male
19	112	64	Male
20	124	70	Female
21	90	60	Female
22	102	64	Female
23	106	70	Male
24	124	74	Male
25	130	72	Male
26	116	70	Female

Source of data: Unpublished findings in a sample of 26 professional persons in Connecticut.

> **BOX 10-1. Calculation of the Standard Error and the 95% Confidence Interval for Systolic Blood Pressure Values of 26 Subjects**
>
> **Part 1. Beginning data (see Table 10-1):**
>
> Number of observations, or $N = 26$
> Mean, or $\bar{x}$ $= 113.1$ mm Hg
> Standard deviation, or SD $= 10.3$ mm Hg
>
> **Part 2. Calculation of the standard error, or SE:**
>
> $$SE = \frac{SD}{\sqrt{N}} = \frac{10.3}{\sqrt{26}} = \frac{10.3}{5.1} = 2.02 \text{ mm Hg}$$
>
> **Part 3. Calculation of the 95% confidence interval, or 95% CI:**
>
> $$\begin{aligned} 95\% \text{ CI} &= \text{mean} \pm 1.96 \text{ SE} \\ &= 113.1 \pm (1.96)(2.02) \\ &= 113.1 \pm 3.96 \\ &= \text{between } 113.1 - 3.96 \text{ and } 113.1 + 3.96 \\ &= 109.1, 117.1 \text{ mm Hg} \end{aligned}$$

> **BOX 10-2. The Idea behind the Degrees of Freedom**
>
> The term "degrees of freedom" refers to the number of observations (N) that are free to vary. A degree of freedom is lost every time a mean is calculated. Why should this be?
>
> Before putting on a pair of gloves, a person has the freedom to decide whether to begin with the left or right glove. However, once the person puts on the first glove, he or she loses the freedom to decide which glove to put on last. If centipedes put on shoes, they would have a choice to make for the first 99 shoes but not for the 100th shoe. Right at the end, the freedom to choose (vary) is restricted.
>
> In statistics, if there are two observed values, only one estimate of the variation between them is possible. Something has to serve as the basis against which other observations are compared. The mean is the most "solid" estimate of the expected value of a variable, so it is assumed to be "fixed." This implies that the numerator of the mean (the sum of individual observations, or the sum of x_i), which is based on N observations, is also fixed. Once $N - 1$ observations (each of which was, presumably, free to vary) have been added up, the last observation is not free to vary, because the total values of the N observations must add up to the sum of x_i. For this reason, 1 degree of freedom is lost each time a mean is calculated. The proper average of a sum of squares when calculated from an observed sample, therefore, is the sum of squares divided by the degrees of freedom ($N - 1$).

or proportions is due to more than random variation (i.e., the more likely it is that the difference can be considered statistically significant and, hence, real).

4. Critical ratios are associated with a particular p value; if the p value of the critical ratio is smaller than alpha (usually 0.05), the null hypothesis is rejected (a true difference between the two measures of interest is asserted to exist).

C. **Degrees of freedom**
1. Degrees of freedom refers to the number of observations that are free to vary (see Box 10-2).
2. For simplicity, the degrees of freedom for any test are considered to be the total sample size minus 1 degree of freedom for each mean that is calculated.

IV. **Use of t-tests**

A. **The purpose of a t-test** is to compare the means of a continuous variable in two samples in order to determine whether or not the difference between the two observed means exceeds the difference that would be expected by chance.

B. **Sample populations and sizes**
1. If the two samples come from two different groups (e.g., a group of men and a group of women), the Student's t-test is used.
2. If the two samples come from the same group (e.g., pretreatment and posttreatment values for the same study subjects), the paired t-test is used.
3. t-Tests require that the sample means of interest be normally distributed. The **central limit theorem** asserts that the distribution of the means of many samples (given a sample size of 20 or more) is normal. Therefore, a t-test may be computed on almost any set of continuous data if the observations can be considered a random sample and the sample size is reasonably large.
4. The t-distribution is similar to the z, or normal, distribution but with wider tails and a lower peak to account for the greater uncertainty (error) introduced when sample sizes are small. For samples larger than 120, the difference between the t and z distributions is negligible.

V. **Student's t-Test**

A. **Calculation of the value of t**
1. In both types of Student's t-test (the one-tailed type and the two-tailed type), t is calculated by taking the observed difference between the means of the two groups (the numerator) and dividing this difference by the standard error of the difference between the means of the two groups (the denominator).
2. The **standard error of the difference between the means** (SED) is the square root of the sum of the respective population variances, each divided by its own sample size.
3. In theoretical terms, the equation for the SED would be as follows:

$$\text{SED of } \mu_E - \mu_C = \sqrt{\frac{\sigma_E^2}{N_E} + \frac{\sigma_C^2}{N_C}}$$

where the Greek symbol μ is the population mean, E is the experimental population, C is the control population, σ^2 is the variance of the population, and N is the number of observations in the population (see Box 10-3).

4. Generally, the population variances (σ^2) are not known; the sample variances (s^2) may be substituted.
5. The assumption is generally made that there is no difference between the variances, so that a pooled estimate of the SED (SED_p) is substituted.
6. The formula for the standard error of the difference becomes the following:

$$\begin{aligned} \text{SED}_p \text{ of } \bar{x}_E - \bar{x}_C &= \sqrt{s_p^2 \left(\frac{1}{N_E} + \frac{1}{N_C} \right)} \\ &= \sqrt{s_p^2 [(1/N_E) + (1/N_C)]} \end{aligned}$$

> **BOX 10-3. The Formula for the Standard Error of the Difference between Means**
>
> The standard error equals the standard deviation (σ) divided by the square root of the sample size (N). Alternatively, this can be expressed as the square root of the variance (σ^2) divided by N:
>
> $$\text{Standard error} = \frac{\sigma}{\sqrt{N}} = \sqrt{\frac{\sigma^2}{N}}$$
>
> As mentioned in Chapter 9, Box 9-1, the variance of a difference is equal to the sum of the individual variances. Therefore, the variance of the difference between the mean of an experimental group (μ_E) and the mean of a control group (μ_C) could be expressed as follows: $\sigma_E^2 + \sigma_C^2$.
>
> As shown above, a standard error can be written as the square root of the variance divided by the sample size, allowing the equation to be expressed as:
>
> $$\text{Standard error of } \mu_E - \mu_C = \sqrt{\frac{\sigma_E^2}{N_E} + \frac{\sigma_C^2}{N_C}}$$

7. The s_p^2, which is the **pooled estimate of the variance**, is a kind of average of s_E^2 and s_C^2.
8. The s_p^2 is calculated as the sum of the two sums of squares, divided by the combined degrees of freedom, as follows:

$$s_p^2 = \frac{\sum(x_E - \bar{x}_E)^2 + \sum(x_C - \bar{x}_C)^2}{N_E + N_C - 2}$$

9. If one sample size is much greater than the other or if the variance of one sample is much greater than the variance of the other, more complex formulas are needed.
10. When the Student's t-test is used to test the null hypothesis in research involving an experimental group and a control group, it usually takes the form of the following equation:

$$t = \frac{\bar{x}_E - \bar{x}_C - 0}{\sqrt{s_p^2[(1/N_E) + (1/N_C)]}}$$

$$df = N_E + N_C - 2$$

11. The 0 in the numerator of the equation for t was added for correctness, because the t-test determines if the difference between the means is significantly different from 0. However, because the 0 does not affect the calculations in any way, it is usually omitted from t-test formulas.
12. The same formula, recast in terms to apply to any two independent samples (e.g., samples of men and women), is as follows:

$$t = \frac{\bar{x}_1 - \bar{x}_2 - 0}{\sqrt{s_p^2[(1/N_1) + (1/N_2)]}}$$

$$df = N_1 + N_2 - 2$$

in which $\bar{x}_1$ is the mean of the first sample, $\bar{x}_2$ is the mean of the second sample, s_p^2 is the pooled estimate of the variance, N_1 is the size of the first sample, N_2 is the size of the second sample, and df is the degrees of freedom.
  a. The 0 in the numerator indicates that the null hypothesis states that the difference between the means will not be significantly different from 0.
  b. The df is needed to enable the investigator to refer to the correct line in the table of values of t and their relationship to p (see the Appendix).
  c. Boxes 10-4 and 10-5 may help in understanding the t-test.
B. **Interpretation of the results**
 1. If the value of t is large, the p value will be small, because it is unlikely that a large t ratio will be obtained by chance alone.
 2. If the p value is 0.05 or less, it is customary to assume that there is a real difference.
C. **One-tailed and two-tailed t-tests** (see Box 10-6)
 1. Two-tailed t-test
  a. In the two-tailed test, alpha is equally divided at the ends of the two tails of the distribution (see Fig. 10-1A).
  b. The two-tailed test is generally recommended, because differences in either direction are usually important to document.
  c. For example, it is obviously important to know if a new treatment is significantly better than a standard or placebo treatment, but it is also important to know if a new treatment is significantly worse and should therefore be avoided. In this situation, a two-tailed test provides an accepted criterion for when a difference shows the new treatment to be either better or worse.
 2. One-tailed t-test
  a. When it is acceptable to look for a difference in only one direction (e.g., to determine whether a new technology is better than an old one before accepting the higher costs associated with its use), a one-tailed t-test is used.
  b. In the one-tailed test, the 5% rejection region for the null hypothesis is all put on one tail of the distribution (see Fig. 10-1B), instead of being evenly divided between the extremes of the two tails.
  c. The null hypothesis nonrejection region extends only to 1.645 standard errors above the "no difference" point in the one-tailed test. In contrast, it extends to 1.96 standard errors above the "no difference" point in the two-tailed test. This makes the one-tailed test more robust (i.e., more able to detect a significant difference).

VI. **Paired t-test**
 A. **Use of the paired t-test**
  1. In many medical studies, individuals are followed over time to see if there is a change in the value of some continuous variable (typically, in a "before and after" experiment). The individual patient thus serves as his or her own control. The appropriate statistical test for this kind of data is the paired t-test.
  2. The paired t-test is more robust than the Student's t-test because it considers the variation from only one group of people, whereas the Student's t-test considers variation from two groups.
  3. Any variation that is detected in the paired t-test is attributable to the intervention or to changes over time in the same person.
 B. **Calculation of the value of t**
  1. To calculate a paired t-test, a new variable is created. This variable, called d, is the difference be-

BOX 10-4. Calculation of the Results of the Student's *t*-Test Comparing the Systolic Blood Pressure Values of 14 Male Subjects with Those of 12 Female Subjects

Part 1. Beginning data (see Table 10-1):

Number of observations, or N = 14 for males, or M
  12 for females, or F

Mean, or $\bar{x}$ = 118.3 mm Hg for males
  107.0 mm Hg for females

Variance, or s^2 = 70.1 mm Hg for males
  82.5 mm Hg for females

Sum of $(x_i - \bar{x})^2$, or TSS = 911.3 mm Hg for males
  907.5 mm Hg for females

Alpha value for the *t*-test = 0.05

Part 2. Calculation of the *t* value based on the pooled variance (s_p^2) and the pooled standard error of the difference (SED$_p$):

$$s_p^2 = \frac{TSS_M + TSS_F}{N_M + N_F - 2} = \frac{911.3 + 907.5}{14 + 12 - 2} = \frac{1818.8}{24} = 75.78 \text{ mm Hg}$$

$$SED_p = \sqrt{s_p^2[(1/N_M) + (1/N_F)]}$$

$$= \sqrt{75.78(1/14 + 1/12)}$$

$$= \sqrt{75.78(0.1548)} = \sqrt{11.73} = 3.42 \text{ mm Hg}$$

$$t = \frac{\bar{x}_M - \bar{x}_F - 0}{\sqrt{s_p^2[(1/N_M) + (1/N_F)]}} = \frac{\bar{x}_M - \bar{x}_F - 0}{SED_p}$$

$$= \frac{118.3 - 107.0}{3.42} = \frac{11.30}{3.42} = \mathbf{3.30}$$

Part 3. Alternative calculation of the *t* value based on the SED equation using the observed variances for males and females, rather than on the SED$_p$ equation using the pooled variance:

$$SED = \sqrt{\frac{s_M^2}{N_M} + \frac{s_F^2}{N_F}} = \sqrt{\frac{70.1}{14} + \frac{82.5}{12}}$$

$$= \sqrt{5.01 + 6.88} = \sqrt{11.89} = 3.45 \text{ mm Hg}$$

$$t = \frac{\bar{x}_M - \bar{x}_F - 0}{SED}$$

$$= \frac{118.3 - 107.0 - 0}{3.45} = \frac{11.30}{3.45} = \mathbf{3.28}$$

Note that the results here (t = 3.28) are almost identical to those above (t = 3.30), even though the sample size is small.

Part 4. Calculation of the degrees of freedom (*df*) for the *t*-test and interpretation of the *t* value:

$$df = N_M + N_F - 2 = 14 + 12 - 2 = 24$$

For a *t* value of 3.30, with 24 degrees of freedom, *p* is less than 0.01, as indicated in the table of the values of *t* (see Appendix). This means that the male subjects have a significantly different (higher) systolic blood pressure than do the female subjects in this data set.

BOX 10-5. Does the Eye Naturally Perform *t*-Tests?

The paired diagrams below show three patterns of overlap between two frequency distributions (e.g., a treatment group and a control group). These distributions can be thought of as the frequency distributions of systolic blood pressure values among hypertensive patients following randomization and treatment either with the experimental drug or with a placebo. The treatment group's distribution is shown in dark gray, the control group's distribution is shown in light gray, and the area of overlap is shown with hatch marks. The means are indicated by the vertical dotted lines. The three different pairs show variation in the spread of systolic blood pressure values.

Take a look at the three diagrams. Then try to guess whether each pair was sampled from the *same* universe (i.e., was not significantly different) or was sampled from two *different* universes (i.e., was significantly different).

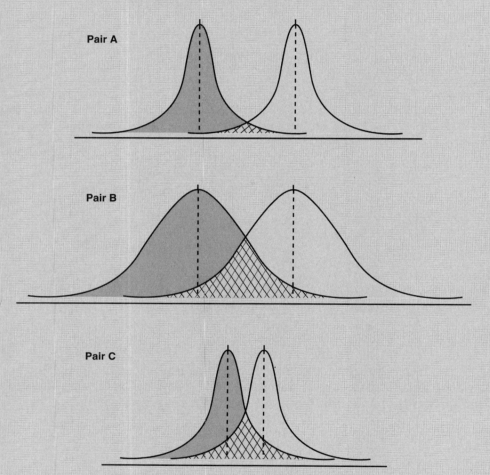

Most observers believe that the distributions in pair A look as though they were sampled from different universes. When asked why they think so, they usually state that there is little overlap between the two frequency distributions. Most observers are not convinced that the distributions in either pair B or pair C were sampled from different universes. They say that there is considerable overlap in each of these pairs, and this makes them doubt that there is a real difference. Their visual impressions are indeed correct.

It is not the absolute distance between the two means which leads most observers to say "different" for pair A and "not different" for pair B, because the distance between the means was drawn to be exactly the same in pairs A and B. Nor is it the absolute amount of dispersion that causes them to say "different" for pair A and "not different" for pair C, because the dispersions were drawn to be exactly the same in pairs A and C. Rather, the essential point, which the eye notices, is the ratio of the distance between the means to the variation around the means. The greater the distance between the means for a given amount of dispersion, the less likely it is that the samples were from the same universe. This ratio is exactly what the *t*-test calculates:

$$t = \frac{\text{Distance between the means}}{\text{Variation around the means}}$$

where the variation around the means is expressed as the standard error of the difference between the means. Therefore, the eye naturally does a *t*-test, although it does not quantify the relationship as precisely as does the *t*-test.

BOX 10-6. The Implications of Choosing a One-Tailed or Two-Tailed Test of Significance

For students who are confused by the implications of choosing a one-tailed or two-tailed test, an analogy may be helpful. The coach of a football team wants to assess the skill of potential quarterbacks. He (or she) is unwilling to allow mere completion of a pass to serve as evidence of throwing accuracy, because a pass could be completed by chance, even if the football did not go where the quarterback intended. Because the coach is something of a Monday-morning statistician, he further infers that if the quarterback were to throw randomly, the ball would often tend to land near the center of the field and less often way off toward one sideline or the other. The distribution of random throws might even be gaussian (thinks the coach).

Therefore, the coach asks quarterback applicants to throw to a receiver along the sideline. The coach announces that each applicant has a choice: He may pick one side ahead of time and complete a pass to that side within 5 feet of the sideline, or he may throw to either side but then must complete the pass within 2.5 feet of the sideline (assuming that the field is 100 feet wide). (If the coach allowed the quarterback who did not specify a particular sideline 5 feet along each side, the coach would "reject" his null hypothesis on the basis of chance 10% of the time, and he is unwilling to take so great a risk of selecting a lucky but unskillful quarterback. The coach's null hypothesis is simply that the quarterback will not be able to complete a pass within the specified zone.) In either case, a completed pass outside the specified zone is attributed to chance, since it is not what was intended.

Clearly, the quarterback has more room to work with if he prefers to throw to one side (one-tailed test) and can count on throwing in only that direction. If he is unsure in which direction he may wish to throw, he can get credit for a completed pass in either direction (two-tailed test) but has only a very narrow zone for which to aim.

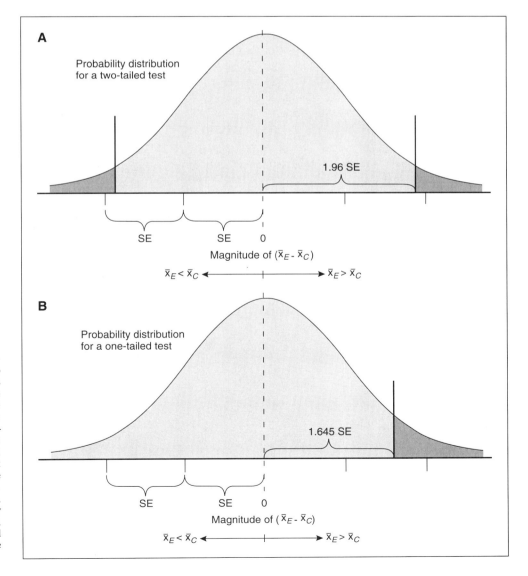

FIGURE 10-1. **Probability distribution of the difference between two means when the null hypothesis is actually true (i.e., when there is no real difference between the two means).** Dark gray indicates the zone for rejecting the null hypothesis, and light gray indicates the zone for failing to reject the null hypothesis. When a two-tailed test is used, there is a rejection zone on each side of the distribution. When a one-tailed test is used, there is a rejection zone on only one side. SE is the standard error, $\bar{x}_E$ is the mean for the experimental group, and $\bar{x}_C$ is the mean for the control group.

tween the values before and after the treatment for each individual studied.
2. The paired t-test is a test of the null hypothesis that, on the average, the difference is equal to 0, which is what would be expected if there were no change over time.
3. Using the symbol $\bar{d}$ to indicate the mean observed difference between the before and after values, the formula for the paired t-test is as follows:

$$t_{\text{paired}} = t_p = \frac{\bar{d} - 0}{\text{Standard error of } \bar{d}}$$

$$= \frac{\bar{d} - 0}{\sqrt{\frac{s_d^2}{N}}}$$

$$df = N - 1$$

4. The numerator contains a 0 because the null hypothesis says that the observed difference will not differ from 0; however, the 0 does not enter into the calculation and therefore can be omitted.
5. In the paired t-test, because only one mean is calculated ($\bar{d}$), only 1 degree of freedom is lost; therefore, the formula for the degrees of freedom is $N - 1$.

C. Interpretation of the results
1. The values of t and their relationship to p are shown in a statistical table in the Appendix.
2. If the value of t is large, the p value will be small, because it is unlikely that a large t ratio will be obtained by chance alone.
3. If the p value is 0.05 or less, it is customary to assume that there is a real difference (i.e., that the null hypothesis of no difference can be rejected).

VII. z-Tests
A. Use of z-tests
1. In contrast to t-tests, which compare differences between means, z-tests compare differences between proportions (e.g., sensitivity, specificity, positive predictive value, risks, percentages of people with a given symptom, percentages of people who are ill, and percentages of ill people who survive their illness).
2. Frequently, the goal of research is to see if the proportion of patients surviving is different in a treated group than in an untreated group.

B. Calculation of the value of z
1. As discussed earlier (see Critical Ratios), a critical ratio is calculated by dividing some parameter by the SE of that parameter. For z, the critical ratio is calculated by taking the observed difference between the two proportions (the numerator) and dividing it by the SE of the difference between the two proportions (the denominator).
2. If p is the proportion of successes (survivals), then $1 - p$ is the proportion of failures (nonsurvivals). If N represents the size of the group on which the proportion is based, the parameters of the proportion are as follows:

$$\text{Variance} = \frac{p(1-p)}{N}$$

$$\text{Standard error} = \text{SE} = \sqrt{\frac{p(1-p)}{N}}$$

95% Confidence interval = 95% CI = $p \pm 1.96$ SE

3. The **standard error of the difference between proportions** can be obtained, and the equation for the z-test can be expressed as follows:

$$z = \frac{p_1 - p_2 - 0}{\sqrt{\bar{p}(1-\bar{p})[(1/N_1) + (1/N_2)]}}$$

in which p_1 is the proportion of the first sample, p_2 is the proportion of the second sample, N_1 is the size of the first sample, N_2 is the size of the second sample, and $\bar{p}$ is the mean proportion of successes in all observations combined. The 0 in the numerator indicates that the null hypothesis states that the difference between the proportions will not be significantly different from 0.

C. Interpretation of the results
1. The value of z and the associated p value appear in a table of z values. If the z ratio is large, the p value will be small, the observed difference is likely to be real, and the null hypothesis will be rejected.

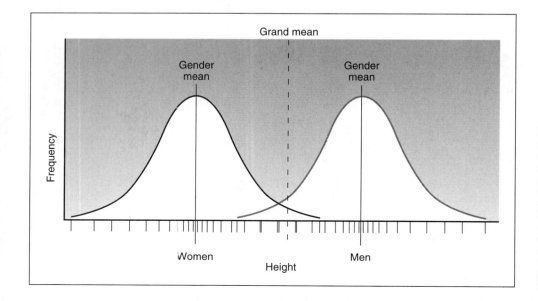

FIGURE 10-2. **Hypothetical frequency distribution of the heights of a sample of women (black marks along the x-axis) and a sample of men (gray marks along the x-axis), indicating the density of observations at the different heights.** An approximate normal curve is drawn over each of the two distributions, and the overall mean (grand mean) is indicated, along with the mean height for women (a gender mean) and the mean height for men (a gender mean).

2. When the data used to compute the z-test are set up as a 2 × 2 table, a **chi-square test** can be performed (see Chapter 11). Technically, the computations for the two tests are identical; however, the chi-square test is generally considered easier to do than a z-test for proportions.

VIII. Other Tests
A. Statistical significance tests used in the analysis of two variables (bivariate analysis) are discussed in Chapter 11.
B. Tests used in the analysis of multiple independent variables (multivariable analysis) are discussed in Chapter 13.

IX. Special Considerations
A. **Variation between groups versus variation within groups**
   1. Biostatistics is used to determine the amount of the **total variation** in a dependent variable (e.g., height) that is explained by an independent variable (e.g., gender) and whether or not the difference ascribable to the independent variable is more than would be expected by chance.
   2. The **total variation** is equal to the sum of the squared deviations, which is usually called the **total sum of squares** (TSS) but is sometimes referred to as the **sum of squares** (SS). Statistics is used to determine how much of this variation is actually due to the independent variable and how much is due to other factors.
   3. If an independent variable (e.g., gender) is responsible for some of the variation in a dependent variable (e.g., height), then the TSS from the overall mean (grand mean) will be greater than the TSS from the means for each group of the independent variable (e.g., the variation in height among women around the mean for women, plus the variation in height among men around the mean for men) (see Fig. 10–2).
      a. From a statistical perspective, *explaining variation implies reducing the unexplained SS.*
      b. If more explanatory variables (e.g., age, height of father, height of mother, nutritional status) are analyzed, the unexplained SS can be reduced still further, and even more of the variation explained.
   4. The greater the distance between the means for groups of the independent variable, the greater is the proportion of the variation that is likely to be explained by **variation between groups** (see Fig. 10–3).
   5. The larger the standard deviations within groups, the greater the proportion of the variation that is likely to be explained by **variation within groups** (see Fig. 10–3).
   6. Statistics is used to divide the total variation into a part that is explained by the independent variables (the model) and a part that is still unexplained. This is called **analyzing variation,** or analyzing the TSS. A specific method for doing this under certain circumstances and testing hypotheses at the same time is called **analysis of variance,** or **ANOVA** (see Chapter 13).

B. **Clinical importance and external validity versus statistical significance**
   1. Even if a finding is statistically significant, it may not be clinically or scientifically important.
   2. Before the findings of a study can be put to general clinical use, **external validity,** or **generalizability,**

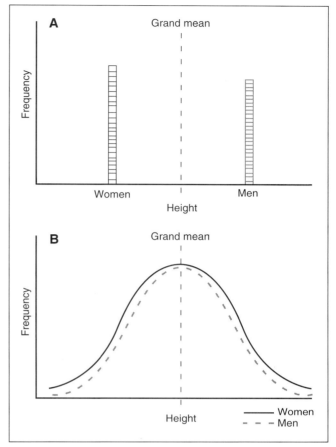

FIGURE 10-3. Two hypothetical frequency distributions of the heights of a sample of women (black lines) and a sample of men (gray lines). Diagram A shows how the distribution would appear if all women were of equal height, all men were of equal height, and men were taller than women. Diagram B shows how the distribution would appear if women varied in height, men varied in height, and the mean heights of the men and women were the same.

must be addressed. For example, whether the **sample** (the patients in the study) is representative of the **universe** (the patients for whom the new intervention might eventually be used) depends on the spectrum of disease and the spectrum of individual characteristics in the sample group.
   a. The **spectrum of disease** in the sample (i.e., the type, stage, and severity of disease) must be clearly defined, including both **inclusion criteria** (requirements for participation in the study) and **exclusion criteria** (characteristics that are not permitted among study subjects).
   b. The **spectrum of individual characteristics** in the sample of patients (e.g., age, gender, income levels, and ethnic backgrounds) can vary. The subject characteristics must be defined, including both inclusion and exclusion criteria.
   c. An appropriate **sampling technique** (see Chapter 12) is needed for the selection of the individual study subjects. The sampling method should always be reported along with the findings, because the generalizability of results will depend on both the sampling techniques and the spectrum of characteristics in the sample of patients.

QUESTIONS

DIRECTIONS. (Items 1–10): Each of the numbered items or incomplete statements in this section is followed by answers or by completions of the statement. Select the ONE lettered answer or completion that is BEST in each case. Correct answers and explanations are given at the end of the chapter.

1. The first three interns you meet feel a lot better since they started to take a commonly prescribed antidepressant. You reluctantly draw the conclusion that internship is associated with depression. (You also strive to identify the least talent you might have to pursue a career in anything else.) This is an example of

   (A) transference
   (B) interpolation
   (C) deductive reasoning
   (D) hypothesis testing
   (E) inductive reasoning

2. The conceptual approach to the formula for a straight line, which is $y = mx + b$, differs between mathematics and statistics. In statistics,

   (A) y is known, and x is to be determined
   (B) x, y, and m are known, and b is to be determined
   (C) this equation is irrelevant
   (D) x and y are known, and m and b are to be determined
   (E) only m is unknown

3. The basic goal of hypothesis testing is

   (A) to confirm the alternative hypothesis
   (B) to distinguish between random and meaningful differences in outcome
   (C) to establish the value of beta
   (D) to establish the value of alpha
   (E) to enhance the predictive value of a particular study design

4. The value of alpha serves as protection against

   (A) type I error
   (B) selection bias
   (C) type II error
   (D) false-negative results
   (E) inadequate sample size

5. Statistical significance is achieved when

   (A) p is greater than alpha
   (B) p is greater than beta
   (C) alpha is greater than p
   (D) beta equals alpha
   (E) the result is two-tailed

Items 6–10

A group of six volunteers have gone on a cholesterol-lowering diet for 3 months. The study diet is very strict: one-third teaspoon of diluted oat bran is sprinkled over the subjects' chocolate mousse every day. Gaunt and haggard from their ascetic experience, the subjects are recovering under the golden arches of the neighborhood dietary deprivation treatment center. You are left in the lab with a stale tunafish sandwich and the following data:

Subject	Pretrial Cholesterol Level (mg/dL)	Posttrial Cholesterol Level (mg/dL)
1	180	182
2	225	220
3	243	241
4	150	140
5	212	222
6	218	216

6. The standard deviation of the pretrial cholesterol values is

   (A) 12.5
   (B) 100
   (C) 42.6
   (D) 210
   (E) 33.8

7. The standard error of the pretrial cholesterol values is

   (A) 13.8
   (B) 27
   (C) 33.8
   (D) 33.6
   (E) 7.4

8. The appropriate test of statistical significance for this trial is

   (A) the critical ratio
   (B) the z-test
   (C) the paired t-test
   (D) the Student's t-test
   (E) the odds ratio

9. By mere inspection of these data, you conclude, regarding the difference in before and after cholesterol values, that

   (A) even if clinically meaningful, the difference is not statistically significant
   (B) if clinically meaningful, the difference must be statistically significant
   (C) if statistically significant, the difference must be clinically meaningful
   (D) the difference cannot be either clinically or statistically significant
   (E) even if statistically significant, the difference cannot be clinically meaningful

10. Before concluding that a lack of statistical significance in this trial proves that oat bran does not lower cholesterol levels, one would want to consider

    (A) type I error
    (B) the p value
    (C) beta error
    (D) the alpha level
    (E) the critical ratio

ANSWERS AND EXPLANATIONS

1. The answer is E: inductive reasoning. As discussed in the outline at the beginning of this chapter, statistics is based on inductive reasoning, a logical progression from the specific to the general. In statistics, conclusions are drawn about general associations based on the available data. In this example, based on a sample of three interns, the general association between internship and depression is drawn. In contrast, mathematics is based on deductive reasoning, in which general associations are known but specific values are unknown. *Epidemiology*, p. 124.
2. The answer is D: x and y are known, and m and b are to be determined. Question 2, like question 1, highlights the distinction between mathematics and statistics. In mathematics, one uses the formula for a linear (straight line) relationship between x and y to calculate the value of y for a given value of x. In statistics, however, the values of both x (the independent variable) and y (the dependent variable, or outcome variable) are known. What is to be established in statistics is the nature of the relationship between x and y. If x and y are related in a linear fashion, then the specific goal of statistics is to estimate values of m (the slope) and b (the y-intercept). *Epidemiology*, p. 124.
3. The answer is B: to distinguish between random and meaningful differences in outcome. A medical study begins with a hypothesis or belief. The belief, for example, that a particular drug effectively lowers blood pressure must be tested before the belief can gain widespread acceptance. The fundamental goals of hypothesis testing are to observe differences in outcome between two groups and to determine whether such differences are the result of random variation or are large enough to be significant (i.e., not likely the result of random variation). Much of statistics is devoted to this one basic task. The values of alpha and beta should be established before hypothesis testing and should influence only the stringency of requirements for statistical significance. Predictive value, discussed in Chapter 7, is not related to hypothesis testing. While rejection of the null hypothesis, and thus acceptance of the alternative hypothesis, might result from hypothesis testing, it cannot be considered the basic goal; hypothesis testing might show a lack of statistical significance and indicate that the null hypothesis should not be rejected. *Epidemiology*, p. 124.
4. The answer is A: type I error. The value of alpha represents the probability of false-positive error, or type I error. By convention, alpha is set at 0.05, indicating that a statistically significant result is one with no more than a 5% chance of occurring due to random variation. The smaller the value of alpha, the more difficult it becomes to achieve statistical significance, and the less likely one is to make a type I error. However, in order to avoid extreme type II error (false-negative error), some risk of type I error is unavoidable. Alpha, therefore, protects against type I error by setting a limit to its likelihood of occurrence. *Epidemiology*, p. 125.
5. The answer is C: alpha is greater than p. The value of p, as discussed in the chapter, is the likelihood that an observed outcome difference is due to random variation, or chance, alone. Alpha is the maximum risk one is willing to take that the observed outcome difference is due to chance when asserting the alternative hypothesis. Therefore, one rejects the null hypothesis, and asserts the alternative hypothesis, whenever p is less than or equal to the preselected value of alpha. By convention, $p \leq 0.05$ indicates statistical significance. *Epidemiology*, p. 125.
6. The answer is E: 33.8. As discussed in Chapter 9, the formula for the standard deviation (SD) is as follows:

$$SD = \sqrt{\frac{\sum(x_i - \bar{x})^2}{N - 1}}$$

The mean for the six pretrial observations is 204.67, rounded to 205. This value is then subtracted from each of the observations, the difference is squared, and these values summed to yield 5712. This figure is divided by ($N - 1$), or 5, to yield 1142.4. The square root of this figure, 33.8, is the standard deviation for the data set. *Epidemiology*, p. 126; Chapter 9.
7. The answer is A: 13.8. As discussed in this chapter, the standard error is the standard deviation divided by the square root of the sample size, or $SE = SD/\sqrt{N}$. The calculation of the standard deviation is explained in the answer to question 6, above. The standard error is then 33.8, divided by the square root of 6. With rounding, the answer is 13.8. Note that the standard error is smaller than the standard deviation. This is to be expected, both mathematically and conceptually. Conceptually, while the standard deviation is a measure of dispersion, or variation, among individual observations, the standard error is a measure of variation among means derived from repeated trials. One would expect that mean outcomes would vary less than their constituent observations. Mathematically, the standard error is the standard deviation divided by the square root of the sample size. Therefore, the larger the sample size, the smaller the standard error, and the greater the difference between the standard deviation and the standard error. *Epidemiology*, p. 127.
8. The answer is C: the paired t-test. A t-test is appropriate whenever two means are being compared and the population data from which the observations are derived are normally distributed. When the two means are from distinct groups, a Student's t-test is appropriate, as discussed in this chapter. When the data represent before and after results for a single group of subjects (i.e., when subjects serve as their own controls), the paired t-test is appropriate. The calculations for the value of t in a paired t-test are provided in the chapter. The paired t-test is more apt to detect a statistically significant difference than the Student's t-test because the variation has been reduced to that from one group rather than two. *Epidemiology*, p. 131.
9. The answer is E: even if statistically significant, the difference cannot be clinically meaningful. As discussed in the chapter, statistical significance and clinical significance are not synonymous. A clinically important intervention might fail to show statistical benefit over another in a trial if the sample size is too small. A statistically significant difference in outcomes in a very large sample might be of no clinical importance. Mere inspection of the data in this example suggests that they are unlikely to result in statistical significance, but one cannot be certain of this without formal hypothesis testing. However, the data clearly do not demonstrate a clinically meaningful effect even if statistical significance is achieved (which it is not, by the way). Unlike statistical significance, which is purely

numerical, clinical significance is the product of judgment. *Epidemiology*, p. 136.

10. The answer is C: beta error. A negative trial result may indicate that the null hypothesis is actually true, or it may be due to beta error (false-negative error). Beta error generally receives less attention in medicine than does alpha error. Consequently, the likelihood of a false-negative outcome is often unknown. A negative result can occur, for example, if the sample size is too small or if an inadequate dosage is administered. See Chapter 12 for further discussion of beta error and the related concept of power. *Epidemiology*, p. 125.

REFERENCE CITED

Jekel, J. F. Epidemiology, Biostatistics, and Preventive Medicine. Philadelphia, W. B. Saunders Company, 1996.

CHAPTER ELEVEN

BIVARIATE ANALYSIS

SYNOPSIS

OBJECTIVES

- To provide guidelines for the selection of an appropriate statistical test.
- To discuss the statistical methods employed to draw inferences from continuous data.
- To introduce the concepts of correlation and regression.
- To discuss statistical means of drawing inferences from ordinal data.
- To discuss statistical means of drawing inferences from dichotomous and nominal data, and to define and discuss the chi-square distribution and tests of significance.
- To discuss the interpretation of data in a 2 × 2 contingency table, with emphasis on the distinction between statistical significance and clinical significance.
- To present information about survival analysis and introduce the actuarial and Kaplan-Meier methods of life table analysis.

I. **Definitions**
  A. **Bivariate analysis** (the focus of this chapter) is the analysis of the relationship between one independent variable and one dependent variable.
  B. **Multivariable analysis** (discussed in Chapter 13) is the analysis of the relationship of more than one independent variable to a single dependent variable.
  C. **Multivariate analysis,** a term that is frequently used incorrectly, refers to methods for analyzing more than one dependent variable as well as more than one independent variable.

II. **Choosing an Appropriate Statistical Test**
  A. A test should be selected according to the **types of variables** being used. Table 11–1 lists tests of statistical significance used in the bivariate analysis of continuous, ordinal, dichotomous, and nominal variables.
  B. A test should also be selected according to the **research design** of the study.
    1. For example, a paired test of statistical significance would be appropriate for before and after comparisons in the same study subjects or comparisons of matched pairs of study subjects.
    2. If the sampling procedure in a study is not random, statistical tests that assume random sampling, such as most of the parametric tests, may not be valid.

III. **Making Inferences from Continuous (Parametric) Data**
  A. **The joint distribution graph**
    1. A plot of two continuous variables (e.g., height and weight) can be used to visualize the relationship between the two variables (if one exists) and to determine the direction (positive or negative) and linearity of such a relationship (see Box 11–1 and Fig. 11–1).
    2. While a joint distribution graph displays the relationship between two continuous variables, it does not indicate whether such a relationship is statistically significant or simply due to chance.
  B. **The Pearson correlation coefficient**
    1. The **Pearson product-moment correlation coefficient,** which is given the symbol r and is referred to as the r **value,** is a measure of strength of the linear relationship between two continuous variables.
    2. This statistic varies from -1 to $+1$, going through 0.
      a. A finding of -1 indicates that the two variables have a perfect negative linear relationship.
      b. A finding of $+1$ indicates that the two variables have a perfect positive linear relationship.
      c. A finding of 0 indicates that the two variables are totally independent of each other.
    3. The formula for the correlation coefficient r is as follows:

$$r = \frac{\sum(x_i - \bar{x})(y_i - \bar{y})}{\sqrt{\sum(x_i - \bar{x})^2 \sum(y_i - \bar{y})^2}}$$

      a. The numerator is the sum of the covariances. The **covariance** is the product of the deviation of an observation from the mean of the x variable multiplied by the same observation's deviation from the mean of the y variable.
      b. The denominator is the square root of the sum of the squared deviations from the mean of the x variable multiplied by the sum of the squared deviations from the mean of the y variable.
    4. The statistical significance of a value of r can be determined either with use of a statistical computer program (most statistical programs provide the p value along with the correlation coefficient) or with use of the following formula to calculate the value of t for a particular value of r, and then use of a t table to determine the associated p value:

$$t = \frac{r\sqrt{N - 2}}{\sqrt{1 - r^2}}$$

$$df = N - 2$$

TABLE 11-1. Choice of an Appropriate Statistical Significance Test To Be Used in Bivariate Analysis (Analysis of One Independent Variable and One Dependent Variable)

Characterization of Variables To Be Tested*		Appropriate Test or Tests of Significance
First Variable	Second Variable	
Continuous	Dichotomous, unpaired	Student's t-test
Continuous	Dichotomous, paired	Paired t-test
Continuous	Nominal	One-way analysis of variance (ANOVA)
Continuous	Continuous	Pearson correlation coefficient (r); linear regression
Ordinal	Dichotomous, unpaired	Mann-Whitney U test; chi-square test for linear trend
Ordinal	Dichotomous, paired	Wilcoxon test
Ordinal	Nominal	Kruskal-Wallis test
Ordinal	Ordinal	Spearman correlation coefficient (rho); Kendall correlation coefficient (tau)
Ordinal	Continuous	Group the continuous variables and calculate Spearman correlation coefficient (rho), Kendall correlation coefficient (tau), or chi-square test
Dichotomous	Dichotomous, unpaired	Chi-square test; Fisher exact probability test
Dichotomous	Dichotomous, paired	McNemar chi-square test
Dichotomous	Nominal	Chi-square test
Nominal	Nominal	Chi-square test

*For tests other than linear regression analysis, it makes no difference whether the first variable or the second variable is the independent variable.

5. An important concept in drawing inferences from continuous variables is the **strength of the association,** measured by the square of the correlation coefficient, or r^2.
   a. The r^2 **value** is the proportion of variation in y explained by x (or vice versa).
   b. The r^2 value may be converted to a percentage to indicate what percent of the variation in y is due to variation in x (see Box 11-1).

C. **Linear regression analysis**
   1. Linear regression seeks to quantify the linear relationship that may exist between an independent variable x and a dependent variable y.
   2. The formula for a regression line is $y = a + bx$.
      a. Linear regression is used to estimate two parameters: the slope of the line (b) and the y-intercept (a).
      b. Most fundamental is the slope, which determines the strength of the impact of variable x on y.
   3. The formulas for the slope (b) and y-intercept (a) are as follows:

$$b = \frac{\sum(x_i - \bar{x})(y_i - \bar{y})}{\sum(x_i - \bar{x})^2}$$

$$a = \bar{y} - b\bar{x}$$

   4. See Box 11-1 for the calculation of the slope (b) for the observed heights and weights of 8 subjects; the graph in Box 11-1 shows the linear relationship between the height and weight data, with the regression line inserted.
   5. The formula for linear regression is a form of statistical modeling, and the adequacy of the model is determined by how closely the value of y can be predicted from the other data in the model.
   6. Confidence intervals around the slope and the intercept may be calculated, using computations based on linear regression formulas (most statistical computer programs perform these computations).
   7. See Chapter 13 for multiple linear regression and other methods involved in the analysis of more than two variables.

IV. **Making Inferences from Ordinal Data**
   A. **Introduction**
      1. Ordinal data are data that are ranked from the lowest value to the highest value but are not measured on an exact scale.
      2. A number of bivariate statistical tests for ordinal data can be used. These are listed in Table 11-1 and described below.
   B. **The Mann-Whitney U test**
      1. The Mann-Whitney U test for ordinal data is similar to the Student's t-test (see Chapter 10); U, like t, designates a probability distribution.
      2. In the Mann-Whitney test, all of the observations in a study of two samples are ranked numerically from the smallest to the largest, without regard to whether the observations came from the first sample (e.g., the control group) or from the second sample (e.g., the experimental group).
      3. Next, the observations from the first sample are identified, the ranks in this sample are summed, and the average rank for the first sample is determined; this process is repeated for the observations from the second sample.
      4. If the null hypothesis is true (i.e., if there is no real difference between the two samples), the average ranks of the two samples should be similar.
      5. If the average rank of one sample is considerably greater or considerably smaller than that of the other sample, the null hypothesis can be rejected.
      6. Looking up the value of U in an appropriate table will indicate the p value associated with this test.
   C. **The Wilcoxon test**
      1. The Wilcoxon test, which is also called the **Wilcoxon matched-pairs signed-ranks test,** is a rank-order test that is comparable to the paired t-test.
      2. All of the observations in a study of two samples are ranked numerically from the largest to the smallest, without regard to whether the observations came from the first sample (e.g., the pretreatment sample) or from the second sample (e.g., the posttreatment sample).
      3. The difference in rank is identified for each pair.
         a. A pair is a matched set of before and after, or pretreatment and posttreatment, values.
         b. If in a given pair the pretreatment observation scored higher than the posttreatment observation, the difference would be noted as negative.
         c. If in a given pair the posttreatment observation scored higher than the pretreatment observation, the difference would be noted as positive.
      4. If the null hypothesis is true (i.e., if there is no real difference between the samples), the sum of the positive scores and negative scores should be close to 0.
      5. If the sum of differences is considerably different from 0, the null hypothesis can be rejected.

BOX 11-1. Analysis of the Relationship between Height and Weight (Two Continuous Variables) in a Study of 8 Subjects

Part 1. Tabular representation of the data:

Subject	Variable x (Height)	Variable y (Weight)
1	182.9 cm	78.5 kg
2	172.7 cm	60.8 kg
3	175.3 cm	68.0 kg
4	172.7 cm	65.8 kg
5	160.0 cm	52.2 kg
6	165.1 cm	54.4 kg
7	172.7 cm	60.3 kg
8	162.6 cm	52.2 kg

Part 2. Graphic representation of the data:

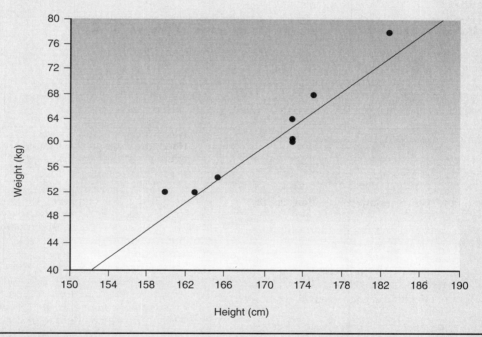

Part 3. Calculation of moments:

$\sum(x_i) = 1364.0$ cm

$\sum(y_i) = 492.2$ kg

$N = 8$

$\bar{x} = 1364.0/8 = 170.50$ cm

$\bar{y} = 492.2/8 = 61.53$ kg

$\sum(x_i - \bar{x})(y_i - \bar{y}) = 456.88$

$\sum(x_i - \bar{x})^2 = 393.1$

$\sum(y_i - \bar{y})^2 = 575.1$

Continued

D. The Kruskal-Wallis test

1. The Kruskal-Wallis test, or the **Kruskal-Wallis one-way ANOVA,** is used to compare three or more groups of ordinal data simultaneously.
2. It is analogous to a one-way analysis of variance (a one-way ANOVA), usually called an F-test, for continuous data.
3. As in the Mann-Whitney U test, in the Kruskal-Wallis test all of the data are ranked numerically, and the rank values are summed in each of the groups to be compared.
4. The Kruskal-Wallis test seeks to determine if the average ranks differ more than would be expected by chance alone.

> **BOX 11-1.** Analysis of the Relationship between Height and Weight (Two Continuous Variables) in a Study of 8 Subjects *(Continued)*

Part 4. Calculation of the Pearson correlation coefficient *(r)* and the strength of the association of the variables *(r^2)*:

$$r = \frac{\sum(x_i - \bar{x})(y_i - \bar{y})}{\sqrt{\sum(x_i - \bar{x})^2 \sum(y_i - \bar{y})^2}}$$

$$= \frac{456.88}{\sqrt{(393.1)(575.1)}} = \frac{456.88}{\sqrt{226{,}071.8}} = \frac{456.88}{475.47} = \mathbf{0.96}$$

$$r^2 = (0.96)^2 = \mathbf{0.92} = 92\%$$

Interpretation: The two variables are highly correlated. The association is strong, with 92% of variation in weight *(y)* explained by variation in height *(x)*.

Part 5. Calculation of the slope *(b)* for a regression of weight *(y)* on height *(x)*:

$$b = \frac{\sum(x_i - \bar{x})(y_i - \bar{y})}{\sum(x_i - \bar{x})^2} = \frac{456.88}{393.1} = \mathbf{1.16}$$

Interpretation: There is a 1.16-kg increase in weight *(y)* for each 1-cm increase in height *(x)*. The *y*-intercept, which indicates the value of *x* when *y* is 0, is not meaningful in the case of these two variables; therefore, it is not calculated here.

Source of data: Unpublished findings in a sample of 8 professional persons in Connecticut.

E. **The Spearman and the Kendall correlation coefficients**
  1. The **Spearman rank test** and the **Kendall rank test** for ordinal data are analogous to correlation for continuous data.
  2. The **Spearman rank correlation coefficient,** which is symbolized by **rho,** is similar to *r*.
  3. The **Kendall rank correlation coefficient** is symbolized by **tau.**
  4. The tests for rho and tau will usually give similar results, but the tau may be slightly preferable because it works better with small sample sizes.

F. **The sign test**
  1. The sign test, which may be used with continuous, ordinal, or dichotomous data, indicates whether or not, on average, one group (experimental group or control group) experienced a better outcome than the other.
  2. If the null hypothesis is true (i.e., if there is no real difference between the groups), then, by chance, for half of the outcome variables the experimental group should perform better, and for half of the outcome variables the control group should perform better.
  3. If the average score in the experimental group is better (by what amount is not important), the result is recorded as a plus sign (+); if the average score in the control group is better, the result is recorded as a minus sign (−); and if the average score in the two groups is exactly the same, no result is recorded and the variable is omitted from the analysis.
  4. Because under the null hypothesis, the expected proportion of plus signs is 0.5 and of minus signs is 0.5, the test compares the observed proportion of successes with the expected value of 0.5.

V. **Making Inferences from Dichotomous and Nominal (Nonparametric) Data**

  A. **Introduction**
   1. As indicated in Table 11–1, the chi-square test, the Fisher exact probability test, and the McNemar chi-square test can be used in the bivariate analysis of dichotomous nonparametric data.
   2. Usually, the data are first arranged in a 2 × 2 table, and the goal is to test the null hypothesis that the variables are independent.

  B. **The 2 × 2 contingency table**
   1. Data arranged as in Box 11–2 form what is known as a contingency table because it is used to determine whether the distribution of one variable is conditionally dependent (contingent) upon the other variable.
   2. Box 11–2 provides an example of a 2 × 2 contingency table, meaning that it has two cells in each direction.
   3. In a contingency table, a **cell** is a specific location in the matrix created by the two variables whose relationship is being studied.
   4. Each cell shows the observed number, the expected number (see below), and the percentage of study subjects in each treatment group with the specified outcome.
   5. The bottom row shows the column totals, and the right-hand column shows the row totals.
   6. If there are more than two cells in each direction of a contingency table, the table is called an *R* × *C* table, where *R* stands for the number of rows and *C* stands for the number of columns.

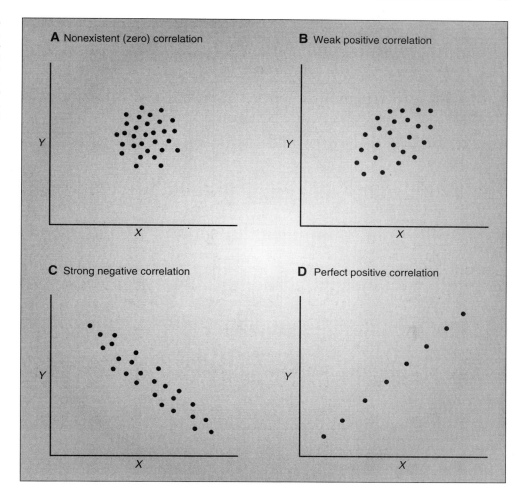

FIGURE 11-1. **Four possible patterns in joint distribution graphs.** As seen in these examples, the correlation between two continuous variables, labeled x and y, can range from nonexistent to perfect. If the value of y increases as x increases, the correlation is positive. If y decreases as x increases, the correlation is negative.

C. The chi-square test of independence
1. The chi-square test is an example of **statistical modeling,** which seeks to develop a statistical expression (the model) that predicts the behavior of a dependent variable on the basis of knowledge of one or more independent variables.
   a. The process of comparing the **observed counts** with the **expected counts** (i.e., of comparing O with E) is called a **goodness-of-fit test,** because the goal is to see how well the observed counts in a contingency table "fit" the counts expected on the basis of the model.
   b. Usually, the model in such a table is the null hypothesis that the two variables are independent of each other.
   c. If the chi-square value is small, the fit is good, but if the chi-square value is large, the data do not fit the hypothesis well.
2. Calculation of percentages
   a. Each of the four cells of Box 11-2 shows an observed count (or O) and a percentage.
   b. The percentage in each cell is calculated as the frequency distribution of the dependent variable (which in Box 11-2 is survival, reflecting the fact that survival is contingent, or dependent, on treatment) within a particular level of the independent variable (treatment assignment).
3. Calculation of expected counts
   a. Each of the four cells in Box 11-2 shows an expected count. Expected counts indicate what *would be expected* in a cell if the null hypothesis were true (i.e., if there were no differences in outcome across levels of the independent variable).
   b. The general formula for calculating the expected count in the top left cell of a contingency table is:

   $$E_{1,1} = \frac{\text{Row}_1 \text{ total}}{\text{Study total}} \times \text{Column}_1 \text{ total}$$

   where $E_{1,1}$ is defined as the cell in row$_1$, column$_1$. The rationale for using this formula is outlined in Box 11-3.
   c. Expected counts may include fractions.
   d. The sum of the expected counts in a given row will equal the sum of the observed counts in that row.
   e. The column totals for expected counts should add up to the column totals for observed counts.
   f. The expected counts in each cell of a 2×2 contingency table should equal five or more. If this is not the case, the assumptions and approximations inherent in the chi-square test may break down. For a study involving a larger contingency table (an $R \times C$ table), 20% of the expected counts may be less than five but none of them may be less than two.

> **BOX 11-2. Chi-Square Analysis of the Relationship between Treatment and Outcome (Two Nonparametric Variables, Unpaired) in a Study of 91 Subjects**
>
> **Part 1. Beginning data, presented in a 2 × 2 contingency table, where O denotes observed counts and E denotes expected counts:**
>
> | | | | OUTCOME | | | | | |
> |---|---|---|---|---|---|---|---|---|
> | | | | Survival for at Least 28 Days | | Death | | Total | |
> | | | | Number | (Percentage) | Number | (Percentage) | Number | (Percentage) |
> | TREATMENT | Propranolol | (O) | 38 | (84) | 7 | (16) | 45 | (100) |
> | | Propranolol | (E) | 33.13 | | 11.87 | | 45 | |
> | | Placebo | (O) | 29 | (63) | 17 | (37) | 46 | (100) |
> | | Placebo | (E) | 33.87 | | 12.13 | | 46 | |
> | | Total | | 67 | (74) | 24 | (26) | 91 | (100) |
>
> **Part 2. Calculation of the chi-square (χ^2) value:**
>
> $$\chi^2 = \sum \left[\frac{(O-E)^2}{E} \right]$$
>
> $$= \frac{(38-33.13)^2}{33.13} + \frac{(7-11.87)^2}{11.87} + \frac{(29-33.87)^2}{33.87} + \frac{(17-12.13)^2}{12.13}$$
>
> $$= \frac{(4.87)^2}{33.13} + \frac{(-4.87)^2}{11.87} + \frac{(-4.87)^2}{33.87} + \frac{(4.87)^2}{12.13}$$
>
> $$= \frac{23.72}{33.13} + \frac{23.72}{11.87} + \frac{23.72}{33.87} + \frac{23.72}{12.13}$$
>
> $$= 0.72 + 2.00 + 0.70 + 1.96 = \mathbf{5.38}$$
>
> **Part 3. Calculation of the degrees of freedom (df) for a contingency table, based on the number of rows (R) and columns (C):**
>
> $$df = (R-1)(C-1) = (2-1)(2-1) = \mathbf{1}$$
>
> **Part 4. Determination of the p value:**
>
> Value from the chi-square table for 5.38 on 1 df: $0.01 < p < 0.025$ (statistically significant)
> Exact p from a computer program: 0.0205 (statistically significant)
> **Interpretation:** The statistically significant result indicates that it is highly probable (only 1 chance in about 50 of being wrong) that the investigator can reject the null hypothesis of independence and accept the alternative hypothesis that propranolol does affect the outcome of myocardial infarction in a positive direction.
>
> Source of data: Snow, P. J. Effect of propranolol in myocardial infarction. Lancet 2:551–553, 1965.

4. Calculation of the **chi-square value**
   a. Method for large numbers
      (1) Begin by calculating the chi-square value for each cell in the table, using the following formula:

      $$\frac{(O-E)^2}{E}$$

      Here, the numerator is the square of the deviation of the observed count in a given cell from the count that would be expected in that cell if the null hypothesis were true. The denominator for chi-square is the expected number (E).

      (2) To obtain the total chi-square value, add up the chi-square values for the separate cells:

      $$\chi^2 = \sum \left[\frac{(O-E)^2}{E} \right]$$

   b. Method for small numbers
      (1) Because the chi-square test is based on the normal approximation of the binomial distribution (which

BOX 11-3. Calculation of Expected Counts in Contingency Tables

As described in the text, the expected count *(E)* for any cell in a contingency table is based on the total for the row in which the cell is located (row total), the total for the column in which the cell is located (column total), and the total for the table (study total). The formula can be expressed as

$$E = \frac{\text{Row total}}{\text{Study total}} \times \text{Column total}$$

Alternatively, the formula can be expressed as

$$E = \frac{\text{Row total} \times \text{Column total}}{\text{Study total}}$$

While the formulas are mathematically straightforward, more details may be helpful. For students who know what to do but do not know the rationale for doing it, calculations for two types of studies are discussed. (Data are fictitious in these examples.)

Study no. 1
The first study was conducted to determine the rates of motion sickness in 400 senior citizens on a cruise, some of whom played shuffleboard and some of whom did not. The following data were gathered:

		MOTION SICKNESS	
		Present	Absent
EXPOSURE TO SHUFFLEBOARD	Present	120	80
	Absent	40	160

These data are representative of data obtained in cohort or case-control studies with a dichotomous exposure and a dichotomous outcome. If the null hypothesis were true for this contingency table, there would be independence between the exposure (shuffleboarding) and the outcome (motion sickness).

In this cohort study, the number of subjects exposed to shuffleboard was 120 + 80 = 200. The number of subjects not exposed was 40 + 160 = 200. The total number of subjects was 120 + 80 + 40 + 160 = 400. Therefore, 200/400, or 50%, of the total number of subjects were exposed.

According to the table, 160 of the subjects had motion sickness. If the null hypothesis were true and 50% of the subjects had been exposed, then what percentage of the subjects with the outcome (motion sickness) would the investigators expect to find in the exposed group (shuffleboarders)? The answer is 50% of 160, or 80. This answer can be calculated for cell *a* by using either of the formulas shown above. With the first formula, (200/400) multiplied by 160 = 80. With the second formula, (200 × 160) divided by 400 = 80.

Study no. 2
The same approach is applicable, although slightly more complicated, when paired data are analyzed. Consider, for example, a referendum in which 100 parents and their 100 young children were asked to vote on whether ravioli and chocolate ice cream seemed like a good breakfast. The investigators know that the voter turnout was 100% and there were no abstentions. They also know that 80% of parents voted "no," 20% of parents voted "yes," 80% of children voted "yes," and 20% of children voted "no." The null hypothesis used to approach a contingency table is that the variables are independent. In this case, the null hypothesis is that the child's tendency to vote in a particular way was not influenced by the parent's vote, or vice versa.

Consider the following table:

		CHILD'S VOTE		
		Yes	No	Row Total
PARENT'S VOTE	Yes			20
	No			80
	Column Total	80	20	100

The table cannot be filled in yet, because the data in the table represent pairs of votes and the investigators do not know yet how the votes were paired. The investigators do know, however, the row and column totals, and this information is sufficient to calculate the expected values. The total for row 1 is the total number of positive votes by parents, or 20. The total for row 2 is the total number of negative votes by parents, or 80. The total for column 1 is the total number of positive votes by children, or 80. The total for column 2 is the total number of negative votes by children, or 20. The study total is 100 pairs of votes.

Continued

> **BOX 11-3.** Calculation of Expected Counts in Contingency Tables (*Continued*)
>
> To calculate the expected value for cell *a*, the investigators must estimate the number of "yes" votes by children that are associated with "yes" votes by parents. The table shows that 80 of 100, or 80%, of votes cast by parents were "no" votes, and it shows that 20% of votes cast by parents were "yes" votes. The converse was true of the children's votes, with 80% of votes positive and 20% negative.
>
> The null hypothesis is that the voting patterns of parents and children are independent. If the voting pattern were truly independent and if 20% of all parental votes were positive, then the following could be expected: 20% of the parental votes coupled to a negative vote by a child would be positive, and 20% of the parental votes coupled to a positive vote by a child would be positive. For cell *a*, the expected count would represent the expected number of positive votes by parents coupled to positive votes by children. Since 80 positive votes were cast by children, 20% of these would be expected to be associated with positive votes by parents if the null hypothesis were true. The expected count for cell *a*, therefore, is 20% of 80, or 16. This is exactly what is calculated when the row total (positive votes by parents = 20) is multiplied by the column total (positive votes by children = 80) and divided by the study total, which in this case is 100 pairs of votes: $(20 \times 80)/100 = 16$.
>
> For cell *b*, the expected count would represent the expected number of positive votes by parents paired with negative votes by children. The row total is again the total number of positive votes by parents, or 20. The column total is the number of negative votes by children, also 20. The expected count is calculated as $(20 \times 20)/100 = 4$. Conceptually, the expected count derives from the assumption that if children's votes are independent of parents' votes and if 20% of the parents' votes are positive, then the following could be expected: 20% of the positive votes by children should be paired with positive votes by parents, and 20% of the negative votes by children should be paired with positive votes by parents. If the investigators find a significant discrepancy between expected and observed counts, they are likely to reject the null hypothesis and conclude that the two variables are related rather than independent. For example, the results of the voting might be the following:
>
> | | | CHILD'S VOTE | |
> |---|---|---|---|
> | | | Yes | No |
> | PARENT'S VOTE | Yes | 20 | 0 |
> | | No | 60 | 20 |
>
> In this extreme example, all of the "yes" votes by parents are coupled to "yes" votes by children, and all of the "no" votes by children are coupled to "no" votes by parents. Even though many more parents than children voted "no," there appears, by mere inspection, to be a relationship between the voting patterns of parents and their children. When the chi-square analysis demonstrates such an apparent relationship to have statistical significance, the null hypothesis of independence is rejected.

is discontinuous), the **Yates correction for continuity** is recommended in tables with small numbers.

(2) The correction results from subtracting 0.5 from the absolute value of the $(O - E)$ in each cell before squaring. The formula is as follows:

$$\text{Yates } \chi^2 = \sum \left[\frac{(|O - E| - 0.5)^2}{E} \right]$$

(3) The use of this formula reduces the size of the chi-square value somewhat and reduces the chance of finding a statistically significant difference, so that correction for continuity makes the test more conservative.

5. Determination of the **degrees of freedom**
   a. The term "degrees of freedom" refers to the number of observations that are free to vary.
   b. According to the null hypothesis, the best estimate of the expected distribution of counts in the cells of a contingency table is provided by the row and column totals; therefore, the row and column totals are considered to be "fixed."
   c. In $R \times C$ contingency tables, the right-hand column and the bottom row are never free to vary, because they must consist of the numbers that make the totals come out right.
   d. In Fig. 11–2, the cells that are free to vary are shown in white, the cells that are not free to vary are shown in light gray, and the fixed row and column totals are shown in dark gray.
   e. The formula for degrees of freedom in a contingency table of any size is as follows:

   $$df = (R - 1)(C - 1)$$

   where *df* denotes degrees of freedom, *R* is the number of rows, and *C* is the number of columns.
6. Interpretation of the results
   a. After the chi-square value and the degrees of freedom are known, a standard table of chi-square values (see the Appendix) can be consulted to determine the corresponding *p* value.
   b. The *p* value indicates the probability that a chi-square value that large would have resulted from chance alone (most computer programs will provide the exact *p* value when calculating a chi-square).
   c. The choice of a one-tailed versus a two-tailed test does not affect the performance of a statistical test but does affect how the critical ratio thus obtained is converted to a *p* value in a statistical table; the direction of difference is obvious by inspection.

D. **The chi-square test for paired data (McNemar test)**
   1. The McNemar chi-square test is used to compare before and after findings in the same individual or to compare findings in a matched analysis for dichotomous variables.

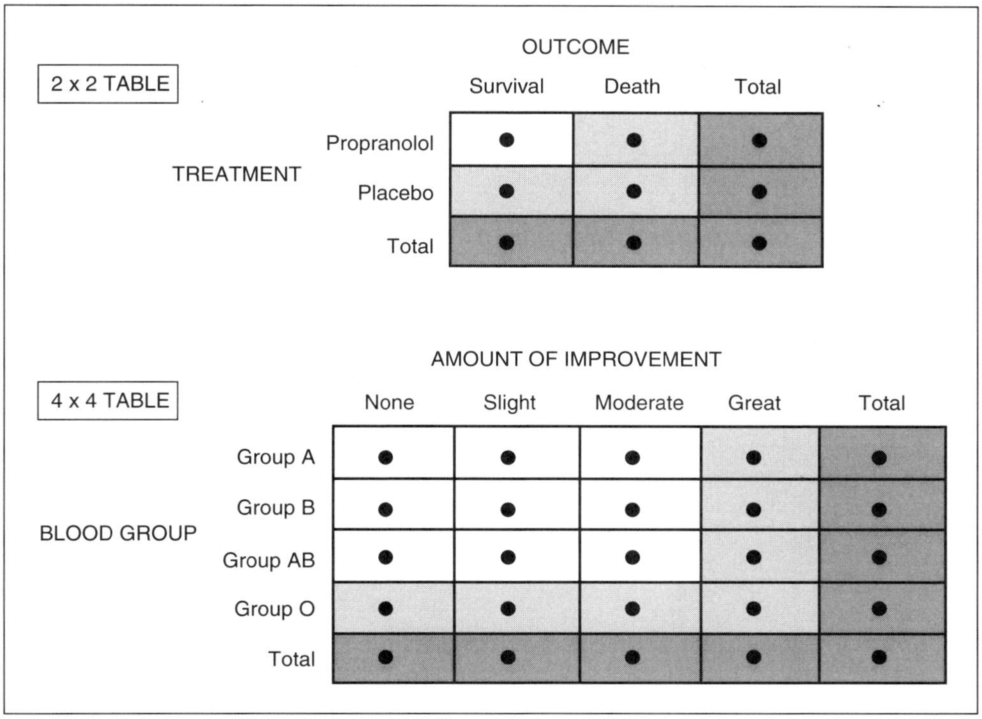

FIGURE 11-2. Conceptualization of the calculation of the degrees of freedom *(df)* in a 2 × 2 contingency table (top) and in a 4 × 4 contingency table (bottom). A white cell is free to vary; a light gray cell is not free to vary; and a dark gray cell is a row or column total. The formula is $df = (R - 1)(C - 1)$, where R denotes the number of rows and C denotes the number of columns. For the 2 × 2 table, $df = 1$. For the 4 × 4 table, $df = 9$.

2. The formula is:

$$\text{McNemar } \chi^2 = \frac{(|b - c| - 1)^2}{b + c}$$

3. Note that the formula uses only cells b and c in the 2 × 2 table; this is because cells a and d do not change and therefore do not contribute to the standard error.
4. Note also that the formula tests data with 1 degree of freedom, using a correction for continuity.
5. Box 11–4 demonstrates use of the McNemar test for before and after data, and Box 11–5 demonstrates its use with matched data.

E. The Fisher exact probability test
1. The Fisher exact probability test is used when one or more of the expected counts in a 2 × 2 table is small (i.e., less than two); the chi-square test cannot be used under these circumstances.
2. The formula is as follows:

$$\text{Fisher } p = \frac{(a + b)! \, (c + d)! \, (a + c)! \, (b + d)!}{N! \, a! \, b! \, c! \, d!}$$

where p is probability; a, b, c, and d denote values in the top left, top right, bottom left, and bottom right cells, respectively, in a 2 × 2 table; N is the total number of observations; and ! is the symbol for factorial. (The factorial of $4 = 4! = 4 \times 3 \times 2 \times 1$.)

3. The Fisher exact probability is tedious to calculate; most commercially available statistical packages calculate the Fisher probability automatically when an appropriate situation arises in a 2 × 2 table.

F. Standard errors for data in 2 × 2 tables
1. **Standard error for a proportion**
   a. Both the standard error for a proportion and the 95% confidence interval for a proportion (percentage) may be calculated by the methods described earlier.
   b. See Use of z-Tests, under Tests of Statistical Significance, in Chapter 10, for further discussion.
2. **Standard error for a risk ratio or an odds ratio**
   a. If a 2 × 2 table is used to compare the proportion of disease in two different exposure groups or is used to compare the proportion of success in two different treatment groups, the relative risk or relative success can be expressed as a risk ratio.
   b. Standard errors can be set around the risk ratio, and if the 95% confidence limits exclude the value of 1.0, there is a statistically significant difference between the risks, at an alpha level of 5%.
   c. If a 2 × 2 table provides data from a case-control study, the odds ratio can be calculated.
   d. Every major statistical computer package includes programs for calculating the standard error of the risk ratio and odds ratio.

G. Strength of association of data in 2 × 2 tables
1. Findings can have statistical significance and at the same time be of no clinical value, especially if the study involves a large number of subjects.
2. In 2 × 2 tables, the strength of association is measured using the **phi coefficient,** which basically adjusts the chi-square value for the sample size and can be considered the correlation coefficient (r) for the data in a 2 × 2 table.
3. The formula is as follows:

$$\text{phi} = \sqrt{\frac{\chi^2}{N}}$$

4. If phi is squared (like r^2), the proportion of explained variation is derived.

> **BOX 11-4.** McNemar Chi-Square Analysis of the Relationship between Data before and Data after an Event (Two Dichotomous Variables, Paired) in a Study of 200 Subjects (Fictitious Data)
>
> **Part 1. Standard 2 × 2 table format on which equations are based:**
>
> | | | FINDINGS AFTER EVENT | | |
> |---|---|---|---|---|
> | | | Positive | Negative | Total |
> | FINDINGS BEFORE EVENT | Positive | a | b | $a + b$ |
> | | Negative | c | d | $c + d$ |
> | | Total | $a + c$ | $b + d$ | $a + b + c + d$ |
>
> **Part 2. Data for a study of the opinions of medical school faculty toward second-year medical students before and after seeing a show presented by the students:**
>
> | | | POSTSHOW OPINION | | |
> |---|---|---|---|---|
> | | | Positive | Negative | Total |
> | PRE-SHOW OPINION | Positive | 150 | 22 | 172 |
> | | Negative | 8 | 20 | 28 |
> | | Total | 158 | 42 | 200 |
>
> **Part 3. Calculation of the McNemar chi-square (χ^2) value:**
>
> $$\text{McNemar } \chi^2 = \frac{(|b - c| - 1)^2}{b + c}$$
>
> $$= \frac{(|22 - 8| - 1)^2}{22 + 8} = \frac{(13)^2}{30} = \frac{169}{30} = 5.63$$
>
> **Part 4. Calculation of the degrees of freedom (df) for a contingency table, based on the number of rows (R) and columns (C):**
>
> $$df = (R - 1)(C - 1) = (2 - 1)(2 - 1) = 1$$
>
> **Part 5. Determination of the p value:**
>
> Value from the chi-square table for 5.63 on 1 df: $p < 0.025$ (statistically significant)
> **Interpretation:** If faculty changed their attitude toward second-year medical students after the show, most of these changes were from a positive attitude to a negative attitude, rather than vice versa.

5. Although phi is not accurate in larger ($R \times C$) tables, a similar test, called **Cramer's V**, can be used in these tables (see Blalock 1972).
6. Every association should be examined for strength of association and clinical utility as well as statistical significance.
   a. Strength of association can be shown by a risk ratio, a risk difference, an odds ratio, an r^2 value, a phi value, or a Cramer's V value.
   b. A **statistically significant association** implies that the association is real (i.e., is not due to chance alone) but not necessarily that it is important.
   c. A **strong association** is likely to be important if it is real.
   d. In some cases, even a relatively weak association can be clinically important if it is real. As discussed in Chapter 6, both the risk ratio (or odds ratio if from a case-control study) and the prevalence of the risk factor determine the population attributable risk. For a prevalent disease (e.g., myocardial infarction), a common risk factor that showed a risk ratio of only 1.3 could be responsible for a large number of preventable cases of disease.

H. **Survival analysis**
   1. **Person-time methods**
      a. In a survival study, some subjects die during the observation period, some subjects may be lost to follow-up (unavailable for examination), and some may be censored (when the time of study of a patient is terminated early because the patient entered late and the study is ending).
      b. To control for the fact that the length of observa-

BOX 11-5. McNemar Chi-Square Analysis of the Relationship between Data from Cases and Data from Controls (Two Dichotomous Variables, Paired) in a Case-Control Study of 54 Subjects

Part 1. Standard 2 × 2 table format on which equations are based:

		CONTROLS		
		Risk Factor Present	Risk Factor Absent	Total
CASES	Risk Factor Present	a	b	a + b
	Risk Factor Absent	c	d	c + d
	Total	a + c	b + d	a + b + c + d

Part 2. Data for a case-control study of the relationship between mycosis fungoides (the disease) and a history of exposure to an industrial environment containing cutting oils (the risk factor):

		CONTROLS		
		History of Industrial Exposure	No History of Industrial Exposure	Total
CASES	History of Industrial Exposure	16	13	29
	No History of Industrial Exposure	3	22	25
	Total	19	35	54

Part 3. Calculation of the McNemar chi-square (χ^2) value:

$$\text{McNemar } \chi^2 = \frac{(|b - c| - 1)^2}{b + c}$$

$$= \frac{(|13 - 3| - 1)^2}{13 + 3} = \frac{(9)^2}{16} = \frac{81}{16} = 5.06$$

Part 4. Calculation of the degrees of freedom (df) for a contingency table, based on the number of rows (R) and columns (C):

$$df = (R - 1)(C - 1) = (2 - 1)(2 - 1) = 1$$

Part 5. Determination of the p value:

Value from the chi-square table for 5.06 on 1 df: $p = 0.021$ (statistically significant)
Interpretation: The cases (subjects with mycosis fungoides) were more likely than expected by chance alone to have been exposed to an industrial environment with cutting oils than were the controls (subjects without mycosis fungoides).

Part 6. Calculation of the odds ratio (OR):

$$OR = b/c = 13 / 3 = 4.33$$

Source of data: Cohen, S. R. Mycosis fungoides: clinicopathologic relationships, survival, and therapy in 54 patients, with observation on occupation as a new prognostic factor. Master's thesis, Yale University School of Medicine, New Haven, Conn., 1977.

tion varies from subject to subject, the person-time methods introduced in an earlier discussion of **incidence density** (see Chapter 2) can be used in calculating risks and rates of death.
   c. Briefly, if one person is observed for 3 years and another for 1 year, the **duration of observation** would be equal to 4 **person-years.** Calculations can be made on the basis of years, months, weeks, or any other unit of time.
   d. The results can then be reported as the number of events (e.g., deaths or remissions) per person-time of observation.
2. **Life table analysis**
   a. The two main methods of life table analysis are the actuarial method and the Kaplan-Meier method. Both require the following information about each patient: (1) the date of entry in the study; (2) the reason for withdrawal (death, loss to follow-up, or censorship); and (3) the date of withdrawal (date of death for those who died, the last time seen alive for those who were lost to follow-up, and the date withdrawn alive for those who were censored).
   b. The **actuarial method**
      (1) The actuarial method is used to calculate the survival rates of patients during *fixed* intervals, such as years.
      (2) First, it determines the number of people surviving to the beginning of each interval.
      (3) Next, it assumes that those who were censored or lost to follow-up during the interval were observed for only half of that interval.
      (4) Then the method calculates the mortality rate for that interval (m_x in life tables) by dividing the number of deaths in the interval by the total person-years for all those who began the interval.
      (5) The survival rate for an interval (p_x) is 1.0 minus the mortality rate.
      (6) The rate of survival of the study group to the end of, say, three of the fixed intervals (P_3) is the product of the survival of each of the three component intervals.
      (7) Thus, if the intervals were years and if the survival to the end of the first interval (p_1) was 0.75, p_2 was 0.80, and p_3 was 0.85, the numbers would be multiplied to arrive at a 3-year survival rate of 0.51, or 51%.
      (8) Fig. 11–3 shows an example of a study in which the actuarial method was used.
   c. The **Kaplan-Meier method**
      (1) The **Kaplan-Meier life table method,** which is the most commonly used approach to survival analysis in medicine, is also referred to as the **product-limit method** because it takes advantage of the fact that the N year survival rate (P_N) is equal to the product of all of the survival rates of the individual intervals (p_1, p_2, and so forth) leading up to year N.
      (2) The Kaplan-Meier method is different from the actuarial method in that it calculates a new line of the life table every time a new death occurs.
      (3) Because deaths occur unevenly over time, the intervals are *uneven* and there are many of them. For this reason, the graph of a Kaplan-Meier life table analysis often looks like uneven stair steps (see Box 11–6).
      (4) In a Kaplan-Meier analysis, a death instantaneously terminates one interval and begins a new interval at a lower survival rate. The periods of time between deaths are death-free periods. Therefore, the proportion surviving between deaths does not change, and the curve of the proportion surviving is flat rather than sloping downward.
      (5) In Box 11–6, p_z is the proportion surviving interval x (i.e., from the time of the previous death to just

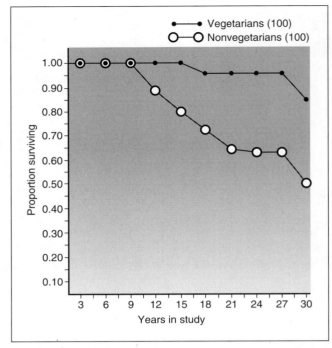

FIGURE 11–3. Graph showing results of survival analysis using the actuarial method in a study of the long-term effects of a vegetarian and a nonvegetarian diet in healthy adults (data are fictitious). The 30-year cumulative survival rates are shown. The subjects were free of known chronic disease when they entered the study. There were 100 vegetarians (indicated by solid circles) and 100 nonvegetarians (indicated by open circles) who participated. The fixed observation interval was 3 years.

before the next death), and P_x is the proportion surviving from the beginning of the study to the end of that interval. P_x is obtained by multiplying together the p_x values of all of the intervals up to and including the row of interest.
      (6) The p_x of the first interval is always 1.0, because the first death ends the first study interval and all of the patients not lost to follow-up survive until the first death.
      (7) The life table method is a powerful tool if the losses are few, if the losses represent a similar percentage of the starting numbers in the groups to be compared, and if the characteristics of those lost to follow-up are similar. Life table methods do not eliminate the bias that occurs if the losses to follow-up happen more frequently in one group than in another, particularly if the characteristics of the patients lost from one group differ greatly from those of the patients lost from the other group.
      (8) The life table method is usually considered the method of choice for describing dichotomous outcomes in longitudinal studies.
3. **Tests of significance for differences in survival**
   a. Significance tests for proportions
      (1) See Chapter 10 for a discussion of *t*-tests and *z*-tests.
      (2) The *t*-test for a difference between actuarial method curves depends on the Greenwood formula for the standard error of a proportion. For details, see Cutler and Ederer (1958) or Dawson-Saunders and Trapp (1994).
   b. The **Logrank test**
      (1) The logrank test is often used to compare data in studies involving treatment and control groups and to test the null hypothesis that each group has the same force of mortality.

BOX 11-6. Survival Analysis by the Kaplan-Meier Method in a Study of 8 Subjects

Part 1. Beginning data:

Timing of deaths in 4 subjects:
  0.8, 3.1, 5.4, and 9.2 months
Timing of loss to follow-up or censorship in 4 subjects:
  1.0, 2.7, 7.0, and 12.1 months

Part 2. Tabular representation of the data:

Number of Months at Time of Subject's Death	Number Living Just before Subject's Death	Number Living Just after Subject's Death	Number Lost to Follow-Up between This and Next Subject's Death	p_x	Interval for p_x	P_x to End of Interval
—	—	—	—	1.000	$0 < 0.8$	1.000
0.8	8	7	2	0.875	$0.8 < 3.1$	0.875
3.1	5	4	0	0.800	$3.1 < 5.4$	0.700
5.4	4	3	1	0.750	$5.4 < 9.2$	0.525
9.2	2	1	0	0.500	$9.2 < 12.1$	0.263
No deaths	1	1	1	1.000	> 12.1	0.263

Note: p_x is the proportion surviving interval x (i.e., from the time of the previous death to just before the next death), and P_x is the proportion surviving from the beginning of the study to the end of that interval.

Part 3. Graphic representation of the data:

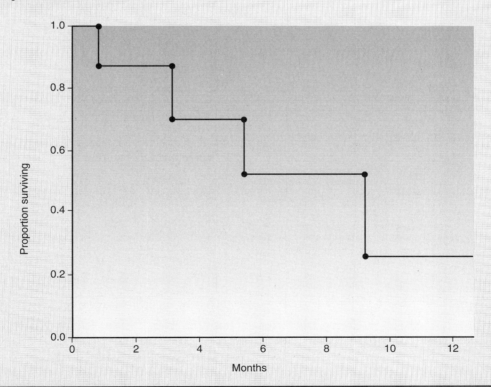

Source of data: Kaplan, E. L., and P. Meier. Nonparametric estimation from incomplete observations. Journal of the American Statistical Association 53:457–481, 1958.

(2) Each time a death occurs, one calculates the probability that the observed death would have occurred in the treatment group and the probability that it would have occurred in the control group.
  (a) These probabilities are proportional to the number of survivors to that point in time in each group.
  (b) For example, if of 100 original subjects in each group 60 survivors were left in the treatment group and 40 in the control group, under the null hypothesis the probability that the next death would occur in the treatment group is 0.6, and the probability that the next death would occur in the control group is 0.4.
(3) Within each study group, the expected probabilities for each death are summed to form the total expected number of deaths (E) for that group; the actual deaths in each group are summed to form the observed number of deaths (O); and then the observed deaths are compared with the expected deaths in the following chi-square test on 1 degree of freedom:

$$\text{logrank } \chi^2 = \frac{(O_T - E_T)^2}{E_T} + \frac{(O_C - E_C)^2}{E_C}$$

where O_T and E_T are the observed and expected deaths in the treatment group, and where O_C and E_C are the observed and expected deaths in the control group.

c. **Proportional hazard models (Cox models)** permit dichotomous outcomes to be used as dependent variables in multiple logistic regression analyses, despite losses to follow-up and censorship of patients. For details, see Dawson-Saunders and Trapp (1994).

QUESTIONS

DIRECTIONS. (Items 1–10): Each of the numbered items or incomplete statements in this section is followed by answers or by completions of the statement. Select the ONE lettered answer or completion that is BEST in each case. Correct answers and explanations are given at the end of the chapter.

1. To employ parametric methods of statistical analysis, the data must be

   (A) age-matched
   (B) normally distributed
   (C) bivariate
   (D) nominal
   (E) linear

2. A joint distribution graph may be used to display

   (A) specificity
   (B) power
   (C) causality
   (D) correlation
   (E) kurtosis

3. A correlation is noted between the frequency of putting on boxing gloves and acts of aggression. The correlation coefficient in this case

   (A) may be greater than 1
   (B) must be less than 0.05
   (C) may be close to 1
   (D) must be statistically significant
   (E) cannot be determined, because the data are dichotomous

4. In the formula for linear regression, the term used to represent the slope is

   (A) b
   (B) a
   (C) x
   (D) y
   (E) m

Items 5–7

Two groups of subjects are assembled on the basis of whether or not they can correctly identify a newt in a pond-life ensemble. The groups are asked to rate the probability that a popular December holiday will be stolen by a member of Congress. They are to use a scale with five choices, ranging from "improbable" to "highly probable."

5. The data in this study are

   (A) parametric
   (B) nominal
   (C) ordinal
   (D) bivariate
   (E) continuous

6. To analyze the responses of the two groups statistically, the data are compared on the basis of

   (A) variance
   (B) means
   (C) standard deviation
   (D) standard error
   (E) ranking

7. The appropriate statistical method for this analysis is

   (A) nonparametric analysis
   (B) the Student's t-test
   (C) the chi-square analysis
   (D) the Fisher exact probability test
   (E) linear regression analysis

8. A distinction between the one-tailed and the two-tailed tests of significance is that a one-tailed test

   (A) does not affect the statistical test but does affect the conversion to a p value
   (B) does not affect statistical significance but does affect power
   (C) requires that the sample size be doubled
   (D) should be made during data analysis
   (E) is based on the number of independent variables

9. A study is conducted to determine the efficacy of influenza vaccine. Subjects agree to participate for 2 years and to receive by random assignment either a placebo or the active vaccine for each of the 2 years. All incident cases of influenza are recorded. Each subject serves as his or her own control. The rate of influenza in the vaccinated group is compared with the rate in the unvaccinated group. The appropriate test of significance for this study is

(A) the McNemar test
(B) the Kaplan-Meier method
(C) the Pearson correlation coefficient
(D) the Wilcoxon test
(E) the Kruskal-Wallis test

10. The phi coefficient is defined as

$$\sqrt{\frac{\chi^2}{N}}$$

This is used to measure

(A) the statistical power of a 2 × 2 table
(B) the standard error of a 2 × 2 table
(C) the strength of association in a 2 × 2 table
(D) the p value of a 2 × 2 table
(E) effect modification in a 2 × 2 table

ANSWERS AND EXPLANATIONS

1. The answer is B: normally distributed. Data are parametric if their distribution is fully described by the mean and the standard deviation, which are the parameters. In order to meet this requirement, the data must be continuous and derived from an underlying set of data that is normally distributed. Appropriate use of parametric methods further requires that the study population be a random sample of the larger population that it is intended to represent. Data need not be age-matched or linear for parametric methods to pertain. Bivariate and nominal data require nonparametric analysis. *Epidemiology*, p. 139.

2. The answer is D: correlation. A joint distribution graph is a plot of the relationship between two continuous variables. The more closely the data points cluster about a line, the greater the correlation between the two variables. If two variables are related in a nonlinear manner, the correlation may not be displayed as a line but rather as a curvilinear distribution of data points. Methods for calculation of the correlation coefficient in such a setting are available but are beyond the scope of this text. Correlation alone does not establish causality. For example, someone who owns a car is also likely to own a television set (or cellular phone), but ownership of one item does not cause ownership of the other item. *Epidemiology*, p. 139.

3. The answer is C: may be close to 1. The Pearson product-moment correlation coefficient, also referred to as the r value, is a measure of the linear relationship between two variables. Its value range is the same as that for the slope of a line, from -1, through 0, to 1. Therefore, it cannot be greater than 1. A strong correlation, such as the one in the question, will produce a correlation coefficient close to 1. Statistical significance is certainly more likely when the correlation is strong but is also dependent on the sample size, about which we are told nothing in this case. *Epidemiology*, p. 140.

4. The answer is A: b. The formula for linear regression is $y = a + bx$, where a is the y-intercept and b is the slope. This formula is derived from the formula for a straight line, which is $y = mx + b$, where m is the slope and b is the y-intercept. Linear regression analysis is used to quantify the linear relationship between an independent variable (x) and a dependent variable (y). *Epidemiology*, p. 141.

5. The answer is C: ordinal. Ordinal data, as defined in Chapter 11, are ranked from lowest to highest but not measured on an exact scale. The scale described here clearly does have directionality and can be ranked, and the data are therefore ordinal rather than nominal. There are more than two choices, and the data are therefore not bivariate. The choices are discrete, and the data are therefore neither continuous nor parametric. *Epidemiology*, p. 143.

6. The answer is E: ranking. Nonparametric methods of statistical analysis are used for ordinal data and based on ranking of the data. Ordinal data do not have a mean or a definable variance, and, therefore, cannot be characterized by a standard deviation or standard error. To analyze these data, which are the ordinal responses provided by the two groups of subjects, the Wilcoxon test is appropriate, as indicated in Table 11–1. *Epidemiology*, p. 144.

7. The answer is A: nonparametric analysis. Nonparametric methods of statistical analysis are used for ordinal data and based on the ranking of the data. To analyze these data, which are the ordinal responses provided by the two groups of subjects, the Wilcoxon test is appropriate, as indicated in Table 11–1. *Epidemiology*, p. 145.

8. The answer is A: does not affect the statistical test but does affect the conversion to a p value. The choice of a one-tailed or two-tailed test of significance should be made before a study is conducted, based on the hypothesis to be tested. If the outcome can differ from the null hypothesis in only one direction (e.g., when an antihypertensive drug you are thoroughly convinced will not *raise* blood pressure is compared with a placebo), a one-tailed test of significance is appropriate. When the outcome may differ from the null hypothesis in either direction, a two-tailed test of significance is warranted.

    The stipulation of a one-tailed test or two-tailed test affects the associated p value. Statistical significance (i.e., a p value less than alpha) is more readily achieved when a one-tailed test is chosen because the extreme 5% of the distribution that differs sufficiently from the null hypothesis to warrant rejection of the null hypothesis (when alpha is set at 0.05) is all to one side. When a two-tailed test of significance is chosen, the rejection region is divided into two areas, with half (or 0.025 of the distribution when alpha = 0.05) at either extreme of the curve. The implications of choosing a one-tailed test or a two-tailed test are discussed in Chapter 10 and Box 10–6. *Epidemiology,* pp. 128–135.

9. The answer is A: the McNemar test. The outcome for each subject in this study is binary: influenza occurs or does not occur. The proportion of vaccinated subjects who acquire influenza will be compared with the proportion of unvaccinated subjects who acquire influenza. A chi-square test is appropriate for this sort of comparison, but, in this study, each subject serves as his or her own control. The McNemar test is a chi-square test modified for matched pairs. *Epidemiology,* p. 149.

10. The answer is C: the strength of association in a 2 × 2 table. A very large sample size is likely to result in statistical significance, even if the association under investigation is weak and trivial. The strength of the association between the two variables under study does not change with the sample size, however. The phi coefficient adjusts the chi-square value for the size of the sample. The phi coefficient squared is the proportion of variation in the dependent variable accounted for by the independent variable. *Epidemiology,* pp. 152–153.

REFERENCES CITED

Blalock, H. M., Jr. Social Statistics, 2nd ed. New York, McGraw-Hill Book Company, 1972.

Cutler, S. J., and F. Ederer. Maximum utilization of the life table method in analyzing survival. Journal of Chronic Disease 8:699–713, 1958.

Dawson-Saunders, B., and R. G. Trapp. Basic and Clinical Biostatistics, 2nd ed. Norwalk, Conn., Appleton and Lange, 1994.

Jekel, J. F. Epidemiology, Biostatistics, and Preventive Medicine. Philadelphia, W. B. Saunders Company, 1996.

CHAPTER TWELVE

Sample Size, Randomization, and Probability Theory

SYNOPSIS

OBJECTIVES

- To discuss issues of importance in the calculation of sample size and to provide commonly used sample size formulas.
- To discuss the goals of randomization and the methods used to achieve random allocation.
- To outline the dangers of data dredging when modern computer techniques are used.
- To define elementary probability theory and delineate the rules of this theory.

I. Sample Size

A. Derivation of the basic sample size formula

1. The basic formula for calculating sample size is derived from the formula for the paired *t*-test:

$$t_\alpha = \frac{\bar{d}}{\frac{s_{\bar{d}}}{\sqrt{N}}}$$

where $\bar{d}$ is the mean difference that was observed, $s_{\bar{d}}$ is the standard error of that mean difference, and N is the sample size.

2. To solve for N, everything can be squared and the equation rearranged so that N is in the numerator:

$$t_\alpha^2 = \frac{(\bar{d})^2}{(s_{\bar{d}}/\sqrt{N})^2} = \frac{(\bar{d})^2 \cdot N}{(s)^2}$$

3. Next, the terms can be rearranged so that the equation for N in a paired (before and after) study becomes:

$$N = \frac{t_\alpha^2 \cdot (s)^2}{(\bar{d})^2}$$

4. The t in the formula must be replaced with z, because determination of the value of t requires that the degrees of freedom (df) be known. However, the df is dependent on N. The value of z is not dependent on df, and z is essentially equal to t when the sample size is large. The formula therefore becomes:

$$N = \frac{z_\alpha^2 \cdot (s)^2}{(\bar{d})^2}$$

5. The above formula is for a study using the paired *t*-test, in which each subject serves as his or her own control. For a study using the Student's *t*-test, such as a randomized controlled trial with an experimental group and a control group, it would be necessary to calculate N for each group.

6. In light of the information provided by the basic formula for N, the following statements can be made:

    a. The larger the *variance* is, the larger the sample size must be, because the variance is in the numerator of the above formula for N; with a large variance (and therefore a large standard error), a bigger N is needed to compensate for the greater uncertainty of the estimate.

    b. To have considerable confidence that a mean difference shown in a study is real, the analysis must produce a small p value for the observed mean difference, which in turn implies that the value for t_α or z_α was large. Because z_α is in the numerator of the sample size formula, the larger z_α is, the larger the N (the sample size needed) will be.

    c. To detect with confidence a very small difference between the mean values of two study groups, a very large N would be needed, because the difference (squared) is in the denominator. The smaller the denominator is, the larger the ratio is and, hence, the larger the N must be. A precise estimate and therefore a large sample size is needed to detect a small difference.

B. The problem of beta error
1. If one observes a mean difference in the data that appears clinically important but is not statistically significant (i.e., the null hypothesis cannot be rejected at the desired level of confidence, such as alpha = 0.05), **beta error** may have occurred because the sample size was small.
2. The importance of beta error is emphasized in a seminal article by Freiman et al. (1978), which reported that most of 71 "negative" randomized controlled trials of new therapies relied on sample sizes that were too small "to provide reasonable assurance that a clinically meaningful 'difference' (i.e., therapeutic effect) would not be missed."
3. A study with a large beta error has a low sensitivity for detecting a mean difference because, as discussed in Chapter 7:

   Sensitivity + False-negative (beta) error = 1.00

   a. When investigators speak of a study as opposed to a clinical test, however, they usually use the term "statistical power" instead of "sensitivity." With this substitution in terms, the equation becomes:

   Statistical power + False-negative (beta) error = 1.00

   b. This means that statistical power is equal to (1 − beta error).
   c. In calculating a sample size, if one accepts a 20% possibility of missing a true finding (beta error = 0.2), the study should have a statistical power of 0.8, or 80%. This provides 80% confidence that a true mean difference of the size specified will be detected with the sample size derived.
   d. Beta error should be incorporated into a study beforehand, in the determination of sample size. Incorporating beta in the sample size calculation is easy, but it is likely to increase the sample size considerably.

C. Steps in the calculation of sample size
1. The appropriate sample size formula should be chosen on the basis of the type of study and the type of error to be considered. Four common formulas for calculating sample size are discussed in this chapter and listed in Table 12–1.
2. The following values must be specified: the variance expected (s^2); the alpha desired (actually, z_α); the smallest clinically important difference ($\bar{d}$); and, usually, beta (actually, z_β).
   a. All but the variance must come from clinical and research judgment.
   b. The estimated variance must be based on knowledge of data.
      (1) If the outcome variable being studied is continuous, such as blood pressure, the estimate of variance can be obtained from the literature or from a small pilot study.
      (2) If the variance of a proportion must be estimated, the formula $\bar{p}(1 - \bar{p})$ can be used.
3. Sample size calculations are demonstrated in Boxes 12–1 through 12–4, and discussed below.

D. Sample size for studies using t-tests
1. The sample size formula and calculations for a be-

TABLE 12-1. Formulas for the Calculation of Sample Size for Studies Commonly Pursued in Medical Research

Type of Study and Type of Errors to Be Considered	Appropriate Formula to Be Used*
Studies using the paired t-test (e.g., before and after studies) and considering alpha (type I) error only	$N = \dfrac{(z_\alpha)^2 \cdot (s)^2}{(\bar{d})^2}$
Studies using the Student's t-test (e.g., randomized controlled trials with one experimental group and one control group) and considering alpha error only	$N = \dfrac{(z_\alpha)^2 \cdot 2 \cdot (s)^2}{(\bar{d})^2}$
Studies using the Student's t-test and considering alpha (type I) and beta (type II) errors	$N = \dfrac{(z_\alpha + z_\beta)^2 \cdot 2 \cdot (s)^2}{(\bar{d})^2}$
Studies using a test of differences in proportions and considering alpha and beta errors	$N = \dfrac{(z_\alpha + z_\beta)^2 \cdot 2 \cdot \bar{p}(1 - \bar{p})}{(\bar{d})^2}$

*The appropriate formula is based on the study design and the type of outcome data. In these formulas, N = sample size; z_α = value for alpha error; z_β = value for beta error; s^2 = variance; $\bar{p}$ = mean proportion of success; and $\bar{d}$ = difference to be detected. See Boxes 12–1, 12–2, 12–3, and 12–4 for examples of calculations using the formulas.

BOX 12–1. Calculation of Sample Size for a Study Using the Paired t-Test and Considering Alpha Error Only

Part 1. Data on which the calculation will be based:

Study Characteristic	Assumptions Made by Investigator
Type of study	Before and after study of an antihypertensive drug
Data sets	Pretreatment and posttreatment observations in the same group of subjects
Variable	Systolic blood pressure
Standard deviation (s)	15 mm Hg
Variance (s^2)	225 mm Hg
Data for alpha (z_α)	p = 0.05; therefore, 95% confidence desired (two-tailed test); z_α = 1.96
Difference to be detected ($\bar{d}$)	10 mm Hg or larger difference between pretreatment and posttreatment blood pressure values

Part 2. Calculation of sample size (N):

$$N = \frac{(z_\alpha)^2 \cdot (s)^2}{(\bar{d})^2} = \frac{(1.96)^2 \cdot (15)^2}{(10)^2}$$

$$= \frac{(3.84)(225)}{100} = \frac{864}{100} = 8.64 = 9 \text{ subjects total}$$

Interpretation: Only 9 subjects would be needed in all, because each subject serves as his or her own control in a before and after study.

BOX 12–2. Calculation of Sample Size for a Study Using the Student's *t*-Test and Considering Alpha Error Only

Part 1. Data on which the calculation will be based:

Study Characteristic	Assumptions Made by Investigator
Type of study	Randomized controlled trial of an antihypertensive drug
Data sets	Observations in one experimental group and one control group of the same size
Variable	Systolic blood pressure
Standard deviation (s)	15 mm Hg
Variance (s^2)	225 mm Hg
Data for alpha (z_α)	$p = 0.05$; therefore, 95% confidence desired (two-tailed test); $z_\alpha = 1.96$
Difference to be detected ($\bar{d}$)	10 mm Hg or larger difference between mean blood pressure values of the experimental group and control group

Part 2. Calculation of sample size (N):

$$N = \frac{(z_\alpha)^2 \cdot 2 \cdot (s)^2}{(\bar{d})^2} = \frac{(1.96)^2 \cdot 2 \cdot (15)^2}{(10)^2}$$

$$= \frac{(3.84)(2)(225)}{100} = \frac{1728}{100} = 17.28$$

= 18 subjects per group × 2 groups = **36 subjects total**

Interpretation: Why is the total number of subjects needed in this box 4 times as large as the total number needed in Box 12–1, even though the values for z_α, s, and $\bar{d}$ are the same in both boxes? First, because two independent groups are being compared in this box, there are two sources of variance. This is why the variance estimate in the numerator of the formula is multiplied by 2. Second, for the type of study depicted in this box, 18 subjects will be needed in the experimental group and 18 in the control group, for a total of 36 study subjects.

BOX 12–3. Calculation of Sample Size for a Study Using the Student's *t*-Test and Considering Alpha and Beta Errors

Part 1. Data on which the calculation will be based:

Study Characteristic	Assumptions Made by Investigator
Type of study	Randomized controlled trial of an antihypertensive drug
Data sets	Observations in one experimental group and one control group of the same size
Variable	Systolic blood pressure
Standard deviation (s)	15 mm Hg
Variance (s^2)	225 mm Hg
Data for alpha (z_α)	$p = 0.05$; therefore, 95% confidence desired (two-tailed test); $z_\alpha = 1.96$
Data for beta (z_β)	20% beta error; therefore, 80% power desired (one-tailed test); $z_\beta = 0.84$
Difference to be detected ($\bar{d}$)	10 mm Hg or larger difference between mean blood pressure values of the experimental group and control group

Part 2. Calculation of sample size (N):

$$N = \frac{(z_\alpha + z_\beta)^2 \cdot 2 \cdot (s)^2}{(\bar{d})}$$

$$= \frac{(1.96 + 0.84)^2 \cdot 2 \cdot (15)^2}{(10)^2}$$

$$= \frac{(7.84)(2)(225)}{100} = \frac{3528}{100} = 35.28$$

= 36 subjects per group × 2 groups = **72 subjects total**

Interpretation: Including z_β in the calculations doubled the sample size in this box, compared with the sample size in Box 12–2. If the investigators insisted on an even smaller beta error, the sample size would have increased even more.

---

fore and after study (i.e., a study for which a **paired *t*-test** is used) are shown in Box 12–1.

2. The sample size formula and calculations for a randomized controlled trial for which the **Student's *t*-test** is used are shown in Box 12–2. This formula differs from the Box 12–1 formula only in that the variance estimate must be multiplied by 2.
3. The randomized controlled trial requires four times the number of subjects as the before and after study described in Box 12–1; the larger sample size is needed for studies using the Student's *t*-test because there are two sources of variance instead of one and because a second person serves as the control for each experimental subject.
4. Box 12–3 considers **beta error** in addition to alpha error. Although there is no complete agreement on the level of beta error acceptable for most studies, usually a beta error of 20% (one-tailed test) is used; this corresponds to a z value of 0.84.
5. If the sample size selected is too small, clinically important differences in outcome may be missed. If the sample size is excessive, costs become prohibitive, and clinically meaningless differences may nonetheless achieve statistical significance.

BOX 12-4. Initial and Subsequent Calculation of Sample Size for a Study Using a Test of Differences in Proportions and Considering Alpha and Beta Errors

Part 1A. Data on which the initial calculation is based:

Study Characteristic	Assumptions Made by Investigator
Type of study	Randomized controlled trial of a drug to reduce the 5-year mortality in patients with a particular form of cancer
Data sets	Observations in one experimental group (E) and one control group (C) of the same size
Variable	Success = 5-year survival after treatment; failure = death within 5 years of treatment
Variance, expressed as $\bar{p}(1-\bar{p})$	$\bar{p} = 0.55$; therefore, $(1-\bar{p}) = 0.45$
Data for alpha (z_α)	$p = 0.05$; therefore, 95% confidence desired (two-tailed test); $z_\alpha = 1.96$
Data for beta (z_β)	20% beta error; therefore, 80% power desired (one-tailed test); $z_\beta = 0.84$
Difference to be detected $(\bar{d})$	0.1 or larger difference between the success (survival) of the experimental group and that of the control group (i.e., 10% difference—because $p_E = 0.6$, and $p_C = 0.5$)

Part 1B. Initial calculation of sample size (N):

$$N = \frac{(z_\alpha + z_\beta)^2 \cdot 2 \cdot \bar{p}(1-\bar{p})}{(\bar{d})^2} = \frac{(1.96 + 0.84)^2 \cdot 2 \cdot (0.55)(0.45)}{(0.1)^2}$$

$$= \frac{(7.84)(2)(0.2475)}{0.01} = \frac{3.88}{0.01} = 388$$

= 388 subjects per group × 2 groups = **776 subjects total**

Part 2A. Changes in data on which the initial calculation was based:

Study Characteristic	Assumptions Made by Investigator
Variance, expressed as $\bar{p}(1-\bar{p})$	$\bar{p} = 0.60$; therefore, $(1-\bar{p}) = 0.40$
Difference to be detected $(\bar{d})$	0.2 or larger difference between the success (survival) of the experimental group and that of the control group (i.e., 20% difference—because $p_E = 0.7$, and $p_C = 0.5$)

Part 2B. Subsequent (revised) calculation of sample size (N):

$$N = \frac{(z_\alpha + z_\beta)^2 \cdot 2 \cdot \bar{p}(1-\bar{p})}{(\bar{d})^2} = \frac{(1.96 + 0.84)^2 \cdot 2 \cdot (0.60)(0.40)}{(0.2)^2}$$

$$= \frac{(7.84)(2)(0.2400)}{0.04} = \frac{3.76}{0.04} = 94$$

= 94 subjects per group × 2 groups = **188 subjects total**

Interpretation: As a result of changes in the data on which the initial calculation was based, the number of subjects needed was reduced from 776 to 188.

E. Sample size for a test of differences in proportions
1. Often a dependent variable is measured as success/failure and is described as the proportion of outcomes that represent some form of success, such as improvement in health, remission of disease, or reduction in mortality.
2. In this case, the formula for sample size must be expressed in terms of proportions, as shown in the formula in Box 12–4.

II. Randomization
A. Definitions
1. **Randomization** entails allocating the available subjects to one or another study group and is often used in clinical studies.
2. **Random sampling** entails selecting a small group for study from a much larger group of potential study subjects.

B. Goals of randomization
1. Randomization is used to establish an **unbiased allocation of study subjects** to the experimental and control groups.
2. Randomization *does not guarantee* that the two (or more) groups created by random allocation are identical in either size or subject characteristics. If properly done, randomization *does guarantee* that the different groups will be free of selection bias and problems due to regression toward the mean.
    a. **Selection bias** can occur if subjects are allowed to choose whether they will be in an experimental group or a control group or if investigators influence the assignment of subjects to one group or another. Differences in outcome may derive from differences in the baseline characteristics of the groups, rather than from the intervention in question.
    b. **Regression toward the mean,** also known as the **statistical regression effect,** affects patients chosen to participate in a study precisely because they had an extreme measurement on some variable (e.g., a high number of throat infections during the past year). They are likely to have a measurement that is closer to average at a later time (e.g., during the subsequent year) for reasons unrelated to the type or efficacy of the treatment they are given.

C. Methods of randomization
1. To prevent selection bias from influencing a particular randomization process, the results of randomization should be concealed until they are required for data analysis. This can be accomplished by **"blinding"** the study subjects (e.g., giving the control group a placebo that looks the same as the treatment for the experimental group) and blinding the persons who record the findings from the study subjects. The result is a **double-blind study.** (Generally, an independent group of monitors uninvolved in the design or interpretation of the study is aware of treatment assignments, so that the study may be terminated early should there be an appreciable difference in outcome referable to the intervention.)
2. **Simple random allocation**
    a. Investigators begin with a **random number table** (see the Appendix) and a stack of sequentially numbered envelopes (e.g., numbered from 1 to 100 in a study with 100 participants).
    b. A number in the table is chosen blindly, and subsequent numbers are chosen by proceeding from that number in a predetermined direction (e.g., up the columns of the table).
    c. If the first number is even, "experimental group" is written on a slip of paper that is put into the first envelope. If the next number is odd, "control group" is written on a slip of paper that is put into the second envelope. The process continues until all of the envelopes contain a random group assignment.
    d. The first patient enrolled in the study is assigned to whatever group is indicated in the first envelope. As each new eligible patient is enrolled, the investigators open the next sequentially numbered envelope to find out the patient's group assignment.
3. **Randomization into groups of two**
    a. Patients can be randomized two at a time. Envelopes are numbered sequentially (e.g., from 1 to 100) and separated into groups of two.
    b. As in the above method, the investigators begin by blindly choosing a number in the random number table and proceeding in a predetermined direction.
    c. If the first number is even, "experimental group" is written on a slip of paper that is put into the first envelope. For the paired envelope, the alternative group assignment (in this case, "control group") is automatically made.
    d. The random number table is used to determine the assignment of the first envelope in each successive pair.
    e. With this method, any time an even number of patients has been admitted into the study, half will be in the experimental group and half in the control group.
4. **Systematic allocation**
    a. Systematic allocation in research studies is equivalent to the old military "Sound off!" The first patient is randomly assigned to a group, and the next patient is automatically assigned to the alternative group.
    b. Subsequent patients are given group assignments on an alternating basis. This will ensure that the experimental and control groups are of equal size if there is an even number of patients entered in the study.
    c. Advantages of this method include an improvement in statistical power because the variance of the data from a systematic allocation is smaller than that from a simple random allocation. The simplicity, convenience, and statistical advantages of systematic sampling make it desirable to use whenever possible.
    d. A disadvantage of this method is that any kind of periodicity in the way patients enter may introduce bias. For example, if the first patient to arrive each day were assigned to the experimental group and the second to the control group, all of the experimental group subjects would be the first patients to arrive, and they might be systematically different (e.g., employed, eager, early risers) from those who arrive later. This danger is easy to avoid, however, if the investigator reverses the sequence frequently, sometimes taking the first person each day into the control group.

e. The systematic allocation method can also be used for allocating study subjects to three, four, or even more groups.
   5. **Stratified allocation**
      a. Stratified allocation, often called **prognostic stratification,** is used to assign patients to different risk groups depending on such baseline variables as the severity of disease (e.g., stage of cancer) and age.
      b. When such risk groups have been created, each stratum can then be allocated randomly to the experimental group or the control group.
      c. This is usually done to ensure homogeneity of the study groups by severity of disease.
      d. If the experimental and control groups have been made prognostically similar, the analysis can be conducted both for the entire group and within the prognostic groups.
D. **Special issues concerning randomization**
   1. Randomization does not guarantee that two or more groups will be identical.
   2. The fact that there are occasional differences between groups that are statistically significant does not mean that the randomization was biased; some differences are expected by chance alone.
   3. Some of the observed differences between the randomized groups could confound the analysis; in that case, the variables of concern can be controlled for in the analysis.
   4. In addition to randomization, other precautions that should be taken in clinical trials include ensuring the accuracy of all of the data by blinding patients and observers, standardizing data collection instruments, and so forth.
   5. Patients may refuse to participate in a study. This means that a particular study is limited to patients who are willing to participate. These patients may not be representative of those who refused to participate. The **generalizability** of a study must be considered in light of this limitation.
   6. Following randomization, if a patient is not doing well, the patient or physician may wish to switch from the experimental treatment to another medication. Ethically, the patient cannot be forced to continue a particular treatment. Once the switch occurs, it will be necessary to choose an alternative way of analyzing the data.
      a. Currently, the popular approach is to analyze the data as if the patient had remained in his or her original group, so that any negative outcomes are assigned to their original treatment. This strategy, called the **"intention to treat" approach,** is based on the belief that if the patient was doing so poorly as to want to switch, a negative outcome should be ascribed to that treatment.
      b. Other approaches include the following:
         (1) The data are analyzed as if the patient had never participated in the study. This can lead to a small, and probably biased, sample, however.
         (2) The patient is reassigned to a third group, and the data are analyzed separately from the original groups. The problem with this approach is that the original groups are changed and it is not clear whom the remaining groups represent.
   7. Another problem in randomized trials of treatment is determining the appropriate starting point for measuring the outcome; most investigators recommend counting from the point of randomization.

III. **Dangers of Data Dredging**
  A. **Analysis of large data sets**
   1. The analysis of large data sets with modern computer techniques permits the assessment of hundreds of possible associations among the study variables. This process is sometimes referred to as **data dredging.**
   2. The search for associations can be appropriate as long as the investigator keeps two points in mind:
      a. Hypothesis development and hypothesis testing must be based on different data sets. One data set is used to develop the hypothesis or model, which is used to make predictions, which are then tested on a new data set.
      b. A correlational study (e.g., the Pearson correlation coefficient or the chi-square test) is only useful for developing hypotheses, not for testing them.
   3. If the $p = 0.05$ cutoff point is used for alpha, then chance would account for at least 1 out of 20 statistically significant associations.
  B. **Multiple hypotheses**
   1. The problem with multiple hypotheses is similar to the problem with multiple associations: the greater the number of hypotheses that are tested, the more likely it is that at least one of them will be found "statistically significant" by chance alone.
   2. One possible way to handle this is to lower the p value required before rejecting the null hypothesis (e.g., make it <0.05).
   3. To keep the risk of a false-positive finding in the entire study to no more than 0.05, the alpha level chosen for rejecting the null hypothesis may be made more stringent by dividing alpha by the number of hypotheses being tested. This method of adjusting for multiple hypotheses is called the **Bonferroni procedure.**
   4. There are other possible adjustments that are less stringent, but they are more complicated and are used in different situations.

IV. **Elementary Probability Theory**
  A. **The independence rule**
   1. The independence rule pertains when the probability of an outcome is unchanged by the outcome of a related event. For example, the probability of heads when flipping a coin ($p = 0.50$) is not altered by the results of one or more prior coin flips.
   2. In statistical terms, this statement can be expressed as follows: $p\{H2+ \mid H1+\} = p\{H2+ \mid H1-\}$. Here, p denotes probability, H2+ denotes heads resulting from the second flip of a coin, the vertical line ($\mid$) means "given that" or "conditional upon" what immediately follows, H1+ denotes heads resulting from the first flip of a coin, and H1− denotes not heads (i.e., tails) resulting from the first flip of the coin.
  B. **The product rule**
   1. The product rule is used to determine the probability of two things being true. The manner of calculation depends on whether the two things are independent.
   2. *If independence is assumed,* the probability that both outcomes will be positive is simply the product of their independent probabilities, or $p\{X+ \text{ and } Y+\} = p\{X+\} \times p\{Y+\}$. The probability that *neither* will be positive is the product of the probabilities of negative outcomes, or $p\{X- \text{ and } Y-\} = p\{X-\} \times p\{Y-\} = (1 - p\{X+\}) \times (1 - p\{Y+\})$.
   3. *If independence cannot be assumed,* a more general product rule must be used. In calculating the probability that *neither* event would occur, the general product

rule says that $p\{X- \text{ and } Y-\} = p\{X- \mid Y-\} \times p\{Y-\}$. The answer would be the same if the rule were expressed as $p\{X- \text{ and } Y-\} = p\{Y- \mid X-\} \times p\{X-\}$.

C. The addition rule
1. According to the addition rule, all of the possible different probabilities in a situation must add up to 1.0 (100%), no more and no less.
2. The addition rule is used to determine the probability of one thing being true under all possible conditions: $p\{X+\} = p\{X+ \mid Y+\} \times p\{Y+\} + p\{X+ \mid Y-\} \times p\{Y-\}$.
3. The numerator and the denominator of Bayes' theorem (see Chapter 8) are based on the general product rule and the addition rule, respectively.

QUESTIONS

DIRECTIONS. (Items 1–10): Each of the numbered items or incomplete statements in this section is followed by answers or by completions of the statement. Select the ONE lettered answer or completion that is BEST in each case. Correct answers and explanations are given at the end of the chapter.

1. The formula for a paired *t*-test is as follows:

$$t_\alpha = \frac{\bar{d}}{\frac{s_{\bar{d}}}{\sqrt{N}}}$$

To calculate for sample size, this formula is often used and rearranged algebraically to solve for N. In the process, z must be substituted for t because

(A) t provides too large a sample
(B) t provides too small a sample
(C) t is dependent on degrees of freedom and z is not
(D) z is dependent on degrees of freedom and t is not
(E) z takes beta error into account

2. It is important to consider beta error when

(A) the null hypothesis is rejected
(B) the null hypothesis is not rejected
(C) the difference under consideration is not clinically meaningful
(D) the difference under investigation is statistically significant
(E) the sample size is excessively large

3. Which one of the following characteristics of a diagnostic test is analogous to the statistical power of a study?

(A) sensitivity
(B) specificity
(C) positive predictive value
(D) negative predictive value
(E) utility

Items 4–7

A study is designed to test the effects of sleep deprivation on academic performance among medical students. Each subject serves as his or her own control. A 10-point difference (10% difference) in test scores is considered meaningful. The standard deviation of test scores in a similar study was 8. Alpha is set at 0.05 (two-tailed test), and beta is set at 0.2.

4. The appropriate sample size is

(A) 6
(B) 16
(C) 60
(D) 100
(E) 34

5. The study is conducted by another group that does not specify beta. The sample size for this study is

(A) 6
(B) 8
(C) 3
(D) 30
(E) 60

6. The study in question 4 is conducted by another group. All the specified parameters remain the same, but the investigators use separate intervention and control groups. The required sample size for this study is

(A) 6
(B) 11
(C) 18
(D) 22
(E) 60

7. The study in question 4 is revised to detect a difference of only 2 points (2%) in test scores. All other parameters of the original study remain unchanged. The required sample size is

(A) 6
(B) 12
(C) 60
(D) 42
(E) 126

8. An investigator studying the health effects of rutabaga in subjects with arthritis tends to assign enthusiastic participants to rutabaga and skeptics to placebo. The best way to avoid this form of bias is

(A) random sampling
(B) statistical regression
(C) intention to treat analysis
(D) randomization
(E) self-selection

9. Statistical methods of adjusting for the testing of multiple hypotheses from a single large data set, such as the Bonferroni procedure, are designed to prevent

   (A) false-negative results
   (B) false-positive results
   (C) bias
   (D) confounding
   (E) effect modification

10. Assume that the risk of myocardial infarction during the next year in a 60-year-old man is 20% but is reduced by half if the man exercises regularly. Assume as well that the risk of the same event in the man's 58-year-old wife is 14% but is reduced by half if she uses hormone replacement therapy. What is the probability that neither husband nor wife will experience a myocardial infarction during the next year if the husband exercises but the wife is unwilling to take supplemental hormones?

    (A) 77%
    (B) 17%
    (C) 20%
    (D) 37%
    (E) 93%

ANSWERS AND EXPLANATIONS

1. The answer is C: t is dependent on degrees of freedom and z is not. Both z and t are used to provide unit-free measures of dispersion about a mean value. Because z is derived from a truly normal distribution, it does not vary with sample size. Because t is an approximation of z for nearly normal distributions, it varies with sample size and is dependent on degrees of freedom. Degrees of freedom derive from the sample size, which is the unknown in a sample size calculation. Therefore, t cannot be used to calculate sample size because the sample size must be known to calculate t. Use of z circumvents this problem and is customary. The substitution confers a slight risk of underestimating the sample size required. *Epidemiology*, p. 160.

2. The answer is B: the null hypothesis is not rejected. Beta, or type II error, is false-negative error: the failure to detect a true difference when one exists. Only a negative result (i.e., failure to reject the null hypothesis) is at risk for being a false-negative result. Therefore, when the null hypothesis is rejected (choice A) or when the difference under investigation is statistically significant (choice D), a difference is being detected, so neither of these choices can be correct. Beta error is the result of inadequate power to detect the difference under investigation and therefore occurs when the sample size is small, not large. Enlarging the sample, or pooling data to increase statistical power, is the appropriate means to correct beta error, provided that the difference is clinically meaningful. *Epidemiology*, p. 160.

3. The answer is A: sensitivity. The sensitivity of a diagnostic test, or $a/(a + c)$, is the ability of the test to detect a condition when it is present. Similarly, statistical power is the ability of a study to detect a difference when it exists. Beta error is $(1 - \text{sensitivity})$, or $(1 - \text{statistical power})$. *Epidemiology*, p. 161.

4. The answer is A: 6. This is a study for which a paired t-test is appropriate. The corresponding sample size formula, as detailed in Box 12–1, is

$$N = \frac{(z_\alpha)^2 \cdot (s)^2}{(\bar{d})^2}$$

In this example, however, the value for beta is also specified, so that the formula becomes

$$N = \frac{(z_\alpha + z_\beta)^2 \cdot (s)^2}{(\bar{d})^2}$$

The value of z_α, when alpha is 0.05 and the test is two-tailed, is 1.96. The value of z_β, when beta is 20%, or 0.2, is 0.84; z_β is one-tailed by convention. The standard deviation (s) is derived from a prior study and is 8. The difference sought is 10 points. The equation becomes the following:

$$N = \frac{(1.96 + 0.84)^2 \cdot (8)^2}{(10)^2}$$

$$= \frac{(7.84)(64)}{100}$$

$$= 501.76/100$$

$$= 5.02$$

By convention, all sample size calculations are rounded up to the nearest whole number. The correct answer is therefore 6. The small sample size required for this study is the result of the large difference sought (a 10% change in test scores is substantial), the small standard deviation, and the use of each subject as his or her own control. The sample required would be even smaller if beta were not specified (see the explanation for question 5, below). *Epidemiology*, p. 161.

5. The answer is C: 3. The sample size calculation here is the same as for question 4, but this time there is no beta term. Therefore, the equation is as follows:

$$N = \frac{(1.96)^2 \cdot (8)^2}{(10)^2}$$

$$= \frac{(3.8416)(64)}{100}$$

$$= 245.86/100$$

$$= 2.46 \text{ rounded to } 3$$

This is obviously a very small sample. A sample this small may make intuitive sense if you consider how many subjects you would need to demonstrate that test scores are higher if the test taker gets 8 hours of sleep the night before the test rather than staying up all night. *Epidemiology*, p. 161.

6. The answer is D: 22. The formula for this calculation is shown in Box 12–2. The basic sample size formula for a trial using separate control and intervention groups is the following:

$$N = \frac{(z_\alpha + z_\beta)^2 \cdot 2 \cdot (s)^2}{(\overline{d})^2}$$

Note that the term for z_β has been added to the numerator. The difference between this equation and the equation for the before-after study in question 4 is the 2 in the numerator, doubling the sample. When the numbers from question 4 are inserted, the equation becomes the following:

$$N = \frac{(1.96 + 0.84)^2 \cdot 2 \cdot (8)^2}{(10)^2}$$

$$= \frac{(7.84)(2)(64)}{100}$$

$$= 1003.52/100$$

$$= 10.04 \text{ rounded to } 11$$

N represents the number of subjects needed per group. As there are now separate control and intervention groups, the total sample is $2N$. The total number of subjects needed is therefore 22. *Epidemiology,* p. 161.

7. The answer is E: 126. The calculation is as shown in the answer to question 4, with one difference: the denominator is now $(2)^2$ instead of $(10)^2$. The numerator is again 501.76. When this is divided by 10^2, or 100, it yields a sample size of 5.02, or 6. When divided by 2^2, or 4, it yields a sample size of 125.44, or 126. This example demonstrates the profound effect that the size of the difference in outcome one is hoping to detect can have on sample size requirements. *Epidemiology,* p. 161.

8. The answer is D: randomization. The bias described is selection bias. Any time an investigator believes in a particular intervention and has the ability to assign to that intervention those subjects most likely to respond well, there is the possibility for selection bias. Self-selection, when subjects choose a particular study or a particular intervention, is another means by which selection bias occurs. Random sampling is a method of drawing a representative study sample from a larger population but does not pertain to assigning treatments. Statistical regression is not a technique but rather a tendency against which one needs to be on guard. Extreme values tend to "regress" toward the mean when they are reassessed over time, simply because they had already become maximally extreme and the only direction they could go is back toward the population "norm." Randomization, as discussed in the chapter, is a method for eliminating selection bias as well as distributing evenly between treatment groups those extreme values susceptible to the statistical regression effect. While randomization cannot ensure that two groups will be comparable in all ways except for the study intervention, it is an unbiased means of allocating subjects to treatment assignments. Randomization has the potential to eliminate both known and unknown confounders.

The intention to treat analysis is the interpretation of study data based on subject assignment to a particular group, regardless of whether or not the subject actually complied with that assignment. *Epidemiology,* p. 165.

9. The answer is B: false-positive results. When alpha is set at 0.05, one is accepting a 5% risk of rejecting the null hypothesis on the basis of a random outcome. This risk is small when a single hypothesis is tested. However, if that same 5% risk pertains to each of many hypotheses being tested, the aggregate risk of rejecting one of the null hypotheses on the basis of chance becomes large. If 20 hypotheses are tested, a single false-positive result can be expected with some confidence. Techniques are used to maintain the aggregate risk of a false-positive result close to the 0.05 level. Such techniques place more stringent requirements on associations before they are deemed significant and the null hypothesis is rejected. *Epidemiology,* pp. 167–168.

10. The answer is A: 77%. The probability that the husband will experience a myocardial infarction is 20% if he does not exercise and 10% if he does. We are told that he is exercising. Therefore, the probability that the husband will not experience a myocardial infarction during the next year is 90%. The wife's risk of myocardial infarction is 14%, and this risk is not reduced because she is disinclined to use hormone replacement therapy. Her probability of not experiencing a myocardial infarction is therefore 86%. The risks for myocardial infarction are independent in the two individuals. Therefore, the probability that neither spouse will experience a myocardial infarction is the product of their separate probabilities of not experiencing a myocardial infarction, or $(0.90)(0.86) = 0.774$. This is approximately 77%. *Epidemiology,* p. 168.

REFERENCES CITED

Freiman, J. A., et al. The importance of beta, the type II error, and sample size in the design and interpretation of the randomized control trial: a survey of 71 "negative" trials. New England Journal of Medicine 299:690–695, 1978.

Jekel, J. F. Epidemiology, Biostatistics, and Preventive Medicine. Philadelphia, W. B. Saunders Company, 1996.

CHAPTER THIRTEEN

MULTIVARIABLE ANALYSIS

SYNOPSIS

OBJECTIVES

- To provide a conceptual understanding of multivariable modeling.
- To discuss the uses of multivariable statistics.
- To introduce basic methods of multivariable analysis, and to provide guidelines for selecting among them.

I. An Overview of Multivariable Statistics

A. Introduction

1. Statistical models often seek to represent only one dimension of reality, such as the effect of a change in one variable (e.g., a nutrient) on another variable (e.g., the growth rate of a rat).
   a. For such models to be meaningful, all factors other than the one being studied must be equalized in the research architecture, usually by randomization.
   b. Often, however, either the other influences cannot be adequately controlled by design or the investigator may actually wish to study the relative simultaneous influence of several independent (possibly causal) variables on a dependent (outcome) variable.
2. Statistical models that have one outcome variable but include more than one independent variable are generally called **multivariable models.**
   a. These models are intuitively attractive to investigators because they seem more true to life than the single-variable models.
   b. Multivariable analysis does not enable an investigator to ignore the basic principles of good research design and analysis, because multivariable analysis also has many limitations.
   c. The methodology and interpretation of findings in this type of analysis are difficult for most physicians, but it is important for health care professionals to understand how to interpret the findings as they are presented in the literature.

B. A conceptual understanding of equations

1. Multivariable analysis may be easier to understand if it is approached conceptually rather than mathematically.
2. For example, suppose that there is a study of the prognosis of patients at the time of diagnosis for a certain cancer for which there is not, as yet, an effective treatment. The physician might surmise that the length of survival for a patient would depend on at least four things: the patient's age, the anatomic stage of the disease at the time of diagnosis, the presence or absence of other diseases (comorbidity), and the degree of systemic symptoms such as weight loss. The relationship could be explained conceptually as follows:

   Cancer prognosis varies with Age and Stage and Comorbidity and Symptoms

   (13–1)

3. This statement could be made to look more mathematical simply by making a few slight changes:

   Cancer prognosis $\approx$ Age + Stage + Comorbidity + Symptoms

   (13–2)

4. The four independent variables on the right side of the equation are not necessarily of equal importance. Expression 13–2 can be improved by giving each independent variable a **coefficient,** which is a **weighting factor** based on its relative importance in predicting prognosis. Thus, the equation becomes:

   Cancer prognosis $\approx$ (Weight$_1$) Age
   + (Weight$_2$) Stage
   + (Weight$_3$) Comorbidity
   + (Weight$_4$) Symptoms

   (13–3)

5. Before equation 13–3 can become useful, two more things are needed. First, an **anchor point** for the equation is needed; it must be something comparable to the a of the formula for simple regression ($y = a + bx$). Second, an **error term** is needed to make the equation true if the prediction is not perfect.
6. By inserting the anchor point and the error term, the $\approx$ symbol (meaning "varies with") can be replaced by an equals sign.
7. Abbreviating the weights with a W, the equation now becomes:

   Cancer prognosis = Anchor point + W$_1$Age
   + W$_2$Stage + W$_3$Comorbidity
   + W$_4$Symptoms + Error term

   (13–4)

8. In common statistical symbols, y is the dependent (outcome) variable (e.g., cancer prognosis) and is customarily placed on the left; $x_1 + x_2 + x_3 + x_4$ are the independent variables 1 (age) through 4 (symptoms), and they are lined up on the right; b_i is the statistical symbol for the weight of the ith independent variable; a is the estimated y-intercept (the anchor point); and e is the error term.
9. In statistical symbols, the equation can be expressed as follows:

$$y = a + b_1x_1 + b_2x_2 + b_3x_3 + b_4x_4 + e \quad (13-5)$$

10. Although equation 13-5 looks complex, it means exactly the same thing as equations 13-1 through 13-4.

C. Best estimates
1. In a multivariable equation, the value for the error term (e) is not known until after the equation has been solved for a and all of the b's. Therefore, the estimated value of y—namely, $\hat{y}$—is used instead of y, and there is no error term. (Because the estimate of y has a circumflex, or "hat," over it, it is usually called y-hat.)
2. If the values of all of the observed y's and all of the x's are inserted, the following equation can be solved:

$$\hat{y} = a + b_1x_1 + b_2x_2 + b_3x_3 + b_4x_4 \quad (13-6)$$

3. When equation 13-6 is subtracted from equation 13-5, the following equation for the error term emerges:

$$(y - \hat{y}) = e \quad (13-7)$$

4. The error term is the observed value of the outcome variable y for a given patient minus the predicted value of y for the same patient.
5. As noted in previous chapters, variation in statistics is measured as the sum of the squares of the observed value (O) minus the expected value (E). In multivariable analysis, the error term e is often called a **residual.**
6. The **best estimate** has been achieved when the sum of the squared error term has been minimized. That sum is expressed as:

$$\sum(y_i - \hat{y})^2 = \sum(y_O - y_E)^2 = \sum e^2 \quad (13-8)$$

7. Those values of a and of the several b_i's which, taken together, produce the smallest total quantity of error (measured as the sum of squares of the error term, or, most simply, e^2) are the **best estimates** that can be obtained from that set of data. This approach is called the **least-squares solution,** because the process is stopped when the sum of squares of the error term is the least.

D. The general linear model
1. The multivariable equation, with one dependent variable and one or more independent variables, as shown in equation 13-6, is usually called the general linear model.
2. The model is "general" because there are many variations regarding the type of variables for y and x_i as well as the number of x variables that can be used.
3. The model is "linear" because it is a linear combination of the x_i terms. For the x_i variables, a variety of transformations (e.g., square of x, cube of x, square root of x, or logarithm of x) could be used and the combination of terms would still be linear, so that the model would remain linear. What cannot happen if the model is to remain linear is for any of the coefficients (the b_i terms) to be a square, a square root, a logarithm, or another transformation.
4. Numerous procedures for multivariable analysis are based on the general linear model. These include analysis of variance (ANOVA), analysis of covariance (ANCOVA), multiple linear regression analysis, multiple logistic regression, the log-linear model, and discriminant function analysis.
5. As discussed below, the choice of which procedure to use depends primarily on whether the dependent and independent variables are continuous, dichotomous, nominal, or ordinal.

E. Uses of multivariable statistics
1. The usual purpose of multivariable analysis is to understand how important, both individually and when acting together, the independent variables are for explaining the variation in the dependent variable y.
2. Multivariable analysis, which uses all of the observations in an analysis, is often more efficient and robust than contingency table analysis.
3. Multivariable techniques make it possible to determine whether there is an **interaction between variables.** Interaction is present when the value of one independent variable influences the way another independent variable explains y.
4. Multivariable techniques can be used to develop a clinical **prediction model** (e.g., a model to predict the likelihood of myocardial infarction given chest pain and a variety of other independent variables), complete with coefficients for use in prediction.
   a. For example, Goldman et al. (1982) used a multivariable technique to develop a protocol to assist in the diagnosis of myocardial infarction in patients presenting to an emergency room with chest pain. Using various combinations of symptoms, signs, laboratory values, and electrocardiographic findings, the authors developed estimates for the probability of myocardial infarction.
   b. Multivariable analysis was also used by investigators in the Framingham Study to develop prediction equations for the 8-year risk of developing cardiovascular disease in people with various combinations of the following factors: smoking, elevated cholesterol levels, hypertension, glucose intolerance, and left ventricular hypertrophy (Breslow 1978).
   c. These prediction equations are now used in various health risk assessment programs.
5. Multivariable analysis is used to determine which of the independent variables are the strongest predictors of y and which of the independent variables overlap with one another or interact in their ability to predict y.

II. Procedures for Multivariable Analysis
A. Introduction
1. The choice of an appropriate statistical method for multivariable analysis depends on whether the dependent and independent variables are continuous, ordinal, dichotomous, or nominal, as shown in Table 13-1.

TABLE 13-1. Choice of an Appropriate Procedure To Be Used in Multivariable Analysis (Analysis of One Dependent Variable and More Than One Independent Variable)

Characterization of Variables To Be Analyzed		
Dependent Variable	Independent Variables*	Appropriate Procedure or Procedures
Continuous	All are categorical.	Analysis of variance (ANOVA).
Continuous	Some are categorical and some are continuous.	Analysis of covariance (ANCOVA).
Continuous	All are continuous.	Multiple linear regression.
Ordinal	—	There is no formal multivariable procedure for ordinal dependent variables. Either treat the variables as if they were continuous (see above procedures) or perform log-linear analysis.
Dichotomous	All are categorical.	Logistic regression; log-linear analysis.
Dichotomous	Some are categorical and some are continuous.	Logistic regression.†
Dichotomous	All are continuous.	Logistic regression; discriminant function analysis.
Nominal	All are categorical.	Log-linear analysis.
Nominal	Some are categorical and some are continuous.	Group the continuous variables and perform log-linear analysis.
Nominal	All are continuous.	Discriminant function analysis; group the continuous variables and perform log-linear analysis.

*Categorical variables include ordinal, dichotomous, and nominal variables.
†If the outcome is a time-related dichotomous variable (such as live/die), then proportional-hazards (Cox) models are best.

2. In cases in which more than one method could be used, the final choice will depend on the investigator's experience, personal preference, and comfort with the methods that are appropriate.

B. Analysis of variance (ANOVA)
1. If the dependent variable is continuous and all of the independent variables are categorical (i.e., nominal, dichotomous, or ordinal), the correct multivariable technique is analysis of variance (ANOVA).
2. **One-way ANOVA (the F-test)**
   a. If the design includes only one independent variable (e.g., treatment), the technique is called one-way analysis, regardless of how many different kinds of treatment (levels of analysis) are under consideration.
   b. The Student's t-test can be used to compare two groups, but when there are more than two groups across levels of a single independent variable, one-way ANOVA is the best approach.
   c. The first step of ANOVA is the **F-test.** (F is derived from the name of the developer of the method, Sir Ronald Fisher.) The F-test is a kind of "super t-test" that compares more than two means simultaneously.
      (1) If there are four groups, the null hypothesis for the F-test is that the mean change in outcome (d) will be the same for all four groups ($\bar{d_1} = \bar{d_2} = \bar{d_3} = \bar{d_4}$), indicating that all samples were from the same population and that any differences between the means are due to chance variation, rather than to true differences.
      (2) Two measures of the variance of the observations are used in the F-test: one is the **between-groups variance,** based on the variation between (or among) the means; the other is the **within-groups variance,** based on the variation within each group (i.e., variation around a single group mean).
      (3) In ANOVA, these two measures of variance are also called the **between-groups mean square** and the **within-groups mean square.** (**Mean square** is simply another name for variance, which is defined as a sum of squares, or SS, divided by the appropriate number of degrees of freedom, or df.)
      (4) The ratio of the two measures can be expressed as follows:

$$F \text{ ratio} = \frac{\text{Between-groups variance}}{\text{Within-groups variance}} = \frac{\text{Between-groups mean square}}{\text{Within-groups mean square}}$$

   (5) If the F ratio is fairly close to 1.0, the two estimates of variance are similar, and the null hypothesis that all of the means came from the same underlying population is not rejected. If the ratio is much larger than 1.0, there must have been some force, attributable to group differences, pushing the means apart, and the null hypothesis of no difference is rejected.
   (6) The assumptions for the F-test are similar to those for the t-test. First, the dependent variable should be normally distributed (although with large samples this assumption can be relaxed because of the central limit theorem). Second, the several samples of the dependent variable should be independent random samples from populations with approximately equal variances. As with the t-test, the F-test requires that an alpha level be specified in advance. After the F statistic has been calculated, its p value can be looked up in a table of the F distribution to determine whether the results are statistically significant.
   d. If the results of the F-test are not statistically significant, either the null hypothesis must be accepted or the study must be repeated using a larger sample.
   e. If the results of the F-test are statistically significant, additional steps are needed to determine which of the differences in means are the "true" differences (i.e., responsible for statistical significance). Most advanced statistical computer packages include a program for ANOVA that performs this task.

3. **N-Way ANOVA**
   a. If the study design includes more than one independent variable (e.g., treatment, age group, and gender), the technique is called N-way ANOVA.
   b. The goal of ANOVA is to explain (to "model") the total variation found in a study.
   c. If only *one* independent variable is tested in a model, the total amount of variation must be explained in terms of how much variation is due to that variable and how much is not. Anything that is not due to the independent variable is considered to be error (residual). See Box 13-1.
   d. If *two* independent variables are tested in a model, the total amount of variation must be explained in terms of how much variation is due to the independent effect of the first independent variable, the independent effect of the second

BOX 13-1. The Analysis of Variance (ANOVA) Table

The goal of ANOVA, stated in the simplest terms, is to explain (i.e., to model) the total variation found in a study. Because the total variation is equal to the sum of squares (SS), the process of explaining total variation is a process that entails partitioning the SS into component parts. The logic behind this process was introduced in Chapter 10 (see Variation between Groups versus Variation within Groups), and the discussion there focused on the example of explaining the difference between the heights of men and women. In the example, the heights of 100 female and 100 male university students were measured, the total variation was found to be 10,000 cm², and 4,000 cm² of the variation was attributed to gender. Because that example is uncomplicated and involves round numbers, it is used here to illustrate the format for an ANOVA table.

Source of Variation	Sum of Squares (SS)	Degrees of Freedom (df)	Mean Square (MS)	F Ratio
TOTAL	10,000	199		
Model (gender)	4,000	1	4,000.0	132.0
Error	6,000	198	30.3	

The model in this example has only one independent variable—namely, gender, a dichotomous variable.

In the column labeled sum of the squares (SS), the figure of 4,000 represents the amount of variation due to gender (i.e., the between-groups variation noted in the ANOVA), while 6,000 represents the amount of variation due to error (i.e., the within-groups variation).

In the column labeled degrees of freedom (df), the total df is listed as 199, reflecting the fact that there were 200 subjects and 1 df was lost in calculating the grand mean for all observations. The df for the model is calculated as the number of categories minus 1. Gender has only two categories (men and women), so 1 df is assigned to it. The df for error is calculated as the total df minus the number of df assigned to the model: $199 - 1 = 198$.

The mean square (MS) is simply another name for variance and is equal to the SS divided by the appropriate df: $4,000/1 = 4,000.0$, and $6,000/198 = 30.3$.

The F ratio is calculated by dividing the model mean square by the error mean square: $4,000.0/30.3 = 132.0$. Because the F ratio is so large, the p value would be extremely small ($p < 0.00001$), and the null hypothesis that there is no true difference between the mean heights of men and women would be rejected.

Note that if there was more than one independent variable in the model being analyzed, there would be more entries under the column labeled source of variation: TOTAL, model, interaction, and error. Interaction would refer to the portion of the variation that is due to interactions between the independent variables in the model. Error would then be defined as the variation not explained by any of the independent variables or their interactions.

---

independent variable, the interaction between the two variables, and error.

e. If *more than two* independent variables are tested, the analysis becomes increasingly complicated, but the underlying logic remains the same.

f. As long as there are equal numbers of observations in all of the study groups, ANOVA can be used to analyze the individual and joint effects of the independent variables and to partition the total variation into the various component parts. Numerous computer programs are available to test significance using the F-test and to perform subsequent calculations of formulas based on the general linear model described earlier in this chapter.

C. Analysis of covariance (ANCOVA)
1. As shown in Table 13-1, ANOVA and ANCOVA are methods for evaluating studies in which the dependent variable is continuous.
2. If the independent variables are all of the categorical type (nominal or dichotomous), then ANOVA is used.
3. If some of the independent variables are categorical and some are continuous, then ANCOVA is appropriate.
4. The ANCOVA procedure adjusts the dependent variable on the basis of the continuous independent variable or variables (covariates), and it then does an N-way ANOVA on the **adjusted dependent variable.**

D. Multiple linear regression
1. If the dependent variable and all of the independent variables are continuous, the correct type of multivariable analysis is multiple linear regression.
2. The formula resembles the general linear model formula shown in equation 13-6. The intercept is really the mean of y, and each of the independent variables improves the prediction somewhat (depending on its strength of association).
3. The most common computerized method is called **stepwise linear regression.**
   a. Variables are entered in the analysis one at a time, and the explanatory strength of each variable entered (i.e., the r^2; see Chapter 11) changes as each new variable is entered.
   b. The "stepping" continues until none of the remaining independent variables meets the predetermined criterion for being entered (e.g., p is ≤ 0.1, or the increase in r^2 is ≥ 0.01) or until all of the variables have been entered. When the stepping stops, the analysis is complete.
4. The analysis reveals the statistical significance of the overall equation and of each variable entered, as well as the overall r^2 for each step, which is the proportion of variation the model has explained so far. In multiple regression equations that are statistically significant, the r^2 indicates how much of the variation is being explained by the variable entered.

E. Other procedures for multivariable analysis
1. Other major multivariable procedures include **logistic regression, log-linear analysis,** and **discriminant function analysis.** All are forms of the general linear model and function in an analogous manner; their uses are outlined in Table 13-1.
2. If the dependent variable in a study is a dichotomous variable, logistic regression is the most powerful technique available.

   a. In medicine, the most commonly used form of logistic regression is the **proportional-hazards (Cox) method.**
   b. This is used to test for differences between Kaplan-Meier survival curves while controlling for other variables. It is also used to determine which variables are associated with better survival.

QUESTIONS

DIRECTIONS. (Items 1–7): Each of the numbered items or incomplete statements in this section is followed by answers or by completions of the statement. Select the ONE lettered answer or completion that is BEST in each case. Correct answers and explanations are given at the end of the chapter.

1. A multivariable analysis is appropriate when

   (A) multiple hypotheses are being tested
   (B) multiple repetitions of an experiment are planned
   (C) more than one independent variable is under investigation
   (D) multiple outcome variables are under investigation
   (E) bivariate analysis fails to reveal statistical significance

2. Which of the following statements can be considered a multivariable model?

   (A) Height and weight vary together.
   (B) Height and weight vary with age and gender.
   (C) Height varies with gender.
   (D) Height varies with age and weight and gender.
   (E) Height varies with gender, and weight varies with age.

3. All of the following are included in the basic equation for a multivariable model EXCEPT

   (A) degrees of freedom
   (B) an error term
   (C) an anchor point, or y-intercept
   (D) the outcome variable
   (E) weights

4. The least-squares solution to a multivariable equation is determined when

   (A) the value of a is minimized
   (B) the residual is maximized
   (C) the model is statistically significant
   (D) the y-intercept is 0
   (E) e^2 is minimized

5. A particular advantage of multivariable analysis over contingency table analysis is that

   (A) statistical significance is more easily achieved with multivariable methods
   (B) multivariable methods permit examination of the interaction between independent variables
   (C) multivariable methods exclude weak associations
   (D) overlapping variables are excluded from multivariable analysis
   (E) type I error is minimized

Items 6–7

You are interested in comparing acetaminophen to ibuprofen and also to placebo in the management of pain due to osteoarthritis. You design a study in which equal numbers of subjects are assigned to groups defined by treatment (acetaminophen, ibuprofen, or placebo); gender; age (dichotomous: < 50 years or ≥ 50 years); and severity of arthritis (categorical: mild, moderate, or severe).

6. The appropriate statistical method for analysis of your data is

   (A) logistic regression
   (B) ANCOVA
   (C) One-way ANOVA
   (D) N-way ANOVA
   (E) Wilcoxon signed-ranks test

7. To analyze these data using the F ratio, you must first establish

   (A) between-groups variance and within-groups variance
   (B) the least squares and the residual
   (C) the degrees of freedom and the value of p
   (D) the between-groups mean square and the degrees of freedom
   (E) the standard error and the mean for each group

ANSWERS AND EXPLANATIONS

1. The answer is C: more than one independent variable is under investigation. As defined in the chapter, multivariable models include multiple independent variables but only one dependent variable. The analysis of multiple dependent variables and multiple independent variables, as noted in Chapter 11, is called multivariate analysis. The testing of multiple hypotheses does not per se indicate which analytic technique will be called for. The appropriate statistical methods depend on the nature of the hypotheses. Similarly, multiple repetitions of an experiment do not dictate the statistical method required. That is always determined by the number and nature of the variables involved and the types of hypotheses being tested. When bivariate analysis fails to reveal statistical significance, use of an alternative analytic technique is generally not indicated (and may be inappropriate). The appropriate responses include increasing the sample size or accepting the null hypothesis. *Epidemiology,* p. 173.
2. The answer is D: Height varies with age and weight and gender. As described in the chapter, multivariable models may be expressed both conceptually and mathematically. A conceptual understanding of multivariable analysis is facilitated by a verbal description of the relationships under study. To conform to the requirements for multivariable analysis, the model must postulate the influence of more than one independent variable on one, and only one, dependent (outcome) variable. In general, such a model, expressed in words, would take the following form: dependent variable y varies with independent variables x_1, x_2, and so on. The only choice provided that fits this pattern is D. *Epidemiology,* p. 173.
3. The answer is A: degrees of freedom. The basic equation for a multivariable model is as follows:

$$y = a + b_1x_1 + b_2x_2 + b_3x_3 + b_4x_4 + e$$

   The outcome variable is y; the y-intercept or anchor point is a; the b terms represent weights; and e is the residual or error term. There is no term for degrees of freedom in this method. *Epidemiology,* p. 173.
4. The answer is E: e^2 is minimized. The goal of the least-squares approach to multivariable analysis is to find the model that produces the smallest sum of squares of the error term, e, for each of the independent variables. The values of a and the b_i's that lead to the smallest error term produce the best statistical model. The least-squares model may or may not be statistically significant, depending on the strength of association between the independent variables and the dependent variable under investigation. *Epidemiology,* p. 174.
5. The answer is B: multivariable methods permit examination of the interaction between independent variables. A contingency table may be used to show that multiple independent variables contribute independently to an outcome. There is no way to use a contingency table to demonstrate the interaction between or among independent variables. For example, the independent contributions of tobacco and alcohol to the risk for head and neck cancers can be shown with contingency table analysis, but the synergistic interaction between the two cannot. To demonstrate interaction such as synergy, multivariable methods are required. The achievement of statistical significance is no easier with multivariable methods. Weak association may be studied by either method. Overlapping or interacting variables are purposefully studied in multivariable methods. With either method, type I error is minimized by setting an appropriately stringent alpha level. *Epidemiology,* p. 174.
6. The answer is D: N-way ANOVA. As discussed in the chapter and shown in Table 13–1, ANOVA is appropriate when the outcome variable is continuous and multiple independent variables are categorical. The study described meets these criteria. The method is N-way ANOVA rather than one-way ANOVA because the model includes several different independent variables. *Epidemiology,* p. 177.
7. The answer is A: between-groups variance and within-groups variance. The F ratio is the ratio of between-groups variance to within-groups variance. Therefore, only these two parameters must be established to calculate the F ratio. Subsequently, a p value for a given F ratio is determined from a table of the F distribution. *Epidemiology,* p. 176.

REFERENCES CITED

Breslow, L. Risk factor intervention for health maintenance. Science 200:908–912, 1978.

Goldman, L., et al. A computer-driven protocol to aid in the diagnosis of emergency room patients with acute chest pain. New England Journal of Medicine 307:588–596, 1982.

Jekel, J. F. Epidemiology, Biostatistics, and Preventive Medicine. Philadelphia, W. B. Saunders Company, 1996.

SECTION III

Preventive Medicine and Public Health

CHAPTER FOURTEEN

Introduction to Preventive Medicine

SYNOPSIS

OBJECTIVES
- To characterize the basic concepts that define preventive medicine.
- To discuss the difficulties in defining the term *health*.
- To discuss the various measures of health status.
- To consider the relationship between the natural history of disease and the activities of preventive medicine.
- To define the levels of prevention of disease.
- To consider the importance of financial constraints in the implementation of preventive services, and to define cost-effectiveness analysis and cost-benefit analysis.

I. Basic Concepts
  A. Definitions
  1. **Preventive medicine** seeks to enhance the lives of individuals by helping them protect and improve their own health.
  2. **Public health** attempts to promote health in populations through the application of organized community efforts.
  3. **Health** is more difficult to define than is disease.
     a. One well-known definition is "Health is a state of complete physical, mental, and social well-being and not merely the absence of disease or infirmity." (From the preamble to the constitution of the World Health Organization.)
     b. This definition has the strengths of recognizing that any meaningful concept of health must include all the dimensions of human life and that such a definition must be positive.
     c. The weaknesses of this definition are that it is too idealistic in its expectations for complete well-being, and it is too static in viewing health as a state rather than as a dynamic process that requires constant effort and activity to maintain.
  B. Successful adaptation
  1. "The states of health or disease are the expressions of the success or failure experienced by the organism in its efforts to respond adaptively to environmental challenges" (Dubos 1965:xvii).
  2. The concept of **stress** addresses the fact that different stressors can induce helpful forms of stress **(eustress)** and harmful forms of stress **(dystress).**
  3. Good health requires the presence of eustress and also the limitation of dystress to a level to which the organism can adapt.
  4. The ongoing level of demand for adaptation in an individual is called the **allostatic load** on that person, and it may be an important contributor to many chronic diseases (McEwen and Stellar 1993).
  C. Satisfactory functioning
  1. Health and disease may be defined by the functional status of an individual rather than by the individual's sense of well-being.
  2. Health derives principally from forces other than medical care (e.g., appropriate nutrition, adequate shelter, a nonthreatening environment, and a prudent life-style).
  3. Medicine contributes to health both directly through patient care and indirectly through the development and dissemination of knowledge.

II. Measures of Health Status
  A. Mortality data have historically been the basis for measures of health status. Researchers assumed that a low age-adjusted death rate and a high life expectancy reflected good health in a population.
  B. As a higher proportion of the population lives to old age than ever before, and this group accumulates various chronic and disabling illnesses, increasing emphasis is being placed on the **health-related quality of life.**
  C. Health status indexes are currently the focus of considerable research effort. Many of these seek to adjust life expectancy on the basis of morbidity or the perceived quality of life, or both.
  1. **Life expectancy** is traditionally defined as the average number of years of life remaining at a given age.
  2. **Quality-adjusted life years** (QALY) incorporates both life expectancy and the perceived impact of illness and disability on the quality of life.
     a. One estimates how many years of life with a disability (limited years) would be equal to 1 year of life with good health (healthy years).

b. For example, a person who has had a stroke and now suffers from hemiparesis might estimate that 2 limited years with this disability is equivalent to 1 healthy year. In this case, a year of life following a stroke might be given a weight of 0.5. If 3 limited years were equivalent to 1 healthy year, each limited year would contribute 0.33 years to the QALY. Severe disability might permit survival but produce a QALY score of 0.0 if the individual felt that his or her quality of life was as bad or worse than no life at all.
3. **Healthy life expectancy** is a measure that attempts to combine both mortality and morbidity into one index. The index reflects the number of years of life remaining that are expected to be free of serious disease. The onset of a serious disease with permanent sequelae reduces the index as much as if the person who has the sequelae had died from the disease.
4. Numerous other measures of the length and quality of life have been devised.

III. **The Natural History of Disease**
A. Disease and illness should be viewed as dynamic, rather than static, concepts. Variations are seen in the natural histories of different diseases, depending on whether the "disease-producing stimuli arise in the environment or within man" (Leavell and Clark 1965).
B. If the disease is an **infectious disease**, there are reasons why the causative organism was present in the patient's environment, why the patient was exposed to the organism, and why the patient was or was not resistant to the organism.
C. If the disease is a **noninfectious disease**, such as atherosclerosis leading to myocardial infarction, there may be multiple **risk factors** present in the patient's genotype, nutrition, life-style, and environment. The risk factors may interact with one another to modify the risk that would be expected by simple addition of the contributions of individual risk factors (see Chapter 4).

IV. **Levels of Prevention**
A. According to **Leavell's levels**, all of the activities of physicians and other health professionals have the goal of prevention (Leavell and Clark 1965). There are three levels of prevention (see below and Table 14–1). What is to be prevented depends on the stage of health or disease in the individual receiving preventive care.
1. **Primary prevention** keeps the disease process from becoming established by eliminating causes of disease or increasing resistance to disease (see Chapters 15 and 16).
2. **Secondary prevention** interrupts the disease process before it becomes symptomatic (see Chapter 17).
3. **Tertiary prevention** limits the physical and social consequences of symptomatic disease (see Chapter 18).

B. **Primary prevention and predisease**
1. Most noninfectious diseases can be seen as having an early stage, during which the causal factors will start to produce physiologic abnormalities. The goal at this time is to modify risk factors in a favorable direction, in order to keep the pathologic process from occurring.
2. **Health promotion** contributes to the prevention of a variety of diseases and the maintenance of well-being by nonmedical changes, such as changes in life-style, nutrition, and the environment.
   a. Health-promoting activities may require structural changes in society to enable the majority of people to take part in them (i.e., changes that make healthy choices easier, such as stores providing low-fat, low-salt, low-sugar, yet appealing and nutrient-rich foods).
   b. Health promotion applies both to noninfectious diseases and to infectious diseases. Infectious diseases are reduced in frequency and seriousness where the water is pure, liquid and solid wastes are disposed of in a sanitary manner, and arthropod and animal vectors of disease are controlled. Crowding promotes the spread of infectious diseases, whereas adequate housing and working environments tend to minimize the spread of disease.
3. **Specific protection** may be necessary if health-promoting changes in environment, nutrition, and behavior are not fully effective. This form of primary prevention is targeted at a specific disease or type of injury. Examples include immunization against poliomyelitis and the use of seat belts in cars.
4. Some measures provide specific protection while also contributing to the more general goal of health promotion (e.g., fluoridation of water supplies, which not only prevents dental caries but is a nutritional intervention that promotes stronger bones and teeth).

C. **Secondary prevention and latent disease**
1. Latent (hidden) disease is present when a pathologic process (e.g., atherosclerosis) is detectable by means of diagnostic testing but has not yet begun to produce symptoms.
2. Presymptomatic diagnosis and treatment through screening programs is referred to as secondary prevention because it is the secondary line of defense against disease. Although it does not prevent the cause from initiating the disease process, it may prevent permanent sequelae.

D. **Tertiary prevention and symptomatic disease**
1. **Disability limitation** consists of medical and surgical measures aimed at correcting the anatomic and physiologic components of disease in symptomatic patients. The majority of care provided by physicians meets this description. It can be considered prevention because its goal is to halt the disease process and thereby prevent or limit the impairment (disability) caused by it (e.g., surgical removal of a tumor may prevent further local invasion).

TABLE 14–1. Modified Version of Leavell's Levels of Prevention*

Stage of Disease	Level of Prevention	Type of Response
Predisease	Primary prevention	Health promotion and specific protection
Latent disease	Secondary prevention	Presymptomatic diagnosis and treatment
Symptomatic disease	Tertiary prevention	Disability limitation for early symptomatic disease
		Rehabilitation for late symptomatic disease

* Modified from Leavell, H. R., and E. G. Clark. Preventive Medicine for the Doctor in His Community, 3rd ed. New York, McGraw-Hill Book Company, 1965. Although Leavell originally categorized disability limitation under secondary prevention, both in Europe and in the USA it has become customary to classify it as tertiary prevention because it involves the management of symptomatic disease.

2. **Rehabilitation** is an attempt to mitigate the effects of disease and thereby prevent it from resulting in total social and functional disability. For example, a person who has suffered a stroke may be taught how to care for himself or herself in the **activities of daily living** (feeding, bathing, and so forth). This may enable him or her to avoid the adverse sequelae associated with prolonged inactivity (e.g., increasing muscle weakness and total disability).

V. The Economics of Prevention

A. Health promotion and disease prevention efforts generally must meet financial as well as medical criteria for utility. Financial feasibility is often determined by means of **cost-benefit analysis** or by means of **cost-effectiveness analysis** (see Box 14–1).

B. **Demonstration of benefits**
   1. Scientific proof of benefits may be difficult because it is often impractical or unethical to undertake randomized trials using people as subjects.
   2. For example, it is not possible to randomly assign people to smoking and nonsmoking groups. Therefore, investigators are limited to observational studies of the effects of smoking, which are usually not as convincing as experiments.
   3. Another problem is that life is filled with risks for one disease or another, and many of these operate together to produce the levels of health observed in a population.

C. **Delay of benefits**
   1. There is often a long delay between the time that the preventive measures are instituted and the time that positive health changes become visible.
   2. For example, because the latent period (incubation period) for lung cancer caused by cigarette smoking is 20 years or more, investments made now in smoking reduction programs may not produce clearly identifiable benefits until 20 or more years have passed.
   3. The delay in positive benefits may obscure or even mask the association between the preventive measure and the resultant outcome.

D. **Accrual of benefits**
   1. Even if a given program could be shown to produce meaningful economic benefit, it is critical to know where the benefits would accrue.
   2. Financial benefits may or may not accrue to the institution (e.g., state government, health maintenance organization) responsible for the cost of the program.
   3. Even a financially beneficial preventive program may not be implemented if the subsidizing institution will not receive the financial benefits, or if such benefits will be long delayed.
   4. Delay in the detection of benefits often limits the political popularity of preventive programs, as the

---

BOX 14–1. Cost-Benefit and Cost-Effectiveness Analysis

Cost-benefit analysis measures the costs and the benefits of a proposed course of action in terms of the same units, usually monetary units such as dollars. For example, a cost-benefit analysis of a poliomyelitis immunization program would determine the number of dollars to be spent toward vaccines, equipment, personnel, and so forth, to immunize a particular population. It then would determine the number of dollars that would be saved by not having to pay for the hospitalizations, medical visits, and lost productivity that would occur if poliomyelitis were not prevented in that population.

It is difficult to incorporate ideas such as the dollar value of life, suffering, and the quality of life into such an analysis. However, cost-benefit analysis is useful if a particular budgetary entity (such as a government or business) is trying to determine whether the investment of resources in health would save money in the long run. It is also useful if a particular entity with a fixed budget is trying to make informed judgments concerning allocations in various sectors (health, transportation, education, and so forth) and to determine the sector in which an investment would produce the greatest economic benefit.

Cost-effectiveness analysis provides a way of comparing different proposed solutions in terms of the most appropriate measurement units. For example, by measuring hepatitis B cases prevented, deaths prevented, and life-years saved per 10,000 population, Bloom et al. (1993) were able to compare the effectiveness of four different strategies of dealing with the hepatitis B virus: (1) no vaccination; (2) universal vaccination; (3) screening followed by vaccination of unprotected individuals; and (4) a combination of the screening of pregnant women at delivery, the vaccination of the newborns of women found to be antibody-positive during screening, and the routine vaccination of all 10-year-old children.

After Bloom et al. estimated the numbers of persons involved in each step of each method, they determined the costs of screening, of purchasing and administering the vaccine, and of medical care for various forms and complications of hepatitis. They calculated that the fourth strategy would have an undiscounted cost of about $367 (or a discounted cost of $1205) per case of hepatitis B prevented, and they concluded that this was clearly the strategy with the lowest cost.

The concept of **discounting**, which is important in business and finance, must also be used in medical cost-benefit and cost-effectiveness analysis when the costs are incurred in the present but the benefits will occur some time in the future. Discounting is a reduction in the present value of delayed benefits (or an increase in their present costs) to account for the time value of money. If the administrators of a prevention program spend $1000 now to save $1000 of expenses in the future, they will take a net loss. This is because they will lose the use of $1000 in the interim and also because of the effects of inflation (the $1000 eventually saved will not be worth as much as the $1000 initially spent). Discounting attempts to adjust for these forces.

To discount a cost-benefit or cost-effectiveness analysis, the easiest way is to increase the present costs by a yearly factor, which can be thought of as the interest that would have to be paid in order to borrow the prevention money until the benefits occurred. For example, if it costs $1000 today to prevent a disease that would have occurred 20 years in the future, the present cost can be multiplied by $(1 + r)^n$, where r is the yearly interest rate for borrowing and n is the number of years until the benefit is realized. If the average yearly interest rate is 5% over 20 years, the formula becomes: $(1 + 0.05)^{20} = (1.05)^{20} = 2.653$. When this is multiplied by the present cost of $1000, the result is $2653. Thus, the expected savings 20 years in the future from a $1000 investment today would have to be greater than $2653 in order for the initial investment to be a financial gain.

advantages of such programs may not be demonstrable within the period between elections.

E. Discounting
1. Discounting (see Box 14–1) is another reason that prevention may show a lower benefit-cost ratio than expected.
2. If a preventive effort is made now, the costs are present costs, but the savings may not be evident until many years later.
3. Even if the savings are expected to accrue to the same budgetary unit that provided funding for the preventive program, the delay in economic return partially devalues the benefits to that unit.
4. The *present* value of the benefits must be discounted as a result of the delay, making it more difficult to demonstrate cost-effectiveness or a positive benefit-cost ratio.

F. Priorities
1. Current, obvious problems usually attract far more attention and concern than future, subtle problems.
2. Even though prevention may be more cost-effective than intervention after a problem has developed, it is more difficult to justify money for medical crises that have not yet appeared.
3. Many individuals may be unwilling to modify risks related to diet, activity levels, or smoking, because the risk of future problems does not speak to them urgently in the present.

QUESTIONS

DIRECTIONS. (Items 1–8): Each of the numbered items or incomplete statements in this section is followed by answers or by completions of the statement. Select the ONE lettered answer or completion that is BEST in each case. Correct answers and explanations are given at the end of the chapter.

1. Measures of health status have traditionally been based on mortality data. The principal reason this is no longer satisfactory is that

   (A) there is less risk of fatal infection than in the past
   (B) the population is older and more subject to chronic illness than in the past
   (C) the infant mortality rate has declined so much that it no longer serves as a useful index
   (D) traditional sources of mortality data have failed to include women
   (E) changes in diagnostic technology permit earlier detection of disease

2. Following the onset of blindness resulting from diabetic retinopathy, a 54-year-old patient seems depressed. You question the patient regarding quality of life, and the patient dejectedly tells you that 10 years "like this" isn't worth 1 year of good health. The patient's statement indicates that

   (A) each year of life contributes less than 0.1 quality-adjusted life years
   (B) each year of life contributes 10 quality-adjusted life years
   (C) the life expectancy is less than 10 years
   (D) the healthy life expectancy is 2 years
   (E) blindness is associated with depression

3. In which one of the following ways is health promotion distinguished from disease prevention?

   (A) Only health promotion is targeted at specific diseases.
   (B) Only health promotion can begin before disease becomes symptomatic.
   (C) Only health promotion is applied once disease has developed.
   (D) Only health promotion involves materials and methods that are generally nonmedical.
   (E) Only health promotion is targeted at populations rather than individuals.

4. Which one of the following is an example of secondary prevention?

   (A) cholesterol reduction in a patient with coronary artery disease
   (B) thrombolysis for acute myocardial infarction
   (C) physical therapy following lumbar disc herniation
   (D) pneumococcal vaccine in a patient who has undergone splenectomy
   (E) hormone replacement therapy for menopause

Items 5–8

You are interested in helping a 45-year-old, perimenopausal female patient in your practice to avoid osteoporosis. The patient is motivated but has a fixed income and is concerned about expense. Assume that the cost of hormone replacement therapy is fixed at $660 per year, that this therapy will prevent the development of a hip fracture in this patient at age 68 years, that the current cost of surgical fixation of the hip fracture is $12,000, and that the yearly rate of inflation is 4%.

5. If this patient had no insurance to offset prescription expenses, how much money would she spend to prevent a hip fracture at age 68 years if she started hormone replacement therapy now?

   (A) $12,000
   (B) $18,000
   (C) $15,180
   (D) $ 4,260
   (E) It cannot be determined from the information provided.

6. Taking inflation into account, what is the cost of surgery to repair a hip fracture in this patient at age 68 years?

   (A) $29,580
   (B) $17,580
   (C) $11,620
   (D) $42,814
   (E) It cannot be determined from the information provided.

7. Based on your answers to questions 5 and 6, above, you decide that hormone replacement therapy

    (A) is cost-effective
    (B) has a favorable cost-benefit ratio
    (C) is not cost-effective
    (D) has an unfavorable cost-benefit ratio
    (E) is appropriate for your patient

8. If the cost of preventing a hip fracture is greater than the cost of surgical repair, then which of the following is true?

    (A) No attempt should be made to prevent hip fracture until a more cost-effective strategy is devised.
    (B) The least costly preventive strategy should be chosen.
    (C) The most effective preventive strategy should be chosen.
    (D) A preventive strategy may still be indicated.
    (E) If a hip fracture occurs, it should be managed nonsurgically.

DIRECTIONS. (Items 9–12): For each numbered item, select the ONE lettered option that is most closely associated with it. Begin by reading the list of options and then, for each item, try to generate the correct answer and locate it in the option list, rather than evaluating each option individually. Each lettered option may be selected once, more than once, or not at all.

(A) performing carotid endarterectomy in a patient with transient ischemic attacks
(B) recommending regular physical activity to a patient with no known medical problems
(C) vaccinating a health care worker against hepatitis B
(D) giving isoniazid for 1 year to a 28-year-old medical student with a newly positive purified protein derivative (PPD) tuberculin skin test

Match the procedure to the corresponding category (level) of prevention.

9. primary prevention
10. secondary prevention
11. tertiary prevention
12. health promotion

ANSWERS AND EXPLANATIONS

1. The answer is B: the population is older and more subject to chronic illness than in the past. When many deaths occur among young people in a population, prolongation of life expectancy is a strong indicator that public health is improving. Until the twentieth century, relatively few individuals in most societies lived long enough to die of processes principally related to senescence, or aging. With the advances in medical technology and hygiene of the past decades, life expectancy has increased to the point where most adult deaths are related to chronic diseases that may potentially compromise the quality of life over a number of years before causing death. Thus far, the state of medical care is better suited to stave off death than to prevent disease. Consequently, the measurement of quality of life has assumed greater importance as the burden of chronic disease on an aging population has increased. *Epidemiology*, p. 184.

2. The answer is A: each year of life contributes less than 0.1 quality-adjusted life years. Quality-adjusted life years (QALY) represents one way to measure the quality as well as the length of life. Each year of life with a disability represents some portion of the quality of that year of life without disability. In this case, if 1 year of good health is worth more to the patient than 10 years with blindness, then the disability results in less than 10% of the quality of life with intact health, and each year of life with disability contributes less than 0.1 quality-adjusted life year. While this scale is useful in comparing the relative impact of various disabilities on quality of life, the absolute measure may not be meaningful. In some instances, the more years spent with a disability, the more miserable a person becomes. In this case, 10 years of life with the disability may not truly add up to the value, or quality, of 1 year of good health. *Epidemiology*, p. 185.

3. The answer is D: Only health promotion involves materials and methods that are generally nonmedical. Both health promotion and disease prevention share the goal of keeping people well. Disease prevention is generally directed specifically at a disease or a related group of diseases; the tools of disease prevention, such as vaccines, are generally medical. Health promotion is not disease oriented but rather is an effort to enhance overall health. The materials and methods of health promotion, such as regular physical activity; a diet rich in vegetables, grains, and fruits; safe sexual practices; and the provision of adequate housing or transportation are generally related to life-style and are therefore nonmedical entities. While health promotion and disease prevention are closely linked in efforts to enhance the public health, they are somewhat disparate in both philosophy and application. *Epidemiology*, p. 186.

4. The answer is A: cholesterol reduction in a patient with coronary artery disease. As defined in the text and discussed in greater detail in Chapter 17, secondary prevention interrupts the disease process before it becomes symptomatic. The implication of this definition is that there must be a disease process in order for secondary prevention to take place. The reduction of an elevated cholesterol level in a patient without coronary artery disease is an example of primary prevention; once the disease process has begun, however, the modification of causal factors to prevent the development of symptoms is secondary prevention. In this case, angina pectoris or myocardial infarction, or both, is the symptomatic state at which preventive efforts are directed. *Epidemiology*, p. 186.

5. The answer is C: $15,180. In reality, the cost of prescription medication is likely to change over time to keep pace with inflation. However, the scenario provided stipulated that the yearly cost of hormone replacement therapy was fixed at $660. Therefore, to prevent a hip fracture

at age 68 years, your 45-year-old patient would need to spend $660 each year for 23 years, or a total of $15,180. Many insurance plans include at least partial coverage of prescription medication, so such an expense is generally not fully borne by an insured patient. *Epidemiology,* p. 187.

6. The answer is A: $29,580. The means for answering this question are provided in Box 14–1. The rate of inflation is analogous to the annual interest rate on money spent, or "borrowed," now to prevent a future event. The current cost of surgical fixation of the hip is provided as $12,000. However, the patient will require this surgery 23 years in the future unless a fracture is avoided. To obtain the cost of the operation for the patient at age 68 years, the current cost is multiplied by (1 + inflation) raised to the number of years, which is $(\$12,000)(1 + 0.04)^{23} = (\$12,000)(2.465) = \$29,580$. *Epidemiology,* p. 187.

7. The answer is B: has a favorable cost-benefit ratio. Determination and then comparison of the cost and financial benefit of an intervention constitutes cost-benefit analysis. In this scenario, the financial benefit of prevention is greater than the cost, so the cost-benefit ratio is favorable. However, the cost-effectiveness of hormone replacement therapy has not been assessed. Cost-effectiveness analysis requires that alternative means of achieving the same goal be compared on the basis of cost; the least costly is the most cost-effective. Hormone replacement therapy would not be cost-effective if an alternative strategy, such as calcium and vitamin D supplementation, were as effective at preventing osteoporotic hip fracture at a lower cost. Finally, one cannot conclude that hormone replacement therapy is indicated for the patient in question on the basis of the information provided. Such a determination is contingent not merely on the patient's risk of osteoporosis and hip fracture but on estimates of risk of cardiovascular disease and breast cancer, as well as the patient's preferences after discussion of alternative preventive strategies. *Epidemiology,* p. 187.

8. The answer is D: A preventive strategy may still be indicated. The scenario provided for questions 5 through 8 is very simplistic. The cost of preventing a hip fracture is compared with the cost of repairing one. However, a great deal of pertinent information is not addressed by this cost-benefit analysis. One of the principal deficiencies of cost-benefit analysis is the expression of human morbidity or mortality in financial terms. The pain and disability associated with a hip fracture are not discussed. The effort required to recover from a fracture is not considered. Even the financial losses that might result from transient or permanent disability, as well as the financial burden associated with rehabilitation, are ignored. A realistic cost-benefit analysis must be far more comprehensive. *Epidemiology,* p. 187.

9. The answer is C: vaccinating a health care worker against hepatitis B. Vaccination is primary prevention because it prevents the initial establishment of the disease process in the host. As opposed to health promotion, which is generally nonmedical, primary prevention is often disease specific, as in this case. *Epidemiology,* p. 186.

10. The answer is D: giving isoniazid for 1 year to a 28-year-old medical student with a newly positive purified protein derivative (PPD) tuberculin skin test. The goal of secondary prevention is the avoidance of symptoms once a disease process has begun. Conversion of the skin test from negative to positive implies that exposure to, and latent infection with, tuberculosis has occurred. Postexposure treatment with isoniazid for a period of up to 1 year is recommended to reduce the risk of developing active (symptomatic) tuberculosis in the future (Barnes and Barrows 1993). *Epidemiology,* p. 186.

11. The answer is A: performing carotid endarterectomy in a patient with transient ischemic attacks. Carotid endarterectomy, while generally *not* recommended in asymptomatic patients with carotid bruits (secondary prevention), is recommended to reduce the risk of cerebrovascular accident in patients with carotid disease and symptoms, such as transient ischemic attacks (American College of Physicians 1989). The limitation of physical consequences of symptomatic disease is tertiary prevention. *Epidemiology,* p. 186.

12. The answer is B: recommending regular physical activity to a patient with no known medical problems. Health promotion is an attempt to preserve good health rather than an effort to avoid or prevent a particular disease. Regular physical activity is associated with better health and greater life expectancy (Lee, Hsieh, and Paffenberger 1995), as well as the prevention of specific diseases (Abbott et al. 1994; Lakka et al. 1994). Recommending regular physical activity is a health-promoting strategy. *Epidemiology,* p. 186.

REFERENCES CITED

Abbott, R. D., et al. Physical activity in older middle-aged men and reduced risk of stroke: The Honolulu Heart Program. American Journal of Epidemiology 139:881–893, 1994.

American College of Physicians. Indications for carotid endarterectomy. Annals of Internal Medicine 111:675–677, 1989.

Barnes, P. F., and S. A. Barrows. Tuberculosis in the 1990s. Annals of Internal Medicine 119:400–410, 1993.

Dubos, Rene. Man Adapting. New Haven, Conn., Yale University Press, 1965.

Jekel, J. F. Epidemiology, Biostatistics, and Preventive Medicine. Philadelphia, W. B. Saunders Company, 1996.

Lakka, T. A., et al. Relation of leisure-time physical activity and cardiorespiratory fitness to the risk of acute myocardial infarction in men. New England Journal of Medicine 330:1549–1554, 1994.

Leavell, H. R., and E. G. Clark. Preventive Medicine for the Doctor in His Community, 3rd ed. New York, McGraw-Hill Book Company, 1965.

Lee, I-M, C. Hsieh, and R. S. Paffenberger. Exercise intensity and longevity in men. Journal of the American Medical Association 273:1179–1184, 1995.

McEwen, B. S., and E. Stellar. Stress and the individual. Archives of Internal Medicine 153:2093–2101, 1993.

Revicki, D. A., et al. Responsiveness and calibration of the General Well Being Adjustment Scale in patients with hypertension. Journal of Clinical Epidemiology 47:1333–1342, 1994.

CHAPTER FIFTEEN

METHODS OF PRIMARY PREVENTION: HEALTH PROMOTION

SYNOPSIS

OBJECTIVES
- To characterize the basic influences of society on health.
- To describe the societal resources necessary for the maintenance and promotion of health.
- To discuss the role of nutritional factors in the promotion of health.
- To note the impact on health of both nutritional deficiencies and nutritional excesses.
- To describe the range of environmental factors that affect health.
- To discuss the routes of entry of environmental hazards into the body.
- To describe the assessment of environmental hazards and the methods employed to reduce associated risks.
- To discuss the management of air pollution, water pollution, and solid waste.
- To note the worldwide importance of food and vector-borne toxins.
- To discuss the role of life-style and behavioral counseling in health promotion.

I. Society's Contribution to Health
   A. The most **fundamental sources of health** are adequate nutrition, a safe environment, and prudent behavior.
   B. Society provides the **basic resources necessary for health** in the form of socioeconomic conditions that include employment opportunity; environmental support systems (e.g., water supply) and sanitation; and the regulation of commerce and public safety.
   C. **Socioeconomic groupings** are somewhat controversial but have been shown to affect health, with better overall health consistently observed in higher socioeconomic strata.

II. Nutritional Factors in Health Promotion
   A. **Undernutrition** is a serious problem in some areas of the world, particularly in developing countries. It may lead to starvation, marasmus, or kwashiorkor.

1. Most **starvation** in recent years has resulted from war or civil unrest.
2. **Marasmus** is a severe wasting syndrome in infants resulting from malnutrition. Marasmus causes almost total growth retardation and occurs when all nutrients in the diet are deficient.
3. **Kwashiorkor** (visceral protein malnutrition) tends to occur in children at the time of weaning, when starchy foods replace breast milk. Kwashiorkor results from protein deficiency despite nearly adequate calorie intake. The development of **ascites** (fluid in the abdominal cavity) produces a distended abdomen, which suggests obesity but is actually due to severe undernutrition.
4. Attempts are being made to alleviate the problem of undernutrition. For example, by mixing starchy crops that are deficient in different **amino acids,** it is possible to obtain a diet that is adequate in all of the essential amino acids. Successful programs based on this principle have been developed, making it possible to reduce the incidence of weaning malnutrition and child mortality resulting from amino acid deficits (King 1969). **Vitamin** supplements alone may markedly reduce mortality in areas where nutrition is poor and infectious disease is common (Rahmathullah et al. 1990).
5. In developing countries, both malnutrition and infection are common in infants and young children, particularly around the time of weaning. The **synergism of malnutrition and infection** (i.e., each condition exacerbates the other) is responsible for a large proportion of the deaths among young children in the developing world and is therefore the target of many programs designed to promote health.
   a. Because the prevention of infection will improve the ratio of caloric intake to caloric need, programs that provide measles vaccinations and programs that provide uncontaminated water supplies will have a positive effect on the nutritional

status of children. Similarly, programs that provide improved nutrition, especially for young children, will help children avoid death or permanent organ damage due to infections.
b. Research on synergism has shown that the approach to nutritional and infectious disease problems in regions of scarcity should be broad-based, attempting to improve all dimensions of life simultaneously.

B. **Overnutrition** is a major health problem in the USA and most of the other industrialized nations. Its effects have become evident in cross-cultural studies and studies of migrants, which demonstrate the differences in the health of populations consuming different diets, and changes in health associated with dietary change (Keyes et al. 1986). Consumption of excessive food, particularly excess fat, represents a threat to the public health in the USA (US Department of Health and Human Services 1988).
1. **High-fat diets** are associated with obesity, a variety of cancers, and coronary artery disease.
   a. Human preference for dietary fat, which is calorically dense (9 kcal/g), is complex, but is likely the result of human evolution in a world of perennial nutritional shortages. Human physiology is better adapted to subsistence than to the relatively recent state of consistent nutritional abundance. The ability of the body to store calories as fat whenever an excess of food is available, the human preference for calorie-rich foods, and the increased metabolic efficiency observed in the use of energy during periods of caloric deficiency all may be biologic mechanisms that served *Homo sapiens* well in the distant past but are maladaptive under conditions of nutritional abundance, particularly when compounded by a sedentary life-style.
   b. The guideline, and the US goal for the year 2000 (US Department of Health and Human Services 1990) for dietary fat intake is no more than 30% of total calories (US Department of Agriculture and US Department of Health and Human Services 1985). Many authorities consider this guideline too high (Butrum, Clifford, and Lanzer 1988). A more optimal level of intake was not established as the guideline because it was thought to be unrealistic in the short term.
   c. Average fat consumption in the USA, until recently reported to be 37% of total calories (US Department of Agriculture 1985 and 1986), is now thought to be approximately 34% of calories (MMWR 1994).
   d. Excessive intake of saturated fat and partially hydrogenated fats with high levels of trans–fatty acids is in particular associated with a variety of chronic diseases.
2. **Low-fiber diets** are common in industrialized nations.
   a. Soluble fiber tends to slow absorption of lipids and glucose, lowering their serum levels. Insoluble fiber increases stool bulk and is beneficial to the health of the gastrointestinal tract.
   b. Many diseases common in industrialized nations, such as appendicitis, are rare in parts of the world where fiber intake is much higher (Burkitt and Temple 1994).
3. **Obesity,** defined as 120% or greater of ideal body weight for height, is extremely prevalent. Data from the US National Health Interview Survey in 1990 indicated that 24% of males and 27% of females were obese by this definition (US Department of Health and Human Services 1990).
   a. Obesity is associated with chronic diseases, including coronary artery disease and at least several cancers, as well as diabetes mellitus and hypertension. Even persons with no discernible pathologic manifestations may suffer from psychologic stress resulting from dissatisfaction with body image and loss of self-esteem.
   b. Weight reduction requires sustainable decreases in caloric intake plus sustainable increases in activity levels. Lifelong management is necessary. To date, most methods that have attempted to achieve sustainable weight loss have demonstrated very limited success.

C. **Nutrition counseling** is usually provided only minimally, or not at all, by physicians, according to available data. This likely stems from lack of training, because only 25% of US medical schools provide formal nutrition education (Winick 1993; Zimmerman and Kretchmer 1993).
1. A diet rich in grain products, fresh vegetables, and fresh fruits is advisable for almost everyone.
2. Patients requiring detailed dietary counseling are best referred to a nutrition expert, such as a registered dietitian.

III. **Environmental Factors in Health Promotion**
A. **Basic concepts**
1. The environment includes microbiologic and chemical agents, as well as physical, social, and psychologic influences that can be beneficial (producing "eustress") or detrimental (producing "dystress").
2. In general, the public tends to show greater concern for environmental hazards which they cannot control (e.g., radon, dioxin) than those which they can control (e.g., ambient noise).
3. Specialists in environmental health are responsible for helping patients interpret the dangers of environmental risks about which they are concerned, exploring the possibility of environmental and occupational causes of acute and chronic disease in patients by performing histories and physical examinations that are environmentally sensitive, and reporting diseases that might have an environmental source to public health agencies (see Chapter 19).

B. **Routes and effects of exposure to environmental hazards**
1. The skin, the gastrointestinal tract, and the lungs represent the usual portals of entry to the body for environmental hazards. Knowledge of the probable **route of entry** of an environmental hazard is essential in devising adequate protection.
2. The **skin** is susceptible to injury by microbes, chemicals, heat and cold, and physical trauma.
3. The **gastrointestinal tract** is susceptible to microbial and chemical ingestion, as well as radiation injury.
4. The **lung** is susceptible to airborne microbes, chemical aerosols, fumes, dusts, and environmental allergens.
5. Exposure to environmental hazards is classified as acute or chronic.
   a. The negative effects of **acute exposure** may result from short-term, high-level exposure to an infec-

tious organism; to certain toxic substances, such as potassium cyanide or carbon monoxide; or to a source of high energy, such as heat, noise, or a heavy blow.

b. The negative effects of **chronic exposure** may result from cumulative contact with or irritation of human tissues. Examples are long-term exposure to asbestos, lead, mercury, and certain types of dust; repetitive motion injuries; or repeated exposure to loud noise of the type faced by airport ground crews or members of a rock band.

6. Chronic exposures often demonstrate a **dose-response relationship,** such as that between the quantity of cigarette smoking and the risk of lung cancer. In research involving a potential carcinogen, the dose-response relationship is measured beyond the usual **latent period** for the carcinogen (i.e., the period between the onset of exposure and the development of cancer).

7. Adverse exposures may or may not have a **threshold level** below which the body can adapt successfully and no harm results. There is such a threshold level for most chemical and physical agents and even for most microbes. Usually, **nonthreshold exposures** are limited to those which alter genetic material, producing cancer or genetic mutations (e.g., ionizing radiation).

C. **Assessment of environmental risks** may be undertaken with the five steps suggested in Table 15-1.

1. The nature of disease or damage (step 3) may be difficult to determine when multiple hazards are present, particularly because experimentation with human subjects is unethical and animal models of exposure to a toxin or other hazard must be used to estimate the type of damage.

2. The **Ames test** is an example of a quick and frequently used test to estimate the mutagenic potential of a chemical substance. *Salmonella typhimurium* bacteria are exposed to the suspected toxin in the presence of mammalian enzymes and then observed for the mutation of a specific gene. A high proportion of the chemicals that cause mutation in this test is also found to be carcinogenic in other, more complicated assays. A negative result is not a guarantee that the tested chemical is not carcinogenic.

3. Assessment of risks (step 5) is particularly difficult for many types of environmental hazards, because much of what is known about them is based on data concerning fairly high levels of exposure and the resulting risk of acute toxic effects.

D. **Methods of environmental modification** for the control of environmental hazards begin with the identification and characterization of the hazards.

1. Once hazards have been adequately characterized, they can be eliminated in certain instances (e.g., eradication of the smallpox virus).

2. If a hazard cannot be eliminated, effort is directed at preventing contact between people and the hazard (e.g., wearing protective clothing or insect repellent in areas of ticks or mosquitoes).

E. **Major sources of environmental hazard** include air pollution, water pollution, solid wastes, contaminated food, and disease vectors.

1. **Air pollution** is caused by a variety of substances, including particulate matter, metal fumes, gases, and dusts. Noise and airborne microorganisms may also be considered types of air pollutants.

a. The lung is the most frequent site of serious contact with environmental hazards, both because the lung tissue is highly absorptive and vulnerable and because the volume of air exchanged by the lungs each day is so great (i.e., approximately 50 pounds per day).

b. **Particulate matter,** whether from cigarette smoking or from fuel combustion, often contains carcinogenic substances.

c. **Metal fumes** are gaseous metal oxides that come primarily from activities in occupational settings, such as welding without adequate ventilation. An acute syndrome called **metal fume fever** may occur a few hours following exposure and is characteristically a flu-like illness. Metals contributing to air pollution include zinc, copper, arsenic, beryllium, cadmium, lead, and mercury.

d. **Gases** such as ozone, nitrogen dioxide, sulfur dioxide, hydrocarbons, and carbon monoxide are produced in abundance and may cause respiratory tract disease.

e. **Dusts** such as coal dust and silica dust are an important cause of chronic lung diseases known as **pneumoconioses.** Occupational asthma may result from exposure to a variety of dusts, including those from grain, wood, cotton, hemp, and flax, as well as certain chemicals, especially isocyanates.

f. **Pollutants produced or released outdoors** tend to be most concentrated outdoors (e.g., pollen, ozone, sulfur and nitrogen oxides, gasoline, and solvents). Air mixture and winds serve to dilute and disperse outdoor pollutants. An **air inversion** occurs when the cooler air is found close to the earth and the warmer air is above, so that natural mixing of air does not occur and pollution is concentrated. Outdoor air pollution can only be controlled by reducing the production and release of pollutants at their source.

g. **Pollutants produced or released indoors** almost always show higher concentrations indoors (e.g., microorganisms, allergens, allergenic molds, aerosols, cooking products, radon, tobacco smoke, and chemicals or fumes released by building or consumer products such as urethane). **Sick building syndrome** refers to a constellation of symptoms such as headache, watery eyes, and wheezing ascribed by people to a building in which they work. The syndrome often, but not

TABLE 15-1. Suggested Steps in the Assessment of an Environmental Risk

(1) Identify the environmental hazard
(2) Characterize the exposure to the hazard in terms of the following:
   (a) Time of occurrence
   (b) Number of exposures
   (c) Duration of each exposure
   (d) Intensity (level) of each exposure
(3) Determine the nature of acute and chronic effects of exposure to the hazard
(4) Determine the relationship between the exposure level and the risk of each effect identified above
(5) Determine the risks for an exposed individual and the probable population impact (population attributable risk and population attributable risk percent)

always, occurs in buildings that are tightly sealed for energy conservation or are not well ventilated. The cause seems to vary from one building to the next but is, in essence, unknown. Tobacco smoke is the most prevalent and most serious indoor air pollutant. A variety of illnesses, including lung cancer, have been ascribed to passive exposure to environmental tobacco smoke.
2. **Water pollution** may affect surface water or ground water.
   a. Water is potable if it is free of harmful substances and is safe for human consumption. The provision of potable water has been one of the major methods of lowering the rates of death and disease throughout the world.
   b. Pollution of oceans, rivers, and lakes may cause human disease when toxins move through the food chain and enter into fish, which are subsequently caught and consumed by people.
   c. **Surface water** includes protected surface reservoirs, lakes, and rivers. **Protected surface reservoirs** are usually the safest water sources. Other sources of surface water are vulnerable to pollution. Water from **rivers** or **lakes** may require considerable treatment to meet potability standards. Rivers are often used for the dumping of sewage and industrial toxins and are further contaminated by agricultural runoff.
   d. **Ground water** is found in underground spaces called **aquifers**. Aquifers may have a low rate of water turnover and thus tend to remain contaminated for years to decades once they become polluted. When adequately protected from surface pollutants, aquifers represent an important source of potable water.
   e. **Water treatment** involves removing some substances and adding others. **Filtration** is the fundamental method for removing unwanted substances from water. Even water that seems pure, such as that from a mountain stream in a pristine area, can be contaminated with the cysts of *Giardia lamblia*, a protozoan spread by animals. **Slow sand filtration** requires the use of a large bed of packed sand, on which an organic layer forms and assists in the filtration process. The organic layer is called a ***Schmutzdecke***, the German term for "dirt layer." Water passes through the bed and is purified. In **rapid sand filtration**, a flocculent (usually aluminum sulfate, called alum) is added to the water before filtration. The flocculent coagulates and traps suspended materials, preventing them from passing through the sand with the filtered water. The flocculent is removed periodically by back flushing, and new flocculent is added to the next batch of water. **Chemical filtration processes** can further purify water, but these methods are generally expensive. **Additions** are often made to drinking water to enhance potability. Chlorine is added to kill vegetative forms of microorganisms. Fluoride is sometimes added to prevent dental caries.
   f. **Sewage treatment** is also important in providing potable water. Human sewage poses a threat to humans and also to marine animals and freshwater fish. Aerobic bacteria in sewage consume oxygen in water, based on their **biologic oxygen demand**. Water deficient in oxygen cannot adequately support aquatic life. The goals of sewage treatment are to remove as much of the organic material as possible from the water (primary treatment) and then to reoxygenate the water (secondary treatment).
   g. In **primary sewage treatment,** water is held in a large tank long enough to allow solids to settle out. If the **sludge** left in the tank does not contain harmful metals, it can sometimes be used as fertilizer. Otherwise, the treated sludge is buried or dumped at sea.
   h. In **secondary sewage treatment,** the water is aerated, usually by means of an activated sludge process or a trickling filter process. Depending on the purity standards in effect, the water may then be filtered and chlorinated.
3. Each year, the average person in the USA generates about 1500 pounds of garbage and related **solid wastes** that must be disposed of by city governments and other governing bodies. The volume of solid waste produced by the industrial sector is even greater than that generated by the private sector.
   a. Most solid waste is disposed of in landfills, but finding locations for new landfills is becoming increasingly difficult. Existing landfills frequently leach toxic chemicals into the ground water.
   b. The Superfund Act was passed in 1980 to expedite the cleanup of toxic waste sites. More than a decade later, the problem is probably worsening, however.
4. **Contaminated food** is a major source of morbidity throughout the world. In developing countries, food contamination with bacteria, viruses, and parasites is a significant cause of mortality among infants who are not being breast-fed. Food can be contaminated during production, transportation, storage, processing, marketing, and preparation. Precautions must be taken at each stage to avoid food-borne illness.
5. **Environmental vectors**
   a. The development of tick-borne diseases in the USA, such as Lyme disease and ehrlichiosis, has served as a reminder of the importance of environmental vectors in disease transmission.
   b. Prevention of vector-borne disease can be achieved by eradicating the vector (e.g., draining swamps to eliminate breeding grounds for mosquitoes) and by reducing the risk of contact with the vector by the use of protective clothing and insect repellent.

IV. Behavioral Factors in Health Promotion
A. Counseling of women before and during pregnancy
1. **Folic acid** supplementation is recommended for all women from at or near the time of conception throughout pregnancy to reduce the risk of neural tube defects. The addition of folic acid to flour to ensure adequate intake for all women who might become pregnant is currently under consideration.
2. **Cigarette smoking** during pregnancy is associated with premature delivery and the delivery of infants with low birth weight or intrauterine growth retardation. All pregnant women who smoke should be counseled to give up cigarettes, at least for the duration of the pregnancy.
3. **Alcohol consumption** during pregnancy, even in small quantities, can have serious negative effects on the developing fetus, and consumption in large

quantities is associated with cognitive and growth impairment of the fetus **(fetal alcohol syndrome)**. Pregnant women should be counseled to abstain from drinking alcohol.
  4. The use of **illegal recreational drugs** during pregnancy is associated with a variety of adverse outcomes, but the association is confounded by medical, psychosocial, and economic factors.

B. Counseling of parents
  1. Parents of infants should receive counseling regarding optimal nutrition; appropriate use of safety devices (e.g., car seats) and child-proofing devices (e.g., cabinet latches, electrical outlet plugs); immunizations; and appropriate scheduling of pediatric examinations.
  2. Parents of adolescents should receive counseling regarding the behavioral risks associated with adolescence (e.g., alcohol and drug use, sexual activity, cigarette smoking) and ways to lower such risks for their children.

C. Counseling of patients with risk factors
  1. Modification of diet, activity levels, and other lifestyle practices (such as tobacco use) represents the first recourse in risk factor management. Only if the risk is urgent, the patient refuses life-style modification, or attempts at life-style modification do not achieve the desired goals, should pharmacotherapy be initiated.
  2. **Smoking cessation** may represent the single most important behavior change for improving health in patients who smoke. Approximately 3.3 million smokers quit smoking each year in the USA. Signs of tobacco dependence include the continuous use of tobacco for at least 1 month plus at least one of the following: a history of unsuccessful attempts to stop smoking on a permanent basis, withdrawal symptoms after an attempt to stop smoking, or continuing use of tobacco despite a serious physical disorder the patient knows is made worse by tobacco use (e.g., obstructive pulmonary disease, heart disease). Smoking cessation counseling is summarized in Box 15–1.
  3. The physician is responsible for monitoring the progress of the patient on a regular basis and for changing the approach if sufficient progress is not being made. The physician should initiate appropriate pharmacotherapy for such risk factors as hypertension and hyperlipidemia, and provide appropriate medical oversight and follow-up assessments.

BOX 15–1. The Approach to Smoking Cessation that the National Heart, Lung, and Blood Institute and the American Lung Association Recommend for Use by Physicians in Counseling Their Patients

Step 1. Act as a role model by not smoking. Measures include adopting a no-smoking policy in the physician's office, posting no-smoking signs, and making pamphlets about smoking cessation available in the office.

Step 2. Provide the patient with information on the risks associated with smoking and the reduction of risks if smoking is stopped. In addition to outlining the patient's risks for specific diseases, the information should outline the risks faced by members of the patient's household, particularly children.

Step 3. Encourage abstinence by direct advice and suggestions. The patient is more likely to stop smoking if the physician recommends it clearly and forcefully. The approach should be brief, direct, unambiguous, and informative. Instead of using a "scare tactic," which is likely to arouse defense mechanisms, the physician should emphasize the benefits that will result from cessation of smoking. The reasons for smoking should be taken into account, and the advice should be tailored to individual needs. For example, an emphasis on the health benefits that will result may be most effective for an individual who has symptoms of heart or lung disease, whereas an emphasis on sports performance or how one smells to others may be more effective for an adolescent. If the patient is willing to try smoking cessation, it is important to set a quit-smoking date. If the patient is unwilling to do this, it may be possible to persuade him or her to take some other positive action, such as contacting a smoking-cessation program, by a defined time. Most successful quitters have tried and failed several times before they finally succeed, and knowing this may help the patient.

Step 4. Refer the patient to a smoking-cessation program. Local hospitals or offices of the American Lung Association or American Cancer Society are good sources to find smoking-cessation programs.

Step 5. Follow up on the use of specific cessation and maintenance strategies. The patient must know that the physician is not abandoning him or her by referral but only enlisting specialized assistance. The physician should schedule specific follow-up visits for physical checkups and for emotional support both during and after the smoking-cessation process. Continual emotional support may be helpful in preventing the patient from smoking again or in encouraging the patient to cease smoking if relapse does occur.

Source of data: National Heart, Lung, and Blood Institute. Clinical Opportunities for Smoking Intervention: A Guide for Busy Physicians. NIH Publication No. 92-2178. Washington, D. C., Government Printing Office, June 1992.

QUESTIONS

DIRECTIONS. (Items 1–10): Each of the numbered items or incomplete statements in this section is followed by answers or by completions of the statement. Select the ONE lettered answer or completion that is BEST in each case. Correct answers and explanations are given at the end of the chapter.

1. The fundamental sources of good health include all of the following EXCEPT

    (A) adequate nutrition
    (B) medical screening and follow-up
    (C) absence of genetic disease
    (D) environmental safety
    (E) prudent life-style

2. In the 1950s, Hollingshead and Redlich demonstrated an association between social class and mental illness, with more mental illness among lower social class individuals. This finding

    (A) indicates that poverty produces mental illness directly
    (B) indicates that poverty is causally related to mental illness but the relationship may be indirect
    (C) indicates that the treatment of mental illness should include social welfare
    (D) indicates that although mental illness and social class may be correlated, the direction of causality is uncertain
    (E) is spurious because it is not biologically plausible

3. Visceral protein malnutrition that tends to occur in developing nations during the weaning of children from breast milk is known as

    (A) kwashiorkor
    (B) anabolism
    (C) marasmus
    (D) anasarca
    (E) cachexia

4. The goal for the year 2000 in the USA for the dietary fat intake of adults is

    (A) 10% of total daily calories
    (B) 20% of total daily calories
    (C) 65 g/d
    (D) 30% of total daily calories
    (E) not yet established

5. The best explanation for the prevalence of obesity in the USA is

    (A) low dietary fiber and excess protein
    (B) a genetic defect resulting in low basal metabolism
    (C) a high-fat diet and a sedentary life-style
    (D) an innate preference for sweet food compounded by insulin resistance
    (E) an aversion to broccoli at the highest levels of American society

6. All of the following are true of dietary counseling EXCEPT

    (A) a patient with hyperglycemia may benefit from increased soluble fiber in the diet
    (B) surveys indicate that a minority of physicians provide dietary counseling
    (C) a diet useful for reducing the risk of coronary artery disease is low in saturated fat, as opposed to a diet useful for managing diabetes, which is low in carbohydrate
    (D) most attempts to achieve sustainable weight loss have been unsuccessful to date
    (E) dietary modification can reduce cancer risk

7. Of the following environmental hazards, the only non-threshold exposure is

    (A) ionizing radiation
    (B) noise
    (C) heat
    (D) lead
    (E) carbon monoxide

8. The Ames test is used to

    (A) determine antibiotic susceptibility
    (B) estimate the mutagenic potential of a chemical
    (C) quantify radiation exposure
    (D) establish environmental standards for heavy metals
    (E) estimate the adequacy of ventilation in a building

9. The site of the most frequent, serious contact with environmental hazards is

    (A) the skin
    (B) the hands
    (C) the eyes
    (D) the gastrointestinal tract
    (E) the lungs

10. The most significant indoor air pollutant is

    (A) radon
    (B) tobacco smoke
    (C) carbon monoxide
    (D) dust
    (E) dioxin

ANSWERS AND EXPLANATIONS

1. The answer is B: medical screening and follow-up. The fundamental sources of good health are those factors which promote and maintain well-being, rather than those that identify or modify existing threats to such well-being. Medical screening and follow-up can be extremely important but are used as an attempt to detect disease or the risk for disease rather than as a fundamental means to maintain health. Provided that genetic disease is absent, adequate nutrition, a prudent life-style (i.e., avoidance of high-risk activities and hazardous substances), and a safe environment are the fundamental sources of good health. This should suggest that in the clinical setting, it is the patient rather than the physician who wields the greater power over health and the risk of disease. *Epidemiology,* p. 191.
2. The answer is D: indicates that although mental illness and social class may be correlated, the direction of causality is uncertain. An association, or correlation, between two variables may suggest, but cannot definitively establish, causality. While there is evidence that lower social class and mental illness are correlated, causality is plausible in either direction. The stresses of poverty may lead to malnutrition, physical abuse, fractured social supports, and, consequently, poor mental health. Alternatively, poor mental health may lead to unemployability and eventually to poverty. There are plausible causes of both low social class and poor mental health, such as racially motivated discrimination. Thus, an association should serve as a basis for further hypothesis generation but not as sufficient evidence for causality. *Epidemiology,* p. 191.
3. The answer is A: kwashiorkor. The question provides the definition of kwashiorkor. Protein malnutrition occurs in the context of adequate, or nearly adequate, total calorie intake when young children are weaned from breast milk to subsist on gruel or porridge. *Epidemiology,* p. 192.
4. The answer is D: 30% of total daily calories. As discussed in the chapter, reduction of dietary fat consumption to no more than 30% of total daily calories represents the goal included in *Healthy People 2000*. While evidence cited in the chapter suggests a decline in the mean fat intake of American adults from 37% to 34% of calories, the rising prevalence of obesity and the notorious difficulty in obtaining accurate dietary intake data perhaps call this trend into question. The goal is a compromise in that most nutrition experts concur that an even lower fat intake would be preferable, but this seems unrealistic and unattainable, at least for the near future. *Epidemiology,* p. 193.
5. The answer is C: a high-fat diet and a sedentary life-style. Obesity is a complex and multifactorial condition, about which a vast and at times controversial literature proliferates continuously. Fundamentally, though, obesity is known to represent an imbalance between the body's need for, and consumption of, fuel. High fat intake is tantamount to high calorie intake, because fat is calorie dense (approximately 9 kcal/g). Low-fat diets may or may not be low-calorie diets, as excessive intake of either carbohydrate or protein can lead to caloric excess. The other side of the equation governing body weight is fuel consumption, due largely to basal metabolism, but substantially modified by the level of physical activity. Having evolved over 4 million years under the pressures of strenuous activity and dietary deficiency, humans are ill prepared to defend themselves against a conveniently available abundance of calorie-dense food and a host of energy-saving devices. The extremely high prevalence of obesity noted in the chapter is the apparent consequence. Obesity and insulin resistance are associated, but at this time, it is unclear which condition precedes the other. *Epidemiology,* p. 193.
6. The answer is C: a diet useful for reducing the risk of coronary artery disease is low in saturated fat, as opposed to a diet useful for managing diabetes, which is low in carbohydrate. A diet low in saturated fat is useful in the management of both coronary artery disease and diabetes. While there is some debate as to whether a diet high in monounsaturated fat or high in complex carbohydrate is the optimal approach to diabetes (Garg et al. 1994), both are known to be preferable to the typical American pattern. The relative simplicity of the dietary message, that a diet rich in vegetables, fruits, and whole grains, and moderate in virtually all other foods, is beneficial to the health of almost everyone, belies the nutritional controversies so frequently forwarded by the media. *Epidemiology,* p. 194.
7. The answer is A: ionizing radiation. In order to produce a harmful effect, most exposures must exceed some threshold below which they are innocuous. This is true of physical and chemical toxins as well as microbial pathogens. Current theory holds that no level of ionizing radiation is entirely innocuous, as the potential for injury to DNA remains as levels fall to anything that can be measured. Injury to even a single gene might be sufficient ultimately to produce cancer, although risk is known to be greater with more substantial exposure and greater genetic injury. Until or unless measurement technology improves and demonstrates otherwise, ionizing radiation is considered a hazard at any level above 0. *Epidemiology,* p. 194.
8. The answer is B: estimate the mutagenic potential of a chemical. The Ames test exposes *Salmonella typhimurium* bacteria to the suspected toxin in the presence of certain enzymes, which are observed for the mutation of a specific gene. A high proportion of the chemicals that cause mutation in this assay can be shown by more complex methods to be carcinogenic. *Epidemiology,* p. 195.
9. The answer is E: the lungs. The principal reasons why the lungs represent the site of greatest vulnerability to environmental hazards are the difficulty in controlling the dispersion of substances in air and the tremendous volume of air exchanged each day. As noted in the chapter, the air exchanged by an adult each day weighs nearly 50 lb. *Epidemiology,* p. 196.
10. The answer is B: tobacco smoke. Significant exposure to carbon monoxide is potentially lethal but is rare compared with tobacco smoke exposure. The intensity and frequency of exposure to radon is small relative to exposures to tobacco smoke. Dust may lead to allergic reactions but is relatively innocuous under most circumstances. There is now a substantial body of evidence that passive exposure to tobacco smoke increases the risk of reactive airway disease (such as asthma), respiratory infections, and lung cancer. Dioxin is a chemical carcinogen introduced into the food or water supply by

industry and is not a component of indoor air. *Epidemiology,* p. 197.

REFERENCES CITED

Burkitt, D. P., and N. J. Temple, eds. Western Diseases: Their Dietary Prevention and Reversibility. Totowa, N.J., Humana Press, 1994.

Butrum, R. R., C. K. Clifford, and E. Lanza. NCI dietary guidelines: rationale. American Journal of Clinical Nutrition 48:888–895, 1988.

Daily Dietary Fat and Total Food-Energy Intakes: Third National Health and Nutrition Examination Survey, Phase 1, 1988–1991. Morbidity and Mortality Weekly Report 43:116–125, 1994.

Garg, A., et al. Effects of Varying Carbohydrate Content of Diet in Patients with Non-Insulin Dependent Diabetes Mellitus. Journal of the American Medical Association 271:1421–1428, 1994.

Hollingshead, A. B., and F. C. Redlich. Social Class and Mental Illness: Appendix 2. New York, John Wiley and Sons, 1958.

Institute of Medicine. Improving America's Diet and Health: From Recommendations to Action: A Report of the Committee on Dietary Guidelines Implementation, Food and Nutrition Board. Paul R. Thomas, ed. Washington, D. C., National Academy Press, 1991.

Jekel, J. F. Epidemiology, Biostatistics, and Preventive Medicine. Philadelphia, W. B. Saunders Company, 1996.

Keyes, A., et al. The Diet and 15-Year Death Rate in the Seven Countries Study. American Journal of Epidemiology 124:903–915, 1986.

King, K. W. The world food crisis: a partial answer. Research/Development, September, pp. 22–25, 1969.

Nestle, M. Promoting health and preventing disease: national nutrition objectives for 1990 and 2000. Nutrition Today, May/June, pp. 26–30, 1988.

Pratt, L. Changes in health care ideology in relation to self-care by families. Paper presented at the annual meeting of the American Public Health Association, Miami Beach, Fla., 1976.

Rahmathullah, L., et al. Reduced mortality among children in southern India receiving a small weekly dose of vitamin A. New England Journal of Medicine 323:929–935, 1990.

US Department of Agriculture. Nationwide Food Consumption Survey: Continuing Survey of Food Intakes by Individuals: Men 19–50 Years. Report 85-3. Washington, D. C., Government Printing Office, 1986. *In:* US Department of Health and Human Services. The Surgeon General's Report on Nutrition and Health. DHHS Publication No. (PHS)88-50210. Washington, D. C., Government Printing Office, 1988.

US Department of Agriculture. Nationwide Food Consumption Survey: Continuing Survey of Food Intakes by Individuals: Women 19–50 Years and Their Children 1–5 Years, 1 Day. NFCS, CSFII Report 85-1, November, 1985. Hyattsville, Md. *In:* US Department of Health and Human Services. The Surgeon General's Report on Nutrition and Health. DHHS Publication No. (PHS)88-50210. Washington, D. C., Government Printing Office, 1988.

US Department of Agriculture and US Department of Health and Human Services. Nutrition and Your Health: Dietary Guidelines for Americans, 2d ed. Home and Garden Bulletin No. 232. Washington, D. C., Government Printing Office, 1985.

US Department of Health and Human Services. Healthy People 2000: National Health Promotion and Disease Prevention Objectives. DHHS Publication No. (PHS)91-50212. Washington, D. C., Government Printing Office, 1990.

US Department of Health and Human Services. Promoting Health, Preventing Disease: Objectives for the Nation. Washington, D. C., Government Printing Office, 1981.

US Department of Health and Human Services. The Surgeon General's Report on Nutrition and Health. DHHS Publication No. (PHS)88-50210. Washington, D. C., Government Printing Office, 1988.

Winick, M. Nutrition education in medical schools. American Journal of Clinical Nutrition 58:825–827, 1993.

Zimmerman, M., and N. Kretchmer. Isn't it time to teach nutrition to medical students? (Editorial.) American Journal of Clinical Nutrition 58:828–829, 1993.

CHAPTER SIXTEEN

Methods of Primary Prevention: Specific Protection

SYNOPSIS

OBJECTIVES
- To define the goals of primary disease prevention.
- To discuss the use of vaccines in the prevention of infectious diseases.
- To distinguish passive from active immunization.
- To summarize the recommended vaccine schedules for children and adults.
- To characterize methods used for vaccine surveillance and testing.
- To discuss the goals of immunization and the reasons for inadequate immunization levels in the USA.
- To describe alternative means of preventing infection.
- To discuss the use of antimicrobial drugs in preventing disease.
- To identify methods for preventing nutrient deficiencies, injuries, toxic exposures, and iatrogenic diseases and injuries.

I. Goals of Primary Prevention
A. One goal is the prevention of **specific diseases** (e.g., by using vaccines and antimicrobial prophylaxis).
B. Another goal is prevention of **specific deficiency states** (e.g., by using iodized salt to prevent iodine deficiency goiter and by using fluoride to prevent dental caries and osteoporosis).
C. A third goal is prevention of **specific injuries and toxic exposures** (e.g., by using helmets to prevent head injuries in construction workers, goggles to prevent eye injuries in machine tool operators, and ventilation systems to control dusts).

II. Prevention of Diseases by Use of Vaccines
A. Intact immunity implies that the immune system has not suffered damage from a disease such as infection with type 1 human immunodeficiency virus (HIV-1) or damage from medications such as certain anticancer drugs or long-term steroid use. There is some evidence that depression and loneliness may also suppress normal immune function.

B. Types of immunity
1. **Passive immunity** is protection against an infectious disease provided by circulating antibodies made in another organism.
   a. Newborn infants are protected by **maternal antibodies** transferred through the placenta before birth and through breast milk after birth.
   b. If a person has recently been exposed to hepatitis B virus and has not been immunized with hepatitis B vaccine, he or she can be given **human immune globulin,** which confers passive immunity and protects against infection with this virus.
   c. In an emergency, a specific type of **antitoxin,** if available, can be used to confer passive immunity. For example, diphtheria antitoxin is used in the presence of clinical diphtheria, and trivalent botulinum antitoxin is used in the presence of botulism.
   d. Passive immunity provides incomplete protection and usually is of short duration.
2. Vaccines confer **active immunity.**
   a. Some vaccines, such as the inactivated polio vaccine (see Chapter 1), confer active immunity by stimulating the production of **humoral (blood) antibody** to the antigen in the vaccine.
   b. Other vaccines, such as the live attenuated polio vaccine, not only elicit a humoral antibody response but also stimulate the body to develop **cell-mediated immunity.** This tissue-based cellular response to foreign antigens involves mobilization of killer T cells.
   c. Active immunity is far superior to passive immunity, because active immunity lasts longer (a lifetime in some cases) and is rapidly stimulated to high levels by a reexposure to the same or closely related antigens.
3. All types of vaccines provide the immunized person with some level of **individual immunity** to a specific disease. Some vaccines also reduce or prevent the shedding (spread) of infectious organisms from an

immunized person to others, and this contributes to **herd immunity** (see Chapter 1 and Fig. 1–1).

C. Types of vaccines
1. As shown in Table 16–1, some vaccines are inactivated (killed), some are live, some are attenuated (altered), and others are referred to as toxoids (inactivated or altered bacterial toxins).
2. **Inactivated (killed) vaccines**
   a. Pertussis vaccine and the older typhoid vaccines are examples of **inactivated bacterial vaccines.**
   b. Influenza vaccine and the inactivated poliomyelitis vaccine are examples of **inactivated viral vaccines.**
3. **Attenuated (altered) vaccines**
   a. The bacillus Calmette-Guérin (BCG) vaccine against tuberculosis is an example of a **live attenuated bacterial vaccine.**
   b. The measles and oral poliomyelitis vaccines are examples of **live attenuated viral vaccines.**
   c. Live attenuated vaccines are created by altering the organisms so that they are no longer pathogenic but still have antigenicity.
4. Diphtheria vaccine and tetanus vaccine are the primary examples of **toxoids** (inactivated or altered bacterial exotoxins).
   a. *Corynebacterium diphtheriae,* the organism that causes diphtheria, produces a potent toxin when it is in the lysogenic state with corynebacteriophage.
   b. *Clostridium tetani,* an organism that is part of the normal flora of many animals and is frequently found in the soil, can cause tetanus in unimmunized people with infected wounds. This is because *C. tetani* produces a potent toxin when it grows under anaerobic conditions, such as those found in wounds with necrotic tissue. Tetanus is almost nonexistent in populations with high immunization levels.

TABLE 16–1. Prevention of Infectious Diseases by Vaccines Available in the USA

Disease	Vaccine
Anthrax	Anthrax vaccine contains inactivated bacteria and is administered subcutaneously.
Cholera	Cholera vaccine contains inactivated bacteria and is administered subcutaneously or intradermally.
Diphtheria	Several combination vaccines are available: DTP is a combined diphtheria, tetanus, and pertussis vaccine; DTaP is a combined diphtheria, tetanus, and acellular pertussis vaccine; DT is a combined diphtheria and tetanus vaccine; Td is like DT but with a reduced amount of diphtheria antigen; and Tetramune is the trade name for a new tetravalent vaccine combining DTP and *Haemophilus* b conjugate vaccine. In all cases, the diphtheria component is a toxoid and the intramuscular route of administration is used.
Haemophilus influenzae infection	The *H. influenzae* type b conjugate vaccine (Hib) contains bacterial polysaccharide conjugated to protein. It is administered intramuscularly. A tetravalent vaccine against diphtheria, tetanus, pertussis, and *H. influenzae* type b is also available and is marketed under the trade name Tetramune.
Hepatitis B	Hepatitis B conjugate vaccine (HBV) contains inactivated viral antigen and is administered intramuscularly.
Influenza	Influenza vaccine contains inactivated virus or viral components and is administered intramuscularly.
Japanese encephalitis	Encephalitis vaccine contains inactivated virus and is administered subcutaneously.
Measles	A vaccine against measles, mumps, and rubella (MMR) is available and contains live viruses. The vaccine is administered subcutaneously.
Meningococcal disease	Meningococcal vaccine contains bacterial polysaccharides of serotypes A, C, Y, and W-135 and is administered subcutaneously.
Mumps	A vaccine against measles, mumps, and rubella (MMR) is available and contains live viruses. A vaccine against mumps alone is also available. The vaccines are administered subcutaneously.
Pertussis	Three combination vaccines are available: DTP, DTaP, and Tetramune (see discussion under diphtheria, above). A vaccine against pertussis alone is also available. In all cases, the pertussis component consists of inactivated bacteria and the intramuscular route of administration is used.
Plague	Plague vaccine contains inactivated bacteria and is administered intramuscularly.
Pneumococcal disease	Pneumococcal vaccine contains bacterial polysaccharides from 23 strains of *Streptococcus pneumoniae.* It is administered intramuscularly or subcutaneously.
Poliomyelitis	Two vaccines are available. The oral polio vaccine (OPV) contains live polioviruses of all three types. The inactivated polio vaccine (IPV) contains inactivated polioviruses of all three types. The OPV, which is administered orally, is also referred to as the Sabin vaccine. The IPV, which is administered subcutaneously, is also referred to as the Salk vaccine.
Rabies	Human diploid cell vaccine (HDCV) contains inactivated virus and can be administered subcutaneously or intramuscularly. The subcutaneous dose is lower than the intramuscular dose and is used only for preexposure vaccination.
Rubella	A vaccine against measles, mumps, and rubella (MMR) is available and contains live viruses. A vaccine against rubella alone is also available and contains live attenuated virus. The vaccines are administered subcutaneously.
Tetanus	Several combination vaccines are available: DTP, DTaP, DT, Td, and Tetramune (see discussion under diphtheria, above). A vaccine against tetanus alone is also available. In all cases, the tetanus component is a toxoid and the intramuscular route of administration is used.
Tuberculosis	The bacillus Calmette-Guérin (BCG) vaccine contains live attenuated mycobacteria and is administered intradermally or subcutaneously.
Typhoid	The older typhoid vaccine contains inactivated bacteria and is administered subcutaneously. Two newer vaccines appear to be as antigenic as the older vaccine and to have fewer side effects. One is called the Ty21A oral vaccine and contains live attenuated bacteria. The other contains capsular polysaccharide and is administered intramuscularly.
Varicella	Varicella (chickenpox) vaccine contains live virus and is administered subcutaneously.
Yellow fever	Yellow fever vaccine contains live virus and is administered subcutaneously.

Sources of data: (1) Centers for Disease Control and Prevention (CDC). Advisory Committee on Immunization Practices (ACIP): general recommendations on immunization. Morbidity and Mortality Weekly Report 43(RR-1), 1994. (2) American Academy of Pediatrics (AAP). Report of the Committee on Infectious Diseases, 22nd ed. Elk Grove Village, Ill., AAP, 1991. (3) CDC. ACIP: recommendations for use of *Haemophilus* b conjugate vaccines and a combined diphtheria, tetanus, pertussis, and *Haemophilus* b vaccine. Morbidity and Mortality Weekly Report 42(RR-13), 1993. (4) CDC. ACIP: typhoid immunization. Morbidity and Mortality Weekly Report 43(RR-14), 1994. (5) CDC. ACIP: diphtheria, tetanus, and pertussis—recommendations for vaccine use and other preventive measures. Morbidity and Mortality Weekly Report 40(RR-10), 1991. (6) CDC. Varicella vaccination. Morbidity and Mortality Weekly Report 44:264, 1995.

TABLE 16-2. Recommended Schedule for Active Immunization of Healthy Infants and Children

Recommended Age	Vaccine and Dose Number*	Comments
Birth	HBV #1	HBV #1 must be given to infant at birth if mother is HBsAg-positive. HBV #1 can be given to infant at birth or at 1 or 2 months if mother is HBsAg-negative.
2 months	DTP #1 OPV #1 Hib #1 HBV #1 or #2	DTP #1 and Hib #1 can be given earlier in areas of high endemicity. HBV #2 is given at 2 months if HBV #1 was given earlier.
4 months	DTP #2 OPV #2 Hib #2 HBV #2	An interval of 6–8 weeks between OPV doses is necessary. HBV #2 is given at 4 months if it was not given sooner.
6 months	DTP #3 OPV #3 Hib #3	DTP #3 should be given at 6 months in areas of high endemicity; otherwise, it can be given at 6 months or any time up to 18 months.
15 months	MMR #1 DTP #4 Hib #4	DTaP may be substituted for DTP.
6–18 months	HBV #3	
4–6 years	DTP #5 OPV #4 MMR #2	DTP #5 is given at or before school entry. DTP #5 and MMR #2 are boosters. DTaP may be substituted for DTP.
14–16 years	Td	Td is given as a booster at 14–16 years of age and every 10 years thereafter.

*Abbreviations are as follows: DTP = diphtheria, tetanus, and pertussis vaccine; DTaP = diphtheria, tetanus, and acellular pertussis vaccine; HBV = hepatitis B conjugate vaccine; Hib = *Haemophilus influenzae* type b conjugate vaccine; MMR = measles, mumps, and rubella vaccine; OPV = oral polio vaccine; Td = tetanus and diphtheria vaccine with a reduced amount of diphtheria antigen. A tetravalent vaccine combining DTP and *H. influenzae* type b conjugate vaccine is now available under the trade name of Tetramune.

Sources of data: (1) Centers for Disease Control and Prevention (CDC). Advisory Committee on Immunization Practices (ACIP): general recommendations on immunization. Morbidity and Mortality Weekly Report 43(RR-1), 1994. (2) ACIP, American Academy of Pediatrics, and American Academy of Family Physicians. Recommended childhood immunization schedule: United States, January 1995. Journal of the American Medical Association 273:693, 1995. (3) CDC. Recommended childhood immunization schedule: United States, 1995. Morbidity and Mortality Weekly Report 44(RR-5), 1995.

D. Immunization recommendations and schedules
1. Active immunization of children
    a. The American Academy of Pediatrics (AAP) publishes periodically updated versions of the *Report of the Committee on Infectious Diseases* (commonly called the "Red Book" because it invariably comes in a bright red cover), which includes an immunization schedule for healthy infants and children whose immunizations start early in life, another immunization schedule for children whose immunizations did not begin during the first year of life, and a discussion of immunization in special clinical circumstances (e.g., in the presence of HIV infection). This publication is essential for physicians providing care to children.
    b. The Advisory Committee on Immunization Practices (ACIP) of the Centers for Disease Control and Prevention (CDC) also publishes immunization schedules and other information in the Recommendations and Reports (RR) supplements to the *Morbidity and Mortality Weekly Report.*
    c. The recommendations of the AAP and the ACIP are similar, but the latter are more frequently updated.
    d. Table 16-2 shows the ACIP recommended immunization schedule for healthy children without specific contraindications (such as immunodeficiency).
        (1) The recommendations for immunizing children who did not start immunizations as infants have similar intervals between vaccine doses, and the most important difference is that the measles, mumps, and rubella (MMR) vaccine should be started immediately in older children.
        (2) Children with **altered immunocompetence,** whether due to HIV-1 infection or another reason, should not be given live attenuated virus vaccines, including oral polio vaccine; measles, mumps, and rubella vaccine; and yellow fever vaccine. Killed vaccines may be given according to clinical judgment.
    e. A tetravalent vaccine combining diphtheria, tetanus, and pertussis (DTP) with Hib has been produced under the trade name of Tetramune.
2. Active immunization of adults
    a. The importance of adequate immunization of adults was revealed by an epidemic of diphtheria that occurred in the newly independent states of the former Soviet Union, where over 50,000 cases were reported between 1990 and 1994. In 70% of the cases, diphtheria occurred in persons 15 years of age or older. Almost 2000 deaths resulted (see Centers for Disease Control and Prevention 1995).
    b. The immunization of adults builds on the foundation of vaccines given during childhood. If an adult is missing polio, diphtheria, and tetanus vaccines, these should be started immediately.
        (1) Many adults need **boosters** because they were immunized as children and their immunity levels have declined since they were immunized.
        (2) For adults, it is better to use the oral polio vaccine (OPV) and to use the combined tetanus and diphtheria (Td) vaccine, which has reduced diphtheria antigen to decrease the number of reactions.
        (3) Some now recommend that adults be immunized with acellular pertussis vaccine because it may provide some herd immunity against pertussis to children.
        (4) The measles, mumps, and rubella (MMR) vaccine should be administered to most adults who were born after 1956 and lack evidence of immunity to measles (a definite history of measles or measles immunization after age 12 months). The exceptions

> **BOX 16-1. Information on Vaccines and Other Protective Measures for International Travelers**
>
> **Basic immunity**
> Several vaccine-preventable diseases that have been largely eliminated from the USA and other highly industrialized countries can still be found in many of the lesser developed countries throughout the world. Plans for international travel to lesser developed regions should therefore serve as a stimulus for updating immunizations. In general, all children and adults who plan to travel to these regions should have had recent boosters for **tetanus, diphtheria,** and **poliomyelitis.** Because exposure to **measles** is a real possibility in many areas of the world, measles vaccination or revaccination is also indicated for those who have never had the disease and have not received measles vaccine at 15 months of age or later. In addition, female travelers of childbearing age should have immunity against **rubella.** Children should also be immunized against **pertussis** and *Haemophilus influenzae* type b. The 23-valent vaccine against **pneumococcal disease** should be given to persons 65 years or older and to persons with chronic illness.
>
> **Required or recommended vaccines**
> Proof of recent administration of **yellow fever** vaccine may be required for entry into some countries if the traveler has come from areas in South America or Africa where the disease is found. Yellow fever was reported in 14 countries in late 1994. Most countries allow travelers to enter from areas where **cholera** is endemic, but a traveler to an area with cholera may wish to obtain the vaccine for self-protection. In late 1994, cholera was reported in 50 countries, most of which were located in Central and South America, central Africa, and the Indian subcontinent.
>
> **Diphtheria** vaccine is recommended for travel to Russia and the other former states of the Soviet Union. **Typhoid fever** might be a threat for a traveler to the Indian subcontinent or Peru, for example, and **meningococcal disease** may be a threat to those visiting the Indian subcontinent, the Arabian peninsula, and sub-Saharan Africa.
> It is wise for travelers to have been immunized against **hepatitis B,** which is a problem not only in developing countries but also in many of the rapidly industrializing countries of the Pacific rim. If there is any question of sanitation being inadequate, particularly in rural areas, the traveler should receive passive immunization (human immune globulin) against **hepatitis A.** Any immune globulin should be given at least 2 weeks after the last dose of all active vaccines has been given. This schedule will prevent neutralization of the antigen.
> Ordinary travel does not justify being immunized against diseases such as tuberculosis, Japanese encephalitis, and rabies.
>
> **Additional information**
> International travelers should contact a state-licensed official vaccination center in their state. Information about the nearest official center can be obtained from the local or state health department.
> Physicians who care for patients before and after they travel internationally may wish to consult special guides for international health travel, such as Jong's *Travel and Tropical Medicine Manual* (1987) and Hill's "Immunizations for Foreign Travel" (1992). Hill's article discusses indications and precautions concerning special groups of travelers, such as pregnant women.

are pregnant women and immunocompromised patients.
   c. Hepatitis B vaccine should be given to susceptible persons who are at high risk for hepatitis B because of their professions (e.g., health care workers, persons with jobs in certain countries overseas); homosexual activity or intravenous drug use; or frequent exposure to blood or blood products.
   d. Pneumococcal polysaccharide vaccine should be given at least once to persons 65 years and older and to those persons with chronic diseases that increase their risk of mortality or serious morbidity from pneumococcal infection, such as chronic pulmonary or cardiac disease, cancer, renal or hepatic disease, asplenia, and immunosuppression.
   e. Experts recommend that influenza vaccine be given annually in the late autumn to the same groups receiving the pneumococcal vaccine, and some believe that the influenza vaccine should be given to the general population, although this is not a national recommendation.
   f. See Box 16-1 for information on vaccines for international travelers.
3. Passive immunization
   a. The medical indications for passive immunization are far more limited than those for active immunization.
   b. Table 16-3 provides information about the biologic agents available in the USA and the indications for their use in immunocompetent persons (those with normal immune systems) and immunocompromised persons (those with impaired immune systems).
   c. For immunocompetent persons who are at high risk for exposure to hepatitis A, usually because of travel to a country where it is common, hepatitis A vaccine can be administered if there is time, or immune globulin can be administered prior to the travel as a method of **preexposure prophylaxis.**
   d. For those who were recently exposed to hepatitis B or rabies, a specific immune globulin can be used as a method of **postexposure prophylaxis.**
   e. For those who lack active immunity to exotoxin-producing bacteria already causing symptoms (such as *Clostridium botulinum,* the organism responsible for botulism), the injection of a specific antitoxin is recommended after tests are performed to rule out hypersensitivity to the antitoxin.
   f. For immunocompromised persons who have been exposed to a common but potentially life-threatening infection such as chickenpox, immune globulin can be lifesaving if given intravenously soon after exposure.
E. **Vaccine surveillance and testing**
   1. The goals of vaccine surveillance are to monitor the effectiveness of vaccines and to detect vaccine failures or adverse effects.
   2. Randomized field trials
      a. The standard way to measure the effectiveness of a new vaccine is through a **randomized field trial** (the public health equivalent of a random-

TABLE 16-3. Indications for Use of Immune Globulins and Antitoxins Available in the USA

Biologic Agent	Type	Indication
Botulinum antitoxin	Specific equine antibody	Treatment of botulism.
Cytomegalovirus immune globulin	Specific human antibody	Prophylaxis for bone marrow and renal transplant recipients.
Diphtheria antitoxin	Specific equine antibody	Treatment of respiratory diphtheria.
Immune globulin (intramuscular)	Pooled human antibody	Hepatitis A preexposure and postexposure prophylaxis.
Immune globulin (intravenous)	Pooled human antibody	Replacement therapy for antibody deficiency disorders.
Hepatitis B immune globulin	Specific human antibody	Hepatitis B postexposure prophylaxis.
Rabies immune globulin	Specific human antibody	Rabies postexposure management of persons not previously immunized with rabies vaccine.
Tetanus immune globulin	Specific human antibody	Treatment of tetanus; postexposure management of persons not previously immunized with tetanus vaccine.
Vaccinia immune globulin	Specific human antibody	Postexposure prophylaxis for susceptible immunocompromised persons and for perinatally exposed newborns.

Source of data: Centers for Disease Control and Prevention (CDC). Advisory Committee on Immunization Practices (ACIP): general recommendations on immunization. Morbidity and Mortality Weekly Report 43(RR-1), 1994.

ized controlled trial, although the level of "control" usually is somewhat less).
   b. In this type of trial, susceptible persons are randomized into two groups and are then given the vaccine or a placebo, usually at the beginning of the high-risk season of the year.
   c. The vaccinated subjects and unvaccinated controls are followed through the high-risk season to determine the **attack rate** (AR) in each group:

$$AR = \frac{\text{Number of persons ill}}{\text{Number of persons exposed to the disease}}$$

   d. Next, the **vaccine effectiveness** (VE) is calculated:

$$VE = \frac{AR_{(unvaccinated)} - AR_{(vaccinated)}}{AR_{(unvaccinated)}} \times 100$$

   In the VE equation, the numerator is the observed reduction in AR due to the vaccination. The denominator represents the total amount of risk that could be reduced by the vaccine.
   e. Testing the efficacy of vaccines by randomized field trials is very costly, but it may be required the first time a new vaccine is introduced.
3. Retrospective cohort studies
   a. The antigenic variability of influenza virus (see Chapter 1) necessitates frequent (often yearly) changes in the constituents of influenza vaccines to keep them up to date with new strains of the virus. There are insufficient resources and time to perform a randomized controlled trial of each new influenza vaccine. **Retrospective cohort studies** are sometimes done during the influenza season to evaluate the protective efficacy of the vaccines.
   b. In retrospective cohort studies, because there is no randomization, investigators cannot be sure that there was no selection bias on the part of the physicians who recommended the vaccine or the individuals who agreed to be immunized.
4. Case-control studies
   a. Because randomized field trials require large sample sizes (often over 100,000) for uncommon diseases, they frequently cannot be performed, because the expense is prohibitive. To overcome this problem, **case-control studies** are often conducted instead.
   b. When the risk of disease in the population is low,

the vaccine efficacy (VE) formula above may be rewritten as follows:

$$VE = 1 - \left[\frac{AR_{(vaccinated)}}{AR_{(unvaccinated)}}\right] = (1 - RR) \cong (1 - OR)$$

   c. The risk ratio (RR) is closely approximated by the odds ratio (OR) when the disease is uncommon.
5. Incidence density measures
   a. Measures of **incidence density** may be used to determine the optimal timing for administration of a new vaccine and the duration of the immunity produced.
   b. The formula for incidence density (ID) is as follows:

$$ID = \frac{\text{Number of new cases of a disease}}{\text{Person-time of exposure}}$$

   c. The denominator (person-time) can be expressed in terms of the number of person-days, person-weeks, person-months, or even person-years of exposure to the risk.
F. Immunization goals
   1. One goal of immunization is **disease eradication.** This is feasible only for diseases in which human beings are the sole reservoir of the infectious organism (e.g., smallpox). Although vaccines are available to prevent some diseases with reservoirs in other animals (e.g., rabies, plague, and encephalitis) and some diseases with reservoirs in the environment (e.g., typhoid fever), they are not candidates for eradication programs.
   2. A second goal of immunization is **regional elimination** of disease (e.g., elimination of poliomyelitis has been achieved in the Western hemisphere).
   3. A third goal of immunization is **control of disease** to reduce morbidity and mortality.
   4. The surveillance systems required to achieve eradication or regional elimination must be excellent. Any eradication or elimination program would require considerably more resources and time, as well as general political and popular support, than would a disease control program.
   5. Defining the goals of an immunization program, and the resources necessary to achieve those goals, often leads to considerable scrutiny and debate.
   6. The **Expanded Program on Immunization**
      a. In May of 1974, the World Health Assembly adopted a global Expanded Program on Immuni-

zation (EPI), with the goal of cooperation between the World Health Organization and the member governments in establishing or expanding existing national immunization programs.
b. A particular emphasis of the EPI has been the surveillance of vaccine-preventable illnesses and the monitoring of immunization levels in the member countries.

7. The **Vaccines for Children Program**
   a. In the USA, because of requirements for complete immunization before children enter school at the age of approximately 5 years, the rate of vaccine-preventable diseases has generally been falling and levels of adequate immunization have been high in school-age children.
   b. In 2-year-old children, however, the immunization rates have remained disappointing.
   c. **Adequate immunization** is defined as receiving the recommended number of doses of vaccines by the ages shown in Table 16–2.
      (1) In 1993, the percentages of 2-year-olds who had received adequate immunization were as follows: 72% for diphtheria, tetanus, and pertussis vaccine (4+ doses of DTP); 79% for oral polio vaccine (3+ doses of OPV); 84% for measles-containing vaccines (1 dose, usually of MMR, the combined measles, mumps, and rubella vaccine); 55% for *Haemophilus influenzae* vaccine; and 16% for hepatitis B vaccine.
      (2) When the basic combination of 4 DTP doses plus 3 OPV doses plus 1 MMR dose is considered, only 67% of children were adequately immunized (see Centers for Disease Control and Prevention 1994a).
   d. The generally poor immunization levels for preschool children led to the establishment of the Vaccines for Children Program (see Centers for Disease Control and Prevention 1994b).
      (1) This program is designed to provide free vaccines to children at participating private and public health care provider sites.
      (2) The eligible groups include American Indians, Alaskan Natives, children on Medicaid, children who lack health insurance, and children whose health insurance does not cover immunizations.

G. Explanations for inadequate immunization levels
1. Health beliefs
   a. According to the **health belief model**, before seeking preventive measures, people generally must believe that the disease at issue is serious, if acquired; that they or their children are personally at risk for the disease; that the preventive measure is effective in warding off the disease; and that there are no serious risks or barriers involved in obtaining the preventive measure.
   b. In addition, there need to be cues to action, consisting of information regarding how and when to obtain the preventive measure, as well as the encouragement from or support of other people.
   c. One of the reasons for inadequate immunization levels in the USA today may be the very success of immunization programs in the past. People who do not remember how widespread or serious a disease was in the past tend to be less concerned with preventing it in the future.
2. Vaccine-related lawsuits
   a. In the USA, the number of lawsuits related to vaccine use increased from 1 lawsuit in 1978 to 219 lawsuits in 1985, and the damages claimed (but not necessarily awarded) increased from $10 million to over $3 billion.
   b. As a consequence of naming vaccine manufacturers in many of these lawsuits, there has been a decrease in the number of companies making vaccines.
   c. In response to the problem, the federal government instituted the **National Vaccine Injury Compensation Program.**
   d. The program essentially protects vaccine manufacturers from liability lawsuits, unless it can be shown that their vaccines differed from the federal requirements. It also simplifies the process and reduces the costs for those people making a claim, and almost all of the costs are borne by the federal government.
3. Missed opportunities
   a. In the USA, the immunization levels are consistently lower in the poorer population groups.
   b. It is not clear whether this is due to less education among these groups, inadequate medical care, or inability to put such things as immunization high on a struggling family's list of priorities.
   c. Opportunities for immunization are often missed in emergency departments because providers lack records, time, and a relationship with the patients.
      (1) Often, physicians do not vaccinate children who have a mild upper respiratory infection without complications, even though part of the reason for the visit may have been to receive a vaccination. More recent guidelines have emphasized that such children should receive the appropriate vaccines.
      (2) Frequently, the siblings of a child who is being evaluated will be brought along by the parent, and these siblings should receive vaccinations if their immunization record is not up to date, although this is seldom done.
   d. Hospitalized children whose immunization records are not up to date should be given the appropriate vaccines unless there are clear contraindications. Clinicians are often ill-informed about contraindications. As discussed in the pediatric "Red Book" (see American Academy of Pediatrics 1991), the following are *not* contraindications to immunizing children: a mild reaction to a previous DTP dose, consisting of redness and swelling at the injection site or a temperature under 40.5°C (105°F), or both; the presence of nonspecific allergies; the presence of a mild illness or diarrhea with low-grade fever in an otherwise healthy child who is scheduled for vaccination; current therapy with an antimicrobial drug in a child who is convalescing well; breast-feeding of an infant scheduled for immunization; and pregnancy of someone else in the household. False contraindications account for part of the reason that US physicians have not adequately immunized children.

III. Prevention of Diseases by Use of Antimicrobial Drugs
A. Another form of specific protection, which can be used for varying lengths of time, is **antimicrobial prophylaxis.**
B. For travelers to countries where malaria is endemic, antimicrobial protection against the causative organism, *Plasmodium*, may be desirable.
C. The chemoprophylaxis of tuberculosis is discussed in detail in Chapter 19.
D. In patients who have had rheumatic fever with valvular disease, a short-term course of bactericidal antibiotics

is recommended before dental or other manipulative medical procedures are performed.
- E. People who have been exposed to virulent meningococcal disease, either meningitis or meningococcal sepsis, should be given prophylactic antibiotics. The most common antibiotic used for this purpose is rifampin.
- F. A high dose of ceftriaxone is sometimes given to prevent syphilis or gonorrhea in people who are known to have had sexual contact with an infected person during the period in which the disease was communicable.

IV. Prevention of Deficiency States
- A. Fortification of food or water with micronutrients has helped to prevent nutritional deficiency syndromes.
  1. Iodine in salt has essentially eliminated goiter.
  2. Vitamin D in milk has largely eliminated rickets.
  3. Fluoride ion in water has markedly reduced the incidence of dental caries in children who grow up in areas with fluoridated water.
- B. The frequent use of vitamin and mineral supplements and the fortification of most breakfast cereals with a number of vitamins and minerals have largely eliminated B vitamin deficiencies in populations with reasonably normal nutrition. Such deficiencies still occur in some elderly persons.

V. Prevention of Injuries and Toxic Exposures
- A. The activities of the **Department of Agriculture** are designed to protect the food sold in the USA.
- B. The regulations of the **Food and Drug Administration** are designed to ensure that prescriptions and over-the-counter drugs are safe and effective and that foods are safe and properly labeled.
- C. Continual efforts are made to protect the public through federal or state laws governing land, sea, and air **transportation** and equipment, ranging from regulations for those who construct highways and automobiles to regulations for those who use them (e.g., laws about seat belts, air bags, speed limits, maximum lengths of air time or driving time in a day or a week for pilots or truck drivers, and penalties for operating equipment under the influence of drugs or alcohol).
- D. Local regulations regarding **building codes** are largely for home and workplace safety and include requirements for hard-wired smoke alarms in new houses and hotels, as well as for properly lighted safety exits and automatic sprinklers in public buildings.
- E. Workplace protection regulations are enforced by the **Occupational Safety and Health Administration (OSHA).**
  1. **Chemical safety** regulations may involve such things as publication of all chemicals used in a manufacturing or research setting.
  2. **Biologic safety** regulations may include protection for laboratory technicians working with hazardous microorganisms and proper immunizations for health care workers.
  3. **Physical safety** regulations may involve protection against repetitive motion injuries and against harmful levels of exposure to noise, heat, and cold.

VI. Prevention of Iatrogenic Diseases and Injuries
- A. Among the most preventable of health problems are those generated during the process of treatment—i.e., diseases and injuries that are **iatrogenic** (from the Greek *iatros,* which means physician, and *gennao,* which means to produce). Examples include infections, falls, medication errors, unnecessary surgery, and surgical and medical errors.
- B. **Nosocomial infections** (hospital-acquired infections) are more common than often supposed. Based on a review of the literature, Inlander et al. (1988) concluded that there were at least 100,000 nosocomial infection–related deaths each year in the USA.
  1. Hand washing between patient contacts is the single most important method of reducing the spread of infections in hospitals.
  2. Proper sterile techniques, appropriate isolation techniques, and proper disposal of needles and other sharp objects are also critical.
- C. **Falls** in hospitals and other institutions may result from improper supervision of patients whose illness or medication causes them to become confused or lose their footing. Falls can be prevented by careful monitoring of patients.
- D. **Medication errors** include incorrect dosages, incorrect medications, and drug interactions. Medication errors can be reduced if careful ward procedures are followed, pharmacists are well informed, and pharmacy information systems that check dosages and drug interactions are used.
- E. **Unnecessary surgery** is being reduced by increasing requirements for second opinions concerning elective surgical procedures.
- F. **Surgical and medical errors** will always occur, but their frequency can be limited by the proper training and evaluation of surgeons and physicians, by the surveillance of complications, and by medical care studies of morbidity and mortality in patients undergoing surgical or medical procedures.

QUESTIONS

Directions. (Items 1–10): Each of the numbered items or incomplete statements in this section is followed by answers or by completions of the statement. Select the ONE lettered answer or completion that is BEST in each case. Correct answers and explanations are given at the end of the chapter.

1. The administration of human immune globulin after exposure to hepatitis B is an example of

   (A) secondary prevention
   (B) health promotion
   (C) passive immunity
   (D) hypersensitivity
   (E) cross-reactivity

2. Of the following vaccines, which would most likely be dangerous to a person with immunodeficiency?

   (A) tetanus vaccine
   (B) typhoid vaccine
   (C) diphtheria vaccine
   (D) hepatitis B vaccine
   (E) measles vaccine

3. The goal of a randomized field trial is to determine vaccine effectiveness (VE). Which of the following equations (in which AR is the attack rate) is a correct expression for VE?

   (A) $VE = AR \times 100$

   (B) $VE = \dfrac{AR_{(unvaccinated)} - AR_{(vaccinated)}}{AR_{(unvaccinated)}} \times 100$

   (C) $VE = \dfrac{AR_{(vaccinated)} - AR_{(unvaccinated)}}{AR_{(vaccinated)}} \times 100$

   (D) $VE = \dfrac{AR_{(vaccinated)} + AR_{(unvaccinated)}}{AR_{(vaccinated)}} \times 100$

   (E) $VE = \dfrac{AR_{(vaccinated)} + AR_{(unvaccinated)}}{AR_{(population)}} \times 100$

4. A modified influenza vaccine must be produced every year because of the antigenic drift of the virus. Which of the following interventions is appropriate to determine the efficacy of the vaccine for a particular year?

   (A) a randomized field trial
   (B) routine surveillance
   (C) a retrospective cohort study
   (D) administration of the vaccine by random assignment
   (E) review of data from previous years

5. Measles vaccine used to be given to children under 1 year of age but is now delayed until children are 15 months of age. The principal reason for this is that

   (A) measles is a live vaccine that caused acute disease in infants
   (B) newborn infants are immune to measles
   (C) herd immunity protects newborn infants from exposure to measles
   (D) measles infection is less severe in newborn infants
   (E) maternal antibody inactivates the vaccine in newborn infants

6. What characteristic must a disease have in order for its eradication to be feasible?

   (A) The disease must be spread by the fecal-oral route.
   (B) The disease must lack an arthropod vector.
   (C) The disease must be geographically isolated.
   (D) The disease must be epidemic rather than endemic.
   (E) The disease must lack an animal reservoir.

7. Controversy regarding approval of a vaccine against varicella resulted from all of the following EXCEPT

   (A) uncertainty about the duration of immunity
   (B) uncertainty about the effectiveness of the vaccine
   (C) uncertainty about the effects of the vaccine on herd immunity
   (D) uncertainty about the effects of the vaccine on the occurrence of zoster
   (E) the relatively mild illness caused by the virus

8. A relatively new acellular vaccine has recently been approved for which one of the following diseases?

   (A) tuberculosis
   (B) pertussis
   (C) measles
   (D) rubella
   (E) Lyme disease

9. Specific protection against malaria is provided by

   (A) vaccination and active immunity
   (B) passive immunization
   (C) antimicrobial prophylaxis
   (D) vitamin A supplementation
   (E) mosquito repellent

10. In the USA, hand washing is thought to be the single most important means of reducing the rate of nosocomial infection. Which of the following statements is correct regarding nosocomial infections?

    (A) Nosocomial infections result in 10,000 deaths each year.
    (B) Nosocomial infections result in 100,000 deaths each year.
    (C) There is currently an epidemic of nosocomial infections.
    (D) Nosocomial infections are systematically reported.
    (E) Nosocomial infections are under active surveillance.

ANSWERS AND EXPLANATIONS

1. The answer is C: passive immunity. Passive immunity is the protection of an individual from an infection by antibodies received passively rather than by antibodies produced by the individual himself or herself. This occurs when newborn infants receive maternal antibodies in breast milk. It also occurs with various immunizations intended to provide postexposure prophylaxis, or prevention of infection after exposure to the pathogen has occurred. The treatment of a wild animal bite often includes postexposure prophylaxis of rabies, a regimen that involves both active and passive immunization. The particular benefit of passive immunization in this setting is the short amount of time needed to achieve prophylaxis. The delivery of preformed antibody confers protective immunity immediately, albeit temporarily, whereas active immunization requires the longer period of time needed for the exposed individual to produce antibodies. Human immune globulin contains preformed antibodies to hepatitis B virus, which protect the exposed individual. Generally, as with rabies, active immunization against hepatitis B is provided at the same time. *Epidemiology,* p. 204.

2. The answer is E: measles vaccine. Of the listed vaccines (tetanus, typhoid, diphtheria, hepatitis B, and measles), only the measles vaccine delivers a live attenuated pathogen. The process of attenuation reduces the pathogenicity of the organism so that infection, but not overt disease, will occur, and this produces lasting immunity. The advantage of this approach is the robustness and longevity of the immunity conferred. The danger of administering a live pathogen is that infection might actually lead to clinical disease, particularly in those with marginal immunocompetency, or immunocompromise. Except in circumstances of compelling need, live attenuated viruses are avoided in individuals known to be immunocompromised, as the risk exceeds the potential benefit for this group. *Epidemiology,* p. 204.

3. The answer is B:

$$VE = \frac{AR_{(unvaccinated)} - AR_{(vaccinated)}}{AR_{(unvaccinated)}} \times 100$$

Vaccine efficacy is an expression of how effectively the vaccine prevents disease. This can be expressed as the percentage of cases of infection (in the unvaccinated population) that will be prevented by vaccination. If the vaccine has no effect (no efficacy), the same attack rate would be expected in vaccinated and unvaccinated groups, and the numerator in the formula above would approximate 0. If no infections occurred in the vaccinated group during the time that infections were occurring in the unvaccinated group, the vaccine efficacy would obviously be high (perfect, in fact). This is reflected in the formula. Subtracting 0 from the attack rate in the unvaccinated group and dividing by the attack rate in the unvaccinated group and then multiplying by 100 provides a vaccine efficacy of 100%. *Epidemiology,* p. 208.

4. The answer is C: a retrospective cohort study. Because the influenza virus usually modifies its surface antigens every year (antigenic drift), the efficacy of the previous year's vaccine cannot be guaranteed. Determination of vaccine efficacy requires more than surveillance because a defined population of vaccinated and unvaccinated subjects must be studied. A randomized field trial conducted prospectively is a hopelessly impractical approach for studying a vaccine that must be reconstituted annually. Administration of the vaccine by random assignment would be an essential component of such a field trial, but this is unethical in any conditions other than a research setting. A retrospective cohort study allows for rapid assembly of vaccinated and unvaccinated groups, which can be followed from the time of vaccination to the present and compared on the basis of disease outcome. As explained in the answer to question 3, above, these data permit calculation of the vaccine efficacy. *Epidemiology,* p. 208.

5. The answer is E: maternal antibody inactivates the vaccine in newborn infants. Measles is a live vaccine but does not cause acute disease in infants because they are protected by maternal antibodies. These maternal antibodies, however, attack the live vaccine so efficiently that the infant's immune system does not have enough exposure to the virus for antibody production to occur. For this reason, the vaccine is delayed until the child is 15 months of age, by which time the maternal antibodies have largely disappeared from the child's circulation (Marks, et al. 1978). *Epidemiology,* p. 208.

6. The answer is E: The disease must lack an animal reservoir. Thus far, only smallpox has been completely eradicated as a human pathogen. Poliomyelitis has been eradicated in the Western hemisphere. In order to eradicate a disease completely, no residual source of the pathogen can remain, or susceptible individuals will eventually become infected. Means are not yet available to eliminate disease in animal populations in the wild; it is difficult enough in human populations, which are not trying to hide. Therefore, disease harbored in animal populations is not eradicable by current technologic methods. *Epidemiology,* p. 209.

7. The answer is B: uncertainty about the effectiveness of the vaccine. The one thing about the varicella vaccine that has been clearly established is that it does effectively protect against chickenpox (Weibel, et al. 1984). Controversy arose because of the unique natural history of the varicella virus, specifically its propensity to lie dormant in the dorsal root ganglia and regroup after years or decades to cause shingles (herpes zoster infection). In order to determine the efficacy of the vaccine against zoster, decades of study would be required; for this reason, uncertainty persists about the duration of immunity with use of the vaccine as well as its effects on herd immunity. Lastly, chickenpox is usually a mild illness, and some have suggested that the risks of vaccination cannot be offset by the benefits of preventing so mild a disease. Arguments in favor of the vaccine have prevailed, however, and it has now been approved by the FDA and is in limited use (Medical Letter 1995). *Epidemiology,* p. 209.

8. The answer is B: pertussis. A relatively new acellular vaccine has been recently approved for pertussis. The acellular vaccine exposes subjects to antigens derived from the pathogen rather than to the whole pathogen.

For further discussion, see Herwaldt 1993, cited below. *Epidemiology*, p. 210.

9. The answer is C: antimicrobial prophylaxis. Specific protection, as the name implies, is the use of a targeted strategy to prevent a particular disease. Mosquito repellent may help protect against malaria but is not specific to malaria and may, in fact, fail to provide protection. Immunization, active or passive, is the archetype of specific protection, but there is, as yet, no effective vaccine against malaria. The consumption of supplemental vitamin A is a nonspecific and generally nonprotective measure. The use of antimalarial drugs in advance of exposure can provide specific protection against the disease. *Epidemiology*, p. 211.

10. The answer is B: Nosocomial infections result in 100,000 deaths each year. Nosocomial infection is unfortunately common and not infrequently fatal (Leape et al. 1991). Reportedly, such infections result in approximately 100,000 deaths each year in this country. The problem of nosocomial infections is certainly not new, and because it is unlikely to represent a marked divergence from precedent, it should not be considered an epidemic. Quality-of-care review in hospitals provides data on iatrogenic complications, including nosocomial infections, but many such infections are not specifically reported. To date, there is no comprehensive system of active surveillance of nosocomial infections. *Epidemiology*, p. 212.

REFERENCES CITED

American Academy of Pediatrics. Report of the Committee on Infectious Diseases, 22nd ed. Elk Grove Village, Ill., American Academy of Pediatrics, 1991.

Centers for Disease Control and Prevention. Diphtheria epidemic: new independent states of the former Soviet Union, 1990–1994. Morbidity and Mortality Weekly Report 44:177–181, 1995.

Centers for Disease Control and Prevention. Vaccination coverage of 2-year-old children: United States, 1993. Morbidity and Mortality Weekly Report 43:705–709, 1994a.

Centers for Disease Control and Prevention. The Vaccines for Children Program, 1994. Morbidity and Mortality Weekly Report 43:705, 1994b.

Herwaldt, L. A. Pertussis and pertussis vaccines in adults. Journal of the American Medical Association 269:93–94, 1993.

Inlander, C. B., et al. Medicine on Trial. New York, Prentice-Hall Press, 1988.

Jekel, J. F. Epidemiology, Biostatistics, and Preventive Medicine. Philadelphia, W. B. Saunders Company, 1996.

Leape, L. L., et al. Adverse events and negligence in hospitalized patients. Iatrogenics 1:17–21, 1991.

Marks, J., T. J. Halpin, and W. A. Orenstein. Measles vaccine efficacy in children previously vaccinated at 12 months of age. Pediatrics 62:955–960, 1978.

Medical Letter on Drugs and Therapeutics. Varicella vaccine. Medical Letter on Drugs and Therapeutics 37:55–57, 1995.

Weibel, R. E., et al. Live attenuated varicella virus vaccine: efficacy trial in healthy children. New England Journal of Medicine 310:1409–1415, 1984.

CHAPTER SEVENTEEN
METHODS OF SECONDARY PREVENTION

SYNOPSIS

OBJECTIVES
- To define secondary prevention.
- To define and discuss screening, and to characterize the appropriate applications of community screening and multiphasic screening.
- To define and discuss case finding, and to characterize the appropriate application of case finding.
- To introduce the concepts of the periodic health examination, lifetime health monitoring, and health risk assessment.
- To define "risk age" as applied to health risk assessment.

I. **Secondary prevention** is aimed at early detection of disease, either through screening or case finding, followed by treatment.

II. **Community Screening**
  A. **Screening** is the process of identifying a subgroup of people who are at high risk for having asymptomatic disease or who have a risk factor that puts them at high risk for developing a disease or becoming injured.
   1. Unlike case finding, which is defined below, screening takes place in a **community setting** and is applied to a community population, such as students in a school or workers in an industry.
   2. A positive screening test result in an individual is not diagnostic of a disease; it merely identifies a person as being at high risk for having that disease.
  B. **Objectives of screening**
   1. Community screening programs seek to test large numbers of persons for one or more diseases or risk factors in a community setting (e.g., in an educational, employment, or recreational setting), on a voluntary basis, usually with little or no direct financial outlay by the persons being screened.
   2. See Table 17–1 for various objectives of community screening.
  C. **Ethical and practical concerns about community screening**
   1. When an apparently well population of individuals who have not sought medical care is screened, the professionals involved in the screening program have a greater obligation to show that the benefits of being screened outweigh the costs than when treatment is indicated for someone known to be ill.
   2. The methods used in performing any public screening program, therefore, should be safe, with minimal side effects.
   3. Test errors are a major concern in screening.
      a. **False-positive test results** lead to extra time and costs and can also cause anxiety and discomfort to those whose results were in error.
      b. **False-negative test results** may lead people with early symptoms to be less concerned and therefore to delay medical visits that they might otherwise have made promptly.
   4. **Lead-time bias** occurs when screening detects disease earlier in its natural history than would otherwise have happened, so that the length of time from diagnosis to death is lengthened. Having additional lead time does not alter the natural history of the disease and, therefore, does not extend the length of life.
   5. **Length bias** occurs when the full spectrum of a particular tumor, such as prostate cancer, is composed of cancers that range from very aggressive to quite indolent. Persons discovered by screening programs are more likely to have a less aggressive tumor (because such persons survive longer to be detected) and therefore are likely to survive longer after detection, regardless of the treatment given.
  D. **Minimum requirements for community screening programs**
   1. **Disease requirements**
      a. The disease must be serious (i.e., produce significant morbidity or mortality), or there is no reason to screen in the first place.
      b. There must be an effective therapy for the disease if it is detected. Screening is of no value unless there is a good chance that detecting the disease in the presymptomatic stage will result in effective therapy.
      c. The natural history of a disease must be understood sufficiently well to know that there is a

TABLE 17-1. Different Possible Objectives of Screening Programs

Target	Objective	Example
Disease	Treatment to cure patients	Cancer
Disease	Treatment to prevent complications	Hypertension
Disease	Measures to eradicate infection and prevent its spread	Gonorrhea, syphilis, or tuberculosis
Disease	Change in diet and lifestyle	Coronary artery disease or type II diabetes mellitus
Behavioral risk factor	Change in life-style or occupation	Cigarette smoking or unsafe sexual practices
Environmental risk factor	Change in occupation	Chronic obstructive pulmonary disease from work in a dusty trade
Metabolic risk factor	Treatment or change in diet and life-style	Coronary artery disease or elevated serum cholesterol levels

period of time during which the disease is detectable and treatable while asymptomatic.
  d. The disease or condition must be common (i.e., have a fairly high prevalence) and yet not be found in most people.
    (1) Screening for a rare disease usually means that many false-positive test results would be expected (see Chapter 7).
    (2) Unless the benefits from discovering one case are very high (such as in the case of treating a newborn child who has phenylketonuria or congenital hypothyroidism), it will seldom be cost-effective to screen general populations for a rare disease.
    (3) There is no reason to do community screening for extremely common conditions, such as dental caries, because most of the screened population will need to return for further diagnosis or treatment.
 2. **Screening test requirements**
   a. The screening test must be reasonably quick, easy, and inexpensive, or the associated costs will be prohibitive.
   b. The screening test must be safe and acceptable both to the persons being screened and to their physicians.
   c. The sensitivity, specificity, positive predictive value, and other operating characteristics of a screening test must be known and be acceptable.
 3. **Health care system requirements**
   a. There must be adequate follow-up for all persons who have positive results in the screening test.
   b. Before a screening program for a particular disease is undertaken, there should already be adequate and accessible treatment for people known to have that disease.
   c. Those who are screened and diagnosed as having the disease in question must have access to treatment, or the process is ethically flawed.
   d. The treatment should be acceptable to those being screened.
   e. The population to be screened should be clearly defined, so that the resulting data will be epidemiologically useful. Although screening at "health fairs" and in shopping centers provides the opportunity to educate the public about health matters, the data obtained are seldom useful because the screening population is not well-defined and tends to be self-selected and highly biased in favor of those concerned about their health (Berwick 1985).
   f. It should be clear who is responsible for the screening, what the cutoff points are for calling a test result positive, and how the findings will become part of a participant's medical record at his or her usual place of care.
 4. **Application of the criteria to examples** is shown in Table 17-2.
E. **Repetition of screening programs**
 1. The frequency of a screening program must be based on knowledge of the availability and cost-effectiveness of the resources necessary for screening; the performance of previous similar screening programs in the same community, including the effectiveness of follow-up diagnosis and treatment and the benefits obtained by those who were diagnosed and treated in this way; the percentage of the community involved in the previous screening efforts and the time elapsed since prior screening; and the incidence and natural history of the disease being screened.
 2. Almost inevitably, a screening effort repeated after a short interval will be quite disappointing, unless the population screened the second time is very different from the one screened the first time. This is because the initial screening will have detected prevalent cases (cases accumulated over many years), whereas the repeated screening will detect only incident cases (new cases since the last screening), making the number of detected cases smaller.
F. **Multiphasic screening**
 1. Multiphasic screening is an attempt to detect a variety of diseases in the same individual.
 2. The principal advantage of multiphasic screening is efficiency.
 3. Disadvantages of multiphasic screening include the following:
   a. In an elderly population, such screening will often detect diseases or abnormal conditions that have been found earlier and are already being treated, in which case funds are being used for unnecessary testing.
   b. Multiphasic screening results in a relatively high frequency of false-positive results, and this requires many participants to return for more expensive follow-up tests.
   c. For each disease-free person screened, the probability that at least one of the screening tests will yield a false-positive finding can be expressed as $[1 - (1 - \text{alpha})^n]$, where alpha is the false-positive error rate (see Chapter 7) and n is the number of screening tests done. If 25 tests are performed, over 70% of disease-free individuals will be brought back for unnecessary but often costly follow-up testing.

III. **Individual Case Finding**
 A. **Case finding** is the process of searching for asymptomatic diseases or risk factors among people while they are in a **clinical setting** (i.e., among people who are under medical care). The distinction between screening and case finding is frequently ignored in the literature and in practice; the distinction is important be-

TABLE 17-2. Requirements for Screening Programs and Ratings of Example Methods to Detect Hypertension, Elevated Cholesterol Levels, Breast Cancer, and Lung Cancer

Requirements	Screening Method and Rating*			
	Sphygmomanometer Reading	Serum Cholesterol Test	Mammogram	Chest X-Ray
Disease requirements				
(1) The disease is serious.	++	++	++	++
(2) Effective treatment exists.	++	+	+	+/−
(3) The natural history of the disease is understood.	++	+	+	+
(4) The disease occurs frequently.	++	++	++	++
(5) Other diseases or conditions may be detected.	−	−	−	+
Screening test requirements				
(1) The test is quick to perform.	++	+	+	++
(2) The test is easy to administer.	++	+	+	+
(3) The test is inexpensive.	++	+	+	+
(4) The test is safe.	++	++	+	+
(5) The test is acceptable to participants.	++	+	+	++
(6) The sensitivity, specificity, and other operating characteristics of the test are acceptable.	++	+	+	−
Health care system requirements				
(1) The method meets the requirements for screening in a community setting.	++	++	+	−
(2) The method meets the requirements for case finding in a medical care setting.	++	++	++	+

* Ratings are applied to four conditions for which community screening has commonly been undertaken: hypertension, tested by a sphygmomanometer reading of blood pressure; elevated cholesterol levels, with total cholesterol measurement based on a rapid screening of blood; breast cancer, tested by mammography; and lung cancer, tested by chest x-ray. Ratings are as follows: ++ means good, + means satisfactory, and − means unsatisfactory.

cause many of the criteria for community screening do not need to be met during the process of case finding.

B. The periodic health examination
1. Historically, the most common method of prevention in clinical medicine, especially in adults, has been the annual physical checkup, which has come to be known as the periodic health examination.
2. During the 1970s, investigators began moving toward the idea of modifying the periodic examination to focus only on those conditions and diseases that would be most likely to be found in a person of a given age, gender, and family history. This approach was given the term "lifetime health monitoring" by Breslow and Somers (1977).
3. The biggest support for a new approach came in 1979, when the Canadian Task Force on the Periodic Physical Examination recommended that the traditional form of periodic checkup be replaced by the use of **health protection packages** that included sex-appropriate and age-appropriate immunizations, screening, and counseling of patients on a periodic basis.

C. US Preventive Services Task Force
1. In an effort to clarify many of the issues concerning screening and case finding and to make well-studied recommendations, the US Department of Health and Human Services created the US Preventive Services Task Force (USPSTF).
2. The first report of the USPSTF was issued in 1989 and consisted of an assessment of the effectiveness of a total of 169 interventions considered to be preventive in nature. This report, which is summarized in Box 17-1, has become a standard reference for investigators and health care workers in the field of preventive medicine.

D. Health risk assessments
1. Health risk assessments (HRAs) use questionnaires or computer programs to elicit and evaluate information concerning individuals in a clinical or industrial medical practice. Each assessed person receives information concerning his or her life expectancy and the types of interventions that are likely to have a positive impact on health or longevity.
2. The **Society for Prospective Medicine** promotes the use of HRAs for assessing the needs of individual patients as they enter a medical care system or of employees in an industrial setting; developing health education information for those who complete the assessment; and developing cost-containment strategies based on better acquisition of health risk information from individuals.
3. A computer is generally used to calculate the patient's "risk age" on the basis of the data entered and an algorithm. Most HRAs use an algorithm based on findings of the Framingham Study.
   a. The **risk age** is defined as the age at which the average individual would have the same risk of dying as the person being assessed.
   b. If the assessed person's risk age is older than his or her chronologic age, he or she has a higher risk of dying than the average individual of the same chronologic age.
   c. If the assessed person's risk age is younger than the chronologic age, the person has a lower risk of dying than the average individual of the same chronologic age.
4. The HRAs usually provide a printed report about the assessed person's relative risk of dying or risk age, combined with an educational message regarding the types of interventions that would have the most positive effect on the person's life expectancy if they were instituted.
5. Criticisms of HRAs have focused on errors or lack of information by the persons entering the data;

BOX 17-1. US Preventive Services Task Force Recommendations Regarding Screening for Specific Diseases, Conditions, and Risk Factors in Asymptomatic Persons*

Disease, Condition, or Risk Factor	Recommendation
Cancers	
Breast cancer	Annual clinical breast examination is recommended for women 40 years or older. Mammography every 1–2 years is recommended for women between 50 and 75 years of age.
Cervical cancer	Papanicolaou testing every 1–3 years is recommended for all women who are or have been sexually active.
Colorectal cancer	There is insufficient evidence to recommend for or against fecal occult blood testing and sigmoidoscopy; these may be of value in persons who are 50 years or older and have a positive family history. Periodic colonoscopy is recommended for persons with a family history of polyposis.
Lung cancer	Routine screening by chest x-ray or sputum cytology is not recommended.
Oral cancer	Routine screening is not recommended, although it may be of value in those with risk factors such as smoking or chewing tobacco. Examination should be incorporated into regular dental visits.
Ovarian cancer	Routine screening is not recommended.
Pancreatic cancer	Routine screening is not recommended.
Prostate cancer	There is insufficient evidence to recommend for or against routine digital rectal examination in asymptomatic men. Transrectal ultrasound and serum tests for tumor markers, such as prostate-specific antigen, are not recommended for routine screening.
Skin cancer	Routine screening is recommended for persons at high risk, such as those with light skin, easy burning, and major sun exposure. Primary prevention by avoiding excessive sunlight and by use of sunscreens is also recommended.
Testicular cancer	Periodic screening is recommended for men with a history of cryptorchidism, orchiopexy, or testicular atrophy. There is insufficient evidence to recommend for or against testicular self-examination or routine screening of other men.
Congenital and perinatal conditions	
Birth defects	Amniocentesis for karyotyping should be offered to any pregnant woman who is 35 years or older. Unless the pregnant woman does not have access to counseling and follow-up services, her serum alpha-fetoprotein level should be measured between week 16 and week 18 of gestation. In normal pregnancies, routine ultrasound examination is not recommended as a screening measure for birth defects.
Fetal distress	In pregnant women, the fetal heart rate should be monitored by auscultation during labor. Electronic fetal monitoring should be used only in women whose fetuses are at high risk for fetal distress.
Intrauterine growth retardation	Ultrasound examination should be performed in women whose fetuses are at high risk for intrauterine growth retardation. All pregnant women should receive counseling regarding the dangers of smoking, alcohol consumption, and use of legal and illegal drugs during pregnancy.
Preeclampsia	In pregnant women, systolic and diastolic blood pressure measurements should be taken during the first prenatal visit and periodically throughout the third trimester.
Hematologic conditions	
Anemia	Testing for anemia is recommended in pregnant women during the first prenatal visit and in infants during the first year of life. Routine testing is not recommended in other asymptomatic persons.
Hemoglobinopathies	Hemoglobin analysis is recommended in newborn infants who are of African, Mediterranean, or Southeast Asian descent and are at risk for hemoglobinopathies. Families of these infants should be counseled regarding the analysis and results.
Rh incompatibility	During the first prenatal visit, pregnant women should undergo ABO and Rh (D) blood typing and testing for anti–Rh (D) antibody. Unsensitized Rh-negative women should be given Rh (D) immune globulin between week 23 and week 29 of gestation and at delivery or termination of the pregnancy.
Infectious diseases	
Bacteriuria	Periodic dipstick testing is indicated in patients with diabetes, and urine culture is indicated in pregnant women.
Genital herpes infection	Cultures are indicated in pregnant women who have had genital herpetic lesions or whose sexual partners have had these lesions.
Gonorrhea and chlamydial infection	Newborn infants should have a topical ophthalmic antibiotic applied immediately after birth to prevent ophthalmia neonatorum. Pregnant women should have an endocervical culture for gonorrhea and chlamydial infection during their first prenatal visit, with a repeat culture during the latter part of pregnancy for those at high risk. Any person at high risk for sexually transmitted diseases should have routine cultures.

Continued

BOX 17-1. US Preventive Services Task Force Recommendations Regarding Screening for Specific Diseases, Conditions, and Risk Factors in Asymptomatic Persons* *(Continued)*

Disease, Condition, or Risk Factor	Recommendation
Infectious diseases *(continued)*	
Hepatitis B	Pregnant women should be tested for hepatitis B surface antigen (HBsAg) during the first prenatal visit, and those at high risk should be tested again during the third trimester. Infants whose mothers are HBsAG-positive should receive hepatitis B immune globulin and hepatitis B vaccine immediately after birth.
HIV-1 infection	Persons whose sexual activity or intravenous drug use places them at high risk for type 1 human immunodeficiency virus (HIV-1) infection should be offered antibody testing and counseling.
Rubella	A serologic test for rubella antibodies is recommended for all women of childbearing age and should be performed at the first clinical opportunity. Pregnant women found to lack rubella antibodies should be immunized immediately after delivery. Nonpregnant women who lack rubella antibodies and are willing to take the necessary precautions to avoid conception during the next 3 months should be immunized.
Syphilis	Pregnant women should be tested for syphilis during the first prenatal visit and at delivery. Persons whose sexual activity places them at high risk for syphilis should be tested periodically, although the optimal screening interval has not been defined.
Tuberculosis	Persons at high risk of acquiring tuberculosis should undergo tuberculin skin testing. Bacillus Calmette-Guérin (BCG) vaccine should be used only in tuberculin-negative children who cannot take isoniazid and are in continuous contact with an infectious person.
Metabolic and genetic diseases	
Diabetes mellitus	Routine screening for hyperglycemia may be of value in high-risk groups and is recommended in pregnant women between week 24 and week 28 of gestation. It is generally discouraged in other persons.
Obesity	Periodic height and weight measurements are recommended in all children and adults.
Osteoporosis	Routine x-rays to detect low bone mineral content are not recommended.
Phenylketonuria	Newborn infants should be screened for phenylketonuria during the first week of life. Those who are tested before 24 hours of age should be retested before the third week of life.
Thyroid disease	Newborn infants should be screened for congenital hypothyroidism during the first week of life. Screening may be indicated for persons with a history of x-ray exposure to the thyroid area. Routine screening for thyroid disease in others is not recommended.
Vascular diseases	
Cerebrovascular disease	Auscultation for carotid bruits may be of value. Eliciting a history of transient ischemic attacks may be of value. Screening for the following risk factors is recommended: diet, exercise, hypertension, and smoking.
Coronary artery disease	Screening for asymptomatic coronary artery disease is not recommended. Screening for the following risk factors is strongly recommended: diet, exercise, hypertension, and smoking.
High blood cholesterol levels	Periodic measurement of total serum cholesterol is most important in middle-aged men, but it also may be wise in young men and women and in the elderly.
Hypertension	Blood pressure should be measured regularly in all persons 3 years of age and older.
Peripheral artery disease	Screening for disease is not recommended. Screening for risk factors, particularly smoking, may be of value.
Other diseases and conditions	
Alcohol and drug use	Adults and adolescents in clinical settings should be asked to describe their use of alcohol and other drugs, but routine drug testing and assays for biochemical markers are not recommended.
Dementia	Routine screening of asymptomatic persons is not recommended.
Depression and suicidal intent	The use of routine screening tests is not recommended.
Diminished visual acuity	Children should be screened for visual acuity once during the preschool period, preferably at the age of 3 or 4 years.
Glaucoma	There is insufficient evidence to recommend for or against routine screening by tonometry. However, persons 65 years or older should probably be screened periodically for glaucoma.
Hearing impairment	There is insufficient evidence to recommend for or against routine screening of asymptomatic children after the age of 3 years. However, children at high risk for hearing impairment should be screened before the age of 3 years.
Lead toxicity	Annual lead screening is recommended for children who are between 9 months and 6 years of age and are at high risk for lead toxicity.
Low back injury	Routine screening is not recommended.

Source of data: US Preventive Services Task Force (USPSTF). Guide to Clinical Preventive Services. Baltimore, Williams & Wilkins, 1989.
* Note: These recommendations are constantly being revised by the USPSTF, based on the latest research.

difficulties in validating the predictions; uncertainties concerning the correct reference population for baseline risks; limitations related to the fact that the instruments focus mainly or exclusively on mortality and not on morbidity or the quality of life; and limitations associated with the middle-class orientation of all of the instruments developed to date.

6. The greatest strength of HRAs may be their ability to clarify how nutritional and life-style factors affect an assessed person's risk of death and to motivate that person to make changes in a positive direction.

QUESTIONS

DIRECTIONS. (Items 1–10): Each of the numbered items or incomplete statements in this section is followed by answers or by completions of the statement. Select the ONE lettered answer or completion that is BEST in each case. Correct answers and explanations are given at the end of the chapter.

1. An example of secondary prevention is

   (A) early treatment of diabetic nephropathy
   (B) vaccination against hepatitis B
   (C) percutaneous transluminal coronary angioplasty
   (D) detection and treatment of hypertension
   (E) hormone replacement therapy at menopause

2. A screening program detects lung cancer early. The survival time in those screened is 3 months longer than in those not screened who present with symptoms. This difference is likely due to

   (A) length bias
   (B) better treatment options for those found through screening
   (C) lead-time bias
   (D) observer bias
   (E) effect modification

3. A screening program designed to find candidates for liver transplantation would be ill-advised because

   (A) the condition is common
   (B) the necessary resources for treatment are in short supply
   (C) the treatment is invasive
   (D) false-negative results might occur
   (E) the population at risk is unknown

4. Data obtained through screening at health fairs are of little epidemiologic value because

   (A) follow-up is not adequate
   (B) self-selection produces a biased sample
   (C) false-positive results are common
   (D) most conditions are rare in random samples
   (E) comorbid conditions may go undetected

5. Screening for which of the following conditions is not recommended?

   (A) lung cancer
   (B) cervical cancer
   (C) hypertension
   (D) breast cancer
   (E) hyperlipidemia

6. There is controversy regarding the use of prostate-specific antigen to screen for prostate cancer because

   (A) prostate cancer is uniformly fatal
   (B) prostate cancer cannot be detected until it is symptomatic
   (C) there is no effective treatment for prostate cancer
   (D) the appropriate management of asymptomatic prostate cancer is uncertain
   (E) the disease is too rare

7. If 25 tests were performed together as part of multiphasic screening and alpha were set at 0.05, the approximate percentage of healthy subjects in whom at least one false-positive result would be found is

   (A) 25%
   (B) 50%
   (C) 33%
   (D) 70%
   (E) 85%

8. The concept of "lifetime health monitoring" refers to

   (A) routine performance of a comprehensive physical examination
   (B) modification of the periodic physical examination to focus on likely conditions in a given individual
   (C) the aggregate use of diagnostic technology during an individual's lifetime
   (D) the completion of periodic health surveys compiled in a national data base
   (E) a comprehensive list of risk factors for chronic disease

9. The US Preventive Services Task Force (USPSTF), which issued its first report in 1989, was created by the US Department of Health and Human Services to

   (A) control the spread of HIV infection
   (B) eradicate poliomyelitis
   (C) recommend the appropriate role for screening and case finding in the periodic physical examination
   (D) devise a strategy for the prevention of antimicrobial resistance
   (E) curtail population growth

10. Health risk assessments (HRAs) are used to determine an individual's "risk age." A correct interpretation of the risk age is as follows:

    (A) Mortality risk is greatest when the risk age equals the chronologic age.
    (B) If chronologic age exceeds risk age, the risk of death is below average.
    (C) If risk age exceeds chronologic age, the risk of death is below average.
    (D) If risk age is low, the risk of death is high.
    (E) The risk age is defined by the onset of risk factors for chronic disease.

ANSWERS AND EXPLANATIONS

1. The answer is D: detection and treatment of hypertension. It is only fair to acknowledge that the distinctions between levels of prevention can at times be vague. The treatment of obesity in a patient without symptoms of any of the diseases with which obesity is associated (e.g., diabetes mellitus) could be considered primary prevention of those diseases, but it could also be considered tertiary prevention of functional limitations that might result from the obesity itself. Distinction between levels of prevention has both clinical and policy implications, however, and a physician should therefore generally be able to make these distinctions. The early treatment of diabetic nephropathy is an attempt to limit the complications of an established disease and is therefore tertiary prevention. Vaccination against hepatitis B, and vaccination in general, is an effort to prevent disease in an unaffected individual and is therefore primary prevention. Percutaneous transluminal coronary angioplasty is tertiary prevention of the complications of established, symptomatic coronary artery disease. Hormone replacement therapy at menopause is usually a factor in the primary prevention of osteoporosis and ischemic heart disease. Hypertension is usually asymptomatic and often detected through screening or case finding. The detection and treatment of an established asymptomatic disease constitutes secondary prevention. *Epidemiology,* p. 215.

2. The answer is C: lead-time bias. Lead-time bias results when a disease is detected by screening at an earlier point in its natural history than it would be if a screening program were not in effect. The patient survives for a longer time after diagnosis only because the diagnosis is made at an earlier stage of the disease and not because detection changes the natural progression of the disease. Length bias is the preferential detection by screening of relatively indolent disease because more fulminant cases produce early deaths that are not in the population at the time of screening. If better treatment options are available for disease found through screening (e.g., early breast cancers detected by mammography as compared with more advanced breast cancers), a longer survival time might be the result of alteration of the natural history of the disease, and this would represent a true treatment effect and not a form of bias. *Epidemiology,* p. 216.

3. The answer is B: the necessary resources for treatment are in short supply. Organs for transplantation are in perennially short supply, and waiting lists of desperately ill recipients are long. Screening is inappropriate whenever the treatment in question is not available in adequate supply to meet and exceed the needs of symptomatic patients. The less symptomatic or asymptomatic patients detected through screening could not be treated. *Epidemiology,* p. 217.

4. The answer is B: self-selection produces a biased sample. To determine the value of a screening program, the relevant population must be identifiable. Community-based health fairs usually appeal only to health-conscious individuals. These individuals are perhaps less likely to have a particular condition (e.g., hypertension) than those less interested in their health, who would consequently be less interested in health fairs. Because of this self-selection bias, the utility of health fairs in screening a community for disease is highly questionable. False-positive results may or may not be common in such screening efforts, depending on the specificity of the diagnostic test used and the prevalence of the condition. In any screening program, lack of adequate follow-up and inattention to comorbid conditions are potential deficiencies. *Epidemiology,* p. 217.

5. The answer is A: lung cancer. A screening program must meet several criteria to be useful. Among these is the ability of the program to identify a disease at a stage where effective treatment can be given. The use of chest radiography to screen for lung cancer fails to meet this criterion. Lung cancer is usually in an advanced stage by the time it can be detected on a chest film and no more treatable than if detected after symptoms have developed. Routine screening for lung cancer is therefore not currently recommended. *Epidemiology,* p. 217.

6. The answer is D: the appropriate management of asymptomatic prostate cancer is uncertain. The best way to detect and treat prostate cancer is the subject of a vigorous ongoing debate in the medical literature (Woolf 1995; Krahn et al. 1994; Catalona 1994). The disease is rapidly progressive and even fatal in some men but indolent in others, and there is no reliable way to distinguish between the two types early enough to affect the outcome. Those with rapidly progressive disease would clearly benefit from early detection and aggressive treatment, whereas for the majority destined to have indolent disease, treatment may do more harm than good (e.g., surgical prostatectomy can result in erectile dysfunction as well as bladder and bowel incontinence). Prostate cancer can be detected while it is asymptomatic through digital rectal examination and perhaps also with the prostate-specific antigen assay. There is effective treatment for prostate cancer, depending on the stage of disease. The disease is by no means rare, as it is the second leading cause of deaths due to cancer among men in the USA (Catalona 1994). *Epidemiology,* p. 218.

7. The answer is D: 70%. For each disease-free person screened, the probability of at least one false-positive outcome is $[1 - (1 - alpha)^n]$, where alpha is the false-positive error rate and n is the number of screening tests done. In this case, alpha is 0.05 and n is 25. Therefore, $[1 - (0.05)^{25}] = 72.3\%$, or approximately 70%. *Epidemiology,* p. 219.

8. The answer is B: modification of the periodic physical examination to focus on likely conditions in a given individual. Lifetime health monitoring is an effort to detect and respond to changes in the status of an individual's health over the course of his or her lifetime. To this end, age-specific and gender-specific examination and diagnostic screening is recommended by the US Preventive Services Task Force (1996). The routine performance of a comprehensive physical examination is essentially what lifetime health monitoring is intended to replace. We know that it is important to be cautious in our use of diagnostic techniques, as they may expose patients to potential harm as well as potential benefit. We also know that it is important to consider the sensitivity and specificity of the tests employed, the disease prevalence, and the implications for the predictive value associated with diagnostic testing. Routine comprehensive physical examinations, which subject patients to arguably the most potent of diagnostic devices, the physician's brain, have not traditionally been grounded in such rational considerations. The concept of lifetime health monitor-

ing treats the physical examination more like other modalities of screening and diagnosis, stipulating that it should be applied not universally but as indicated by the circumstances and attributes characterizing a particular patient. *Epidemiology,* p. 219.

9. The answer is C: recommend the appropriate role for screening and case finding in the periodic physical examination. The annual physical examination became popular in the USA in the 1940s, as advances in diagnostic and therapeutic technology raised expectations of medical care. The actual utility of an annual examination was first assessed in the 1970s. In the late 1970s, the Canadian Task Force on the Periodic Physical Examination (1979) advised that screening, counseling, and interventions be age-specific and gender-specific rather than comprehensive. The USPSTF was convened to make recommendations regarding appropriate use of the periodic physical examination in the USA. The comprehensive examination became a fixture of the medical landscape before its value was established. In unmodified application, it is a low-yield procedure. *Epidemiology,* p. 219.

10. The answer is B: If chronologic age exceeds risk age, the risk of death is below average. The risk age is the age of the average individual from a population whose risk of death equals that of the patient. Because the risk of death rises with advancing age, if risk age is below chronologic age, the implication is that the patient has the same risk of death as an average, younger person. The lower the risk age, the lower the risk of death. If risk and chronologic age are equivalent, the patient in question has the average risk of death for his or her age group. *Epidemiology,* p. 220.

REFERENCES CITED

Berwick, D. M. Screening in health fairs: a critical review of benefits, risks, and costs. Journal of the American Medical Association 254:1492–1498, 1985.

Breslow, L., and A. R. Somers. The lifetime health monitoring program: a practical approach to preventive medicine. New England Journal of Medicine 296:601–608, 1977.

Canadian Task Force on the Periodic Physical Examination. The periodic health examination. Canadian Medical Association Journal 121:1193–1254, 1979.

Catalona, W. J. Management of cancer of the prostate. New England Journal of Medicine 331:996–1004, 1994.

Jekel, J. F. Epidemiology, Biostatistics, and Preventive Medicine. Philadelphia, W. B. Saunders Company, 1996.

Krahn, M. D., et al. Screening for prostate cancer: a decision-analytic view. Journal of the American Medical Association 272:773–780, 1994.

US Preventive Services Task Force. Guide to Clinical Preventive Services, 2nd ed. Baltimore, Williams & Wilkins, 1996.

Woolf, S. H. Screening for prostate cancer with prostate-specific antigen: an examination of the evidence. New England Journal of Medicine 333:1401–1405, 1995.

CHAPTER EIGHTEEN

Methods of Tertiary Prevention

SYNOPSIS

OBJECTIVES

- To define tertiary prevention and its two basic categories, disability limitation and rehabilitation.
- To characterize situations in which tertiary prevention is indicated.
- To discuss the prevention of disability due to coronary artery disease by the assessment and reduction of the risk factors for this disease.
- To define and characterize hyperlipidemia, introduce various measures of lipemia, and discuss briefly the management of hyperlipidemia.
- To describe the public health importance of hypertension and characterize strategies for detecting, categorizing, and managing hypertension.
- To discuss the prevention of complications of diabetes mellitus and emphasize the importance of strict glycemic control.
- To discuss the role of rehabilitation in preventing social disability and describe the resources used in medical rehabilitation.
- To describe the four formal categories of disability used in most state workers' compensation programs.

I. **Categories of Tertiary Prevention**
   A. **Disability limitation** has the goal of halting the progress of disease or limiting the damage caused by an injury.
   B. **Rehabilitation** is directed at reducing the social disability produced by a given level of impairment.

II. **Opportunities for Prevention**
   A. The onset of **symptomatic disease** may provide a window of opportunity for health promotion to prevent progression of the disease.
   B. **Primary health care personnel** are in the most strategic position to initiate preventive measures in symptomatic patients.
   C. Symptoms may be useful for defining a cohort with particular need for preventive as well as therapeutic health care, and for enhancing the motivation of an individual to adopt a healthful life-style.

   D. Treatment of a symptomatic infection in one individual may constitute primary prevention in another, by **preventing transmission** of the infection.

III. **Disability Limitation**
   A. **Therapy** seeks to undo the threat or damage from an existing disease.
   B. **Symptomatic stage prevention** attempts to halt or limit the future progression of disease.
   C. **Coronary artery disease**
      1. Risk factors
         a. **Male gender** is a risk factor for coronary artery disease, because men are much more likely than women to show signs of atherosclerotic heart disease before the age of 60 years.
         b. A **family history of myocardial infarction** confers an increased risk.
         c. **Cigarette smoking** essentially doubles the age-related risk of myocardial infarction.
         d. **Diabetes mellitus** essentially doubles the age-related risk of myocardial infarction.
         e. **Hypertension** increases the risk of myocardial infarction to varying degrees, depending on its severity.
         f. A **sedentary life-style** is associated with an increased risk of myocardial infarction, but the relationship is difficult to quantify.
         g. **Excess body weight,** and in particular truncal adiposity, is associated with an increased risk of coronary artery disease.
         h. The risk of coronary artery disease is increased in patients with **hyperlipidemia.**
      2. Interaction of risk factors is significant in coronary artery disease. Data from the Framingham Study suggest that multiple risk factors interact in a synergistic manner, so that their effects are more than additive.
      3. Appropriate **therapy** for symptomatic coronary artery disease depends on the stage to which the disease process has progressed when diagnosed. Myocardial infarction may be the first symptomatic manifestation of advanced coronary artery disease.

4. Symptomatic stage prevention
   a. Behavior modification
      (1) Patients who smoke should be counseled to stop smoking.
      (2) All patients should receive counseling about appropriate levels of physical activity and healthful nutrition.
      (3) Counseling should be tailored to facilitate modification of any behaviors contributing to the risk of disease progression in a given patient.
   b. Risk factors amenable to medical interventions, such as hyperlipidemia and hypertension, should be managed by the physician (health care provider) to lower the risk of disease progression.

D. **Hyperlipidemia**
   1. Hyperlipidemia refers to an abnormal elevation of one or more of the lipids found in blood.
   2. The **complete lipid profile** includes the total cholesterol (TC), high-density lipoprotein (HDL), low-density lipoprotein (LDL), very low density lipoprotein (VLDL), and triglycerides.
   3. The **TC level** is equal to the sum of the HDL, LDL, and VLDL levels. The VLDL level is approximated by dividing the triglyceride level by 5. The following equation can be used:

      TC = HDL + LDL + (triglycerides/5)

   4. Assessment of serum lipids is done to determine the need for intervention and to monitor the success of preventive measures.
      a. In general, a **TC level** under 200 mg/dL is desirable.
      b. **HDL** is apparently protective against coronary artery disease. An HDL level greater than 50 mg/dL is considered desirable for women, while a level greater than 35 mg/dL is desirable for men.
      c. **LDL** is the most atherogenic lipid moiety, particularly if oxidized. The desirable LDL level is under 160 mg/dL in people without symptoms of coronary artery disease; under 130 mg/dL in patients with two or more risk factors for coronary artery disease; and under 100 mg/dL in patients with myocardial infarction or known coronary artery disease.
      d. A **triglyceride** level under 150 mg/dL is desirable, although higher levels may occur postprandially.
      e. The **VLDL** level is derived from the triglyceride level, as shown above.
      f. The **total non-HDL cholesterol level** represents all of the potentially harmful serum lipids and is a measure advocated by some. A level under 200 mg/dL is desirable.
      g. The **LDL/HDL ratio** is used to assess the risk of developing coronary artery disease, because a high HDL cholesterol level may compensate for a high LDL cholesterol level. A ratio under 3.5 is considered normal.
         (1) Physical activity may help to lower the LDL level and raise the HDL level.
         (2) Moderate dietary alcohol may raise levels of HDL.
         (3) Monounsaturated fat in the diet (e.g., olive oil) may help lower the LDL level but maintain the HDL level.
      h. The **TC/HDL ratio** may be used instead of the LDL/HDL ratio to estimate risk.
      i. The **triglyceride-HDL relationship** may also be important. The combination of an HDL level below 35 mg/dL and a triglyceride level above 200 mg/dL confers increased risk for coronary artery disease.

   5. Therapy and symptomatic stage prevention
      a. Moderate hyperlipidemia should be managed initially with life-style modification (i.e., increased physical activity and reduced dietary fat intake) if the patient is willing; pharmacotherapy should generally be reserved for situations where a life-style approach is ineffective.
      b. The choice of lipid-lowering medication depends on multiple patient characteristics, as well as the lipid pattern.
         (1) The statins, medications that block a hepatic enzyme required for cholesterol biosynthesis, are generally the most effective drugs for hyperlipidemia.
         (2) Lipid patterns associated with very high levels of triglycerides are best treated with niacin or with medications known as fibrates (e.g., gemfibrozil).

E. **Hypertension**
   1. Hypertension is defined as an average systolic blood pressure of 140 mm Hg or higher or an average diastolic blood pressure of 90 mm Hg or higher, or both. According to this definition, approximately 50 million people in the USA have hypertension.
   2. Assessment
      a. Hypertension may be detected while asymptomatic by screening or case finding. Alternatively, hypertension may be detected only after symptomatic manifestations, such as renal insufficiency or congestive heart failure, have developed.
      b. The mortality risk from coronary artery disease and stroke among hypertensives has dropped by 50% over the past 20 years as a result of the treatment of hypertension.
      c. Staging of hypertension is summarized in Table 18–1.
   3. Therapy and symptomatic stage prevention
      a. Individuals with **optimal or normal blood pressure** (see Table 18–1) should be monitored at 2-year intervals.
      b. Individuals with **high normal blood pressure** should be monitored at 1-year intervals.
      c. **Stage 1 (mild) hypertension** should be confirmed, evaluated, and treated within 2 months.
      d. **Stage 2 (moderate) hypertension** should be evaluated and treated within 1 month.

TABLE 18–1. Evaluation of Blood Pressure and Staging of Hypertension, Based on Average Systolic and Diastolic Blood Pressures in Persons Who Are Not Acutely Ill and Are Not Taking Antihypertensive Medications*

Systolic Blood Pressure (mm Hg)	Diastolic Blood Pressure (mm Hg)	Interpretation
<120	<80	Optimal blood pressure
120–129	80–84	Normal blood pressure
130–139	85–89	High normal blood pressure
140–159	90–99	Stage 1 (mild) hypertension
160–179	100–109	Stage 2 (moderate) hypertension
180–209	110–119	Stage 3 (severe) hypertension
≥210	≥120	Stage 4 (very severe) hypertension

Source of data: National Institutes of Health. The Fifth Report of the National Committee on Detection, Evaluation, and Treatment of High Blood Pressure. Publication No. (NIH)93-1088. Washington, D. C., Government Printing Office, 1993.

* The highest stage for which either part of the blood pressure qualifies is taken as the stage of hypertension. For example, if the systolic blood pressure is 165 mm Hg and the diastolic blood pressure is 115 mm Hg, the stage is 3 (severe).

- e. **Stage 3 (severe) hypertension** should be evaluated and treated within 1 week.
- f. **Stage 4 (very severe) hypertension** should be evaluated and treated immediately.
- g. Any stage of hypertension is more severe if evidence of target organ injury is discernible.
- h. **Essential hypertension** is high blood pressure without a detectable cause; this accounts for most cases of hypertension. **Nonessential hypertension** is due to specific, treatable causes, such as renal artery stenosis or tumors of the adrenal medulla. Underlying causes of hypertension should usually be sought and treated before antihypertensive medication is prescribed or treated concurrently while antihypertensive medication is being given.
- i. The goals of **antihypertensive therapy** are the reduction of blood pressure to normal levels and the prevention of injury to target organs (e.g., eyes, heart, kidneys, brain).
  - (1) **Life-style modification** is indicated to reduce high blood pressure and prevent its complications.
  - (2) Potential means of reducing blood pressure include weight reduction; increased physical activity; decreased intake of dietary sodium and fat; increased intake of potassium, calcium, and magnesium; and moderation of alcohol intake.
  - (3) The importance of smoking cessation is even greater in hypertensives because smoking and hypertension interact synergistically to increase the risk of coronary artery disease.
  - (4) Pharmacotherapy is indicated when life-style approaches are ineffective or incompletely effective. Among the major classes of antihypertensive medications are diuretics, beta blockers, calcium channel blockers, angiotensin-converting enzyme (ACE) inhibitors, alpha blockers, vasodilators, and the recently added class of angiotensin-receptor antagonists (see Medical Letter on Drugs and Therapeutics 1995). The criteria for selecting a particular antihypertensive for a given patient are complex, subtle, somewhat subjective, and constantly changing. For (nearly) current consensus, see National Institutes of Health 1995.

F. Diabetes mellitus
1. Approximately 700,000 people in the USA have insulin-dependent diabetes mellitus (IDDM), and a great many more have non–insulin-dependent diabetes mellitus (NIDDM).
2. Routine screening for diabetes is not cost-effective, except in pregnancy, but early detection through health monitoring is advantageous.
3. The Diabetes Control and Complications Trial (DCCT) in patients with IDDM demonstrated that microvascular and macrovascular complications of diabetes could be prevented or delayed by tight glycemic control.
4. Optimal control of diabetes depends on patient motivation to engage in regular physical activity, monitor the serum glucose and adjust the dosages of glucose-lowering medication accordingly, and adhere to a prudent diet. Patients must also be committed to regular medical follow-up.

IV. Rehabilitation
A. The goal of rehabilitation is to reduce (or prevent) the social disability produced by a given level of physical impairment, both by strengthening the patient's remaining functions and by helping the patient learn to function in alternative ways.

B. General approach to rehabilitation
1. Rehabilitation must begin in the early phases of treatment if it is to be maximally effective.
2. **Rehabilitation counselors** coordinate the efforts of a team of specialists.
3. **Physical therapists** provide counseling and conditioning to facilitate independent performance of the **activities of daily living.**
4. **Speech therapists** help restore communication skills following a stroke or other insult.
5. **Occupational therapists** provide job-related training or retraining.
6. **Psychologic or spiritual counseling,** or both, may be important to the rehabilitation of some patients.

C. Categories of disability
1. Disability is a socially defined concept rather than a medically defined concept.
2. There are four formal categories of disability used in most states for reimbursement of workers who have job-related injuries or illnesses covered under a workers' compensation program:
   - a. **Permanent total disability** (e.g., the loss of two limbs or of vision in both eyes).
   - b. **Permanent partial disability** (e.g., the loss of one limb or of vision in one eye).
   - c. **Temporary total disability** (e.g., a fractured arm or leg in a truck driver).
   - d. **Temporary partial disability** (e.g., a fractured arm in an elementary school teacher).
3. Temporary disability results in compensation for the costs of medical care and for a portion of the lost wages (see Chapters 19 and 21).
4. A person with a permanent disability is reimbursed at a fixed rate for the rest of his or her life.

QUESTIONS

DIRECTIONS. (Items 1–10): Each of the numbered items or incomplete statements in this section is followed by answers or by completions of the statement. Select the ONE lettered answer or completion that is BEST in each case. Correct answers and explanations are given at the end of the chapter.

1. The following is an example of tertiary prevention:

    (A) the treatment of essential hypertension
    (B) postexposure prophylaxis for rabies
    (C) occupational therapy following a stroke (cerebrovascular accident)
    (D) the use of nasal decongestants
    (E) hospice care

2. Primary and tertiary prevention may be achieved concurrently when

    (A) a patient is treated for myocardial infarction
    (B) a patient is treated for cystitis
    (C) a patient is treated for a hip fracture
    (D) a patient is treated for active tuberculosis
    (E) never, because primary and tertiary prevention are mutually exclusive

3. The age-specific risk of myocardial infarction is greater in smokers than nonsmokers by a factor of

    (A) 2
    (B) 3
    (C) 4
    (D) 5
    (E) 7

4. The desirable level of LDL cholesterol is most influenced by which of the following?

    (A) the level of HDL cholesterol
    (B) the presence or absence of symptomatic coronary artery disease
    (C) gender
    (D) age
    (E) the body-mass index

5. A disadvantage of using only the total cholesterol level to predict cardiovascular risk is that

    (A) unlike triglyceride levels, total cholesterol levels vary with meals
    (B) total cholesterol levels do not correlate with risk
    (C) the total cholesterol level is estimated rather than measured
    (D) the ratio of LDL to VLDL is unknown
    (E) HDL is included in the measure

6. Dietary intake of which one of the following substances is inversely associated with blood pressure?

    (A) calcium
    (B) insoluble fiber
    (C) sodium
    (D) polyunsaturated fat
    (E) alcohol

7. Which one of the following antihypertensive medications has been shown to reduce the risk of cardiovascular disease?

    (A) vasodilators
    (B) calcium channel blockers
    (C) angiotensin receptor blockers
    (D) thiazide diuretics
    (E) alpha blockers

8. The Diabetes Control and Complications Trial (DCCT) demonstrated that

    (A) microvascular complications of diabetes are independent of glycemic control
    (B) tight glycemic control delays the onset of microvascular complications
    (C) the risk of hypoglycemia outweighs the benefit of tight glycemic control
    (D) only macrovascular complications of diabetes are preventable
    (E) monitoring urine glucose is more cost-effective than monitoring blood glucose

9. Microalbuminuria of diabetes is best treated with

    (A) a sulfonylurea
    (B) insulin
    (C) dialysis
    (D) an ACE inhibitor
    (E) life-style modification

10. Which of the following statements is true regarding cerebrovascular accidents?

    (A) The incidence of cerebrovascular accidents is rising in the USA.
    (B) Patients require treatment but are no longer candidates for preventive services.
    (C) The restoration of functional ability through physical therapy is a form of prevention.
    (D) Risk factors for cerebrovascular disease are largely unidentified.
    (E) Speech is unlikely to improve following injury to the dominant cerebral hemisphere.

ANSWERS AND EXPLANATIONS

1. The answer is C: occupational therapy following a stroke (cerebrovascular accident). Tertiary prevention is the prevention of disease progression and complications that might result in further impairment and is also rehabilitation to reverse impairment and disability. Occupational therapy is a form of rehabilitation directed at preventing disability and therefore an example of tertiary prevention. The treatment of essential hypertension is best considered a form of secondary prevention, although one might argue it is primary prevention of ischemic heart disease or cardiomyopathy. Postexposure prophylaxis for rabies is secondary prevention. The use of nasal decongestants is a way of treating symptoms and not truly a preventive measure, although it may play a role in preventing sinusitis. Hospice care is intended to provide comfort during the latter stages of terminal illness and does not specifically have the goal of preventing disease progression. *Epidemiology*, p. 226.

2. The answer is D: a patient is treated for active tuberculosis. Communicable disease, such as tuberculosis, offers a unique opportunity for prevention. By treating active disease, progression of disease and impairment are prevented in the infected individual; this is tertiary prevention. The spread of disease to the patient's various social contacts is also prevented; this is primary prevention. *Epidemiology*, p. 226.

3. The answer is A: 2. Multiple data sources suggest that smoking raises the risk of age-specific cardiovascular mortality by a factor of approximately 2 (Bartecchi, Mackenzie, and Schrier 1994). Cigarette smoking is generally considered the most important cause of preventable death and disease in the USA. *Epidemiology*, p. 226.

4. The answer is B: the presence or absence of symptomatic coronary artery disease. According to the National Cholesterol Education Program's Adult Treatment Panel (1994), LDL levels should be less than 160 mg/dL in all adults, even in the absence of other risk factors for coronary artery disease. In those without coronary disease but with two or more other risk factors, LDL levels should be maintained below 130 mg/dL. The panel recommends that LDL measures be maintained below 100 mg/dL in patients known to have coronary disease. *Epidemiology*, p. 228.

5. The answer is E: HDL is included in the measure. As discussed in the chapter, HDL levels are inversely associated with cardiovascular disease risk. All other moieties contributing to the total cholesterol level are associated directly, to varying degrees, with such risk. Therefore, the total cholesterol level includes several positive correlates and one negative correlate of heart disease risk, and this reduces its utility. The ratio of total cholesterol to HDL "purifies" the measure, so that a positive correlate of cardiovascular disease risk is produced (National Cholesterol Education Program 1994). *Epidemiology*, p. 228.

6. The answer is A: calcium. Intake of alcohol and sodium is positively associated with hypertension, but intake of insoluble fiber and polyunsaturated fat does not seem to have any appreciable association with the disease. Calcium intake has been inversely associated with hypertension in some studies, as has intake of potassium, magnesium, and soluble fiber (Stamler et al. 1987; National High Blood Pressure Education Program Working Group 1993). *Epidemiology*, p. 230.

7. The answer is D: thiazide diuretics. As of early 1996, three classes of antihypertensive medication have been shown to reduce the rate of cardiovascular or cerebrovascular morbidity and mortality: thiazide diuretics, beta-adrenergic blockers, and angiotensin-converting enzyme inhibitors (National Institutes of Health 1995). Of these, only thiazide diuretics appear among the choices for this question. The other medications listed are effective in treating hypertension but have not been shown in clinical trials to date to reduce complications and mortality rates. There was recent discussion in the lay press regarding an association between the use of short-acting dihydropyridine calcium channel antagonists and increased mortality rates (Altman 1995; Furberg, Psaty, and Meyer 1995). There is no evidence of such an association with the longer-acting calcium channel antagonists generally preferred in the management of hypertension. *Epidemiology*, p. 230.

8. The answer is B: tight glycemic control delays the onset of microvascular complications. The Diabetes Control and Complications Trial (1993) was designed specifically to test the hypothesis that tight glycemic control could forestall microvascular complications of diabetes. The trial demonstrated that the closer to normal serum glucose levels and glycohemoglobin levels were maintained, the less progression was observed in such microvascular complications as retinopathy and nephropathy. Greater effort is now devoted to achieving nearly normal control of serum glucose levels whenever this is feasible, although the risk of hypoglycemia is raised by such an effort. *Epidemiology*, p. 231.

9. The answer is D: an ACE inhibitor. Before renal function begins to decline as a result of diabetic nephropathy, there is a period during which glomerular filtration actually increases because a high osmotic load is delivered to the glomerulus. The renal hyperfunction is associated with microalbuminuria (microscopic albumin spillage in urine) and presages a decline in creatinine clearance. Evidence is now substantial that ACE inhibitors attenuate renal hyperfunction in diabetes, mitigate the associated microscopic proteinuria, and slow the subsequent decline in glomerular filtration (Ravid et al. 1993; Lewis et al. 1993; Viberti et al. 1994). Sulfonylureas and insulin may indirectly slow the progression of nephropathy by providing tight glycemic control. Dialysis is life-sustaining once renal failure has occurred. Life-style modification may contribute to better glycemic control and thereby indirectly help in the preservation of renal function in diabetes. *Epidemiology*, p. 231.

10. The answer is C: The restoration of functional ability through physical therapy is a form of prevention. As noted in the beginning of the chapter, tertiary prevention consists of efforts to prevent disease progression as well as rehabilitation. Physical therapy is a form of rehabilitation and a means of preventing further impairment and disability. The incidence of stroke is falling in the USA. Risk factors for cerebrovascular disease are known and discussed in the chapter. Speech is often affected by strokes involving the dominant cerebral hemisphere but often improves with time and therapy. *Epidemiology*, p. 231.

REFERENCES CITED

Altman, L. K. Agency issues warning for drug widely used for heart disease. New York Times, September 1, 1995.

Bartecchi, C. E., T. D. Mackenzie, and R. W. Schrier. The human costs of tobacco use. New England Journal of Medicine 330:907–914, 1994.

Dawber, T. R., G. F. Meadors, and F. E. Moore, Jr. Epidemiologic approaches to heart disease: the Framingham Study. American Journal of Public Health 41:279–286, 1951.

Diabetes Control and Complications Trial (DCCT) Research Group. The Diabetes Control and Complications Trial. New England Journal of Medicine 329:683–689, 1993.

Furberg, C. D., B. M. Psaty, and J. V. Meyer. Nifedipine: dose-related increase in mortality in patients with coronary heart disease. Circulation 92:1326–1331, 1995.

Gordon D. J., and B. M. Rifkind. High-density lipoprotein: the clinical implications of recent studies. New England Journal of Medicine 321:1311–1316, 1989.

Jekel, J. F. Epidemiology, Biostatistics, and Preventive Medicine. Philadelphia, W. B. Saunders Company, 1996.

Kuhn, F., and C. E. Rackley. Coronary artery disease in women: risk factors, evaluation, treatment, and prevention. Archives of Internal Medicine 153:2626–2636, 1993.

Lemieux, S., et al. Sex differences in the relation of visceral adipose tissue accumulation to total body fatness. American Journal of Clinical Nutrition 58:463–467, 1993.

Lewis, E. J., et al. The effect of angiotensin-converting enzyme inhibition on diabetic nephropathy. New England Journal of Medicine 329:1456–1462, 1993.

Medical Letter on Drugs and Therapeutics. Drugs for Hypertension. Medical Letter on Drugs and Therapeutics 37:45–50, 1995.

National Cholesterol Education Program. Second report of the expert panel on detection, evaluation, and treatment of high blood cholesterol in adults. Circulation 89:1329–1445, 1994.

National High Blood Pressure Education Program Working Group. Primary prevention of hypertension. Archives of Internal Medicine 153:186–208, 1993.

National Institutes of Health. The Fifth Report of the Joint National Committee on Detection, Evaluation, and Treatment of High Blood Pressure. Publication No. (NIH)95-1088. Washington, D. C., Government Printing Office, 1995.

Ravid, M., et al. Long-term stabilizing effect of angiotensin-converting enzyme inhibition on plasma creatinine and on proteinuria in normotensive type II diabetic patients. Annals of Internal Medicine 118:577–581, 1993.

Scandinavian Simvastatin Survival Study Group. Randomised trial of cholesterol lowering in 4444 patients with coronary heart disease. Lancet 344:1383–1389, 1994.

Shepherd, J., et al. Prevention of coronary heart disease with pravastatin in men with hypercholesterolemia. New England Journal of Medicine 333:1301–1307, 1995.

Stamler, R., et al. Nutritional therapy for high blood pressure: final report of a four-year randomized controlled trial: the Hypertension Control Program. Journal of the American Medical Association 257:1484–1491, 1987.

Viberti, G., et al. Effect of captopril on progression to clinical proteinuria in patients with insulin-dependent diabetes mellitus and microalbuminuria. Journal of the American Medical Association 271:275–279, 1994.

CHAPTER NINETEEN

SPECIAL TOPICS IN PREVENTION

SYNOPSIS

OBJECTIVES
- To introduce public health concerns for which there are important applications of primary, secondary, and tertiary prevention.
- To describe the applications of preventive medicine to maternal and child health.
- To describe the applications of prevention to the acquired immunodeficiency syndrome (AIDS).
- To discuss the incidence and natural history of tuberculosis, and the applications of preventive strategies to control the disease.
- To outline the applications of prevention to chemical substance abuse.
- To discuss the role of preventive services in mental health care.
- To describe the prevention of injuries and of occupational disease.
- To discuss strategies for the avoidance of toxic exposures.

I. Maternal and Child Health
A. Family planning
1. Opportunities for prevention begin in the preconception stage, when primary health care workers can answer questions and provide information about contraception to those who wish to plan the number of children and timing of childbirth.
2. The preconception period is also the time to ensure that potential mothers are in good health, are eating a healthy diet, and are taking folic acid supplements to reduce the risk for neural tube defects in the fetus (see Chapter 15).
3. Family planning can be considered **health promotion,** by enabling good spacing of children, but because of the technical aspects of contraception, it can also be considered **specific protection.**
4. Family planning should not be equated with population control, although family planning may contribute to population control. Such control is achieved in a country only when the policy goals of that country are in agreement with the fertility goals of its people.

B. Prenatal care
1. Primary prevention includes counseling to promote good nutrition and healthful behavior, as well as referral to appropriate programs and specialists if necessary.
   a. Some women may be referred, for example, to the **Women, Infants, and Children (WIC) Program,** which provides food vouchers to those with low income.
   b. Some women may require referral for specific programs concerning cigarette smoking, alcohol use, illegal drug use, or other behaviors that affect the health of the pregnant woman and her fetus.
2. Secondary prevention
   a. Pregnant women are screened for syphilis, type 1 human immunodeficiency virus (HIV-1), and other infectious diseases; a positive screening is followed by confirmatory diagnosis and treatment.
   b. The prevention of isoimmunization and erythroblastosis fetalis is now possible by screening for maternal antibodies to Rh (D) proteins on the fetal red blood cells, followed by the timely use of Rh (D) immune globulin when levels of these antibodies begin to rise.
3. Tertiary prevention includes the control of existing disease (e.g., heart disease, hypertension, and diabetes) and the control of toxemia during pregnancy.

C. Labor and delivery care
1. Many complications of pregnancy and delivery that can cause death or long-term damage to mother or fetus can be treated at the time of labor and delivery by expert obstetric and pediatric management.
2. Infants who are unwell at birth or whose mothers had significant risk factors can be monitored for a time and treated when necessary in a newborn special care nursery.
3. There is debate about whether certain interventions, such as caesarean section, are being overused and causing unnecessary morbidity for the mother and child.

D. Well child care
1. Primary prevention in well child care includes immunizations (see Chapter 16) and review and counseling regarding the child's nutrition, growth, and development.
2. Secondary prevention includes screening for visual and hearing problems. According to the **US Preventive Services Task Force (USPSTF),** 2–5% of children in the USA have amblyopia ("lazy eye") and strabismus (ocular misalignment), and almost 20% have refractive errors by the time they are 16 years old.

Children should be screened for these before entering school, usually by their primary care physicians.

E. Day-care and preschool programs
1. Day-care and preschool programs offer many children the opportunity to learn social skills, interact with other children and adults, and receive stimulation outside the home environment.
2. For children enrolled in **Headstart** programs, there is the added benefit of involvement in a medical screening program called **Healthstart,** which can supplement well child visits.
3. Most states have regulations to promote health and safety in day-care facilities, including regulations concerning environmental sanitation, the qualifications of personnel caring for children, and the ratio of personnel to children.

F. School health
1. Schools can promote health by offering health education courses.
2. Health promotion in children requires that the school setting be regularly inspected to ensure a safe water supply, sanitary food service facilities, and compliance with fire safety regulations.
3. Specific protection is achieved through laws that make evidence of adequate immunization a prerequisite for enrolling in schools in the USA.
4. The newest role for schools in health is to provide school-based health centers that offer counseling, immunizations, and health care; many students, particularly adolescents, apparently prefer school-based facilities to a practitioner or clinic outside the school.

II. Acquired Immunodeficiency Syndrome (AIDS)
A. Worldwide impact of AIDS
1. The number of persons infected with HIV is estimated to be more than 1 million in the USA and 19 million worldwide. All of these infected persons are expected to die of AIDS eventually, if they do not die of other causes in the meantime.
2. In the mid-1990s, between 300,000 and 600,000 persons worldwide were expected to die of AIDS per year. By the turn of the century, the number of AIDS deaths is expected to be between 1.5 and 3 million per year.
3. AIDS, a disease which has only been recognized since 1981 and for which there is no effective treatment to date, is now the leading cause of death in the USA for men between the ages of 25 and 44 years.
4. In some regions of central Africa, the prevalence of HIV infection is greater than 50% and the most productive age groups are being decimated, leaving large numbers of orphaned children. The situation is becoming very grave in Southeast Asia, with rapid increases also noted in South America and on the Indian subcontinent.

B. The spread of HIV infection
1. HIV-1 is spread primarily by sexual contact (both heterosexual and homosexual) and by intravenous drug use (IVDU).
2. The spread of HIV among drug users results from the sharing of the equipment ("works") for injecting drugs. These "works" include needles, syringes, cookers (for heating the drug to dissolve it), cottons (to filter the drug to be injected), and water (used for mixing the drug and cleaning the needle and syringe).
3. HIV can also be spread by transfusions of blood and blood products, although with modern testing, this risk is vanishingly small where all of the tests are done and done correctly.
4. Rarely, HIV is spread by accidental punctures of the skin with contaminated needles or other medical equipment.
5. HIV is not spread by ordinary household contact that does not involve one of the above risk behaviors.
6. In the USA, homosexual intercourse is the most common route and IVDU is the second most common route of spread of HIV, but spread via IVDU is rising rapidly.
7. In some countries with high rates of IVDU, sharing of drug equipment is the leading route of spread.
8. In central Africa and Southeast Asia, heterosexual intercourse is the predominant route of spread.

C. Primary, secondary, and tertiary prevention
1. The best means of preventing the spread of HIV include restricting sexual activity to a monogamous relationship and avoiding IVDU. If these two practices were followed, exposure to HIV would be limited to extremely rare events, such as exposure due to blood transfusion.
2. If a person chooses to have multiple sexual partners, the next best prevention is to use condoms for every act of sexual intercourse. Condoms, if used consistently, reduce the risk of exposure considerably.
3. If a person chooses to use intravenous drugs, exposure can be prevented if only new, clean "works" are used for every injection. Sharing any part of the "works" with another intravenous drug user is extremely hazardous. Needle-exchange programs have been shown to reduce the rate of spread of HIV in urban areas (Kaplan 1992), but they do not eliminate it.
4. Other means to prevent the spread of AIDS include testing donated blood for HIV antibodies and discarding units that are infected and treating HIV-positive pregnant women with azidothymidine (AZT), which reduces the proportion of infants who become infected.

III. Tuberculosis
A. Stages and natural history of tuberculosis
1. The natural history of mycobacterial infection makes the control of tuberculosis considerably more complex than the control of other bacterial diseases.
2. In a small percentage of individuals who are newly infected with mycobacteria, the infection proceeds fairly rapidly either to invade lung tissue or to cause a generalized systemic disease such as miliary tuberculosis.
3. In most persons with normal immune systems, lesions develop in the lung and become contained as cell-mediated immunity develops. The presence of cell-mediated immunity is revealed by a positive reaction in the tuberculin skin test using purified protein derivative (PPD).
4. The initial infection with tuberculosis, when it is successfully resolved, is called **primary tuberculosis,** and it often produces a characteristic abnormality on chest radiography called a **primary (Gohn) complex.**
5. The mycobacteria remain alive—albeit isolated—in the body after the primary infection is resolved. The person is considered to have **inactive tuberculosis** rather than to be completely healed. Inactive tuber-

culosis, which is noninfectious, will ultimately take one of three courses:
  a. The tuberculosis may remain inactive for the rest of the infected person's life. In Europe and the USA, this is by far the most common course.
  b. The infected person's own disease may reactivate later in life to become **active tuberculosis.** This occurs in 4–8% of infected persons and is called **reactivation tuberculosis** or endogenous ("from within") tuberculosis. Reactivation tuberculosis is usually infectious.
  c. The infected person may be exposed to a new tuberculosis infection, which may or may not become active infectious pulmonary tuberculosis. If a new exposure results in active disease, it is called **reinfection tuberculosis** or exogenous ("from without") tuberculosis.

B. The incidence of tuberculosis in the USA
1. In the USA, the incidence of tuberculosis declined from historically high levels until the mid-1980s. In 1985, the decline stopped, and a resurgence of tuberculosis was noted.
2. Although the incidence has dropped somewhat since 1993, tuberculosis still represents a formidable threat, especially in immunocompromised patients.
3. From the 1960s to the mid-1980s, most new active cases of tuberculosis in the USA were due to endogenous tuberculosis, the reactivation of long-standing infection.
   a. During this period, the central goal for the US Public Health Service tuberculosis program was to minimize the spread of tuberculosis from the older population to the younger population.
   b. Because of its central goal, the program was called the "child-centered" program to prevent tuberculosis (see Centers for Disease Control 1965).
4. The incidence of tuberculosis (often called the new active case rate) declined at approximately 5% per year from the 1950s to the mid-1980s.
5. In 1985, the incidence of tuberculosis leveled off and then began to rise for the following reasons:
   a. An increasing proportion of newly discovered cases of tuberculosis were found to be resistant to more than one antimicrobial agent as a result of incomplete courses of antituberculous therapy. This is known as **multiple drug–resistant tuberculosis (MDRTB).**
   b. There were changes in the patterns of tuberculosis because of the presence of type 1 human immunodeficiency virus (HIV-1) infection.
      (1) In HIV-positive individuals, tuberculosis is frequently the first sign of AIDS (Selwyn et al. 1992).
      (2) When HIV-positive individuals are exposed to mycobacteria, the result is often severe and sometimes overwhelming tuberculosis.
      (3) Reactivation of inactive tuberculosis in HIV-positive persons tends to occur as the immune deficiency progresses.
      (4) Persons with HIV infection are often in a position to give their infections to other persons with immunodeficiency, thus continuing the cycle.
6. The problem of MDRTB has been especially severe in prisons, general hospitals, and homeless shelters.

C. Primary, secondary, and tertiary prevention
1. Primary prevention
   a. A vaccine derived from a live, attenuated mycobacterium called the bacillus Calmette-Guérin (BCG) vaccine after its developers is applied to a scratch in the skin of a previously uninfected child or adult to stimulate the production of cell-mediated immunity. This provides some protection against a first infection with *Mycobacterium tuberculosis.*
   b. Immunization with BCG is the least expensive approach to tuberculosis control, but there has been considerable debate regarding its efficacy (see Clemens, Chuong, and Feinstein 1983).
   c. BCG vaccine is widely used in developing nations that have high rates of tuberculosis.
   d. Immunization with BCG has the disadvantage of making subsequent PPD skin test results positive, at least temporarily.
2. Secondary prevention
   a. A 6-month course of isoniazid (INH) reduces the risk of endogenous (reactivation) tuberculosis by more than 50% in people with inactive primary tuberculosis (Mount and Ferebee 1961).
   b. The US Public Health Service has chosen not to recommend the use of BCG vaccine but instead to emphasize the identification of those who had positive results in the tuberculin skin test (particularly recent skin test converters) and the use of INH to reduce their risk of reactivation tuberculosis.
   c. In the presence of immunodeficiency, the tuberculin skin test using PPD often yields false-negative results (Selwyn et al. 1992).
      (1) To prevent this problem in an individual who might be immunodeficient, the PPD skin test can be performed on one arm while an **anergy panel** is tested on the other arm.
      (2) The anergy panel consists of a number of common allergens, at least one of which will elicit a reaction if the immune system is not impaired.
      (3) Immunodeficient patients may show no reaction, and this tells the clinician that a negative result in the PPD skin test cannot be used to exclude the presence of tuberculosis.
3. Tertiary prevention
   a. Increasingly, tuberculosis control depends on the early identification and appropriate treatment of patients with MDRTB or immunodeficiency, or both.
   b. Patients with MDRTB must be treated with a combination of antituberculosis agents (Iseman 1993).
   c. To ensure compliance, therapy may be given in a setting in which patients can be directly observed while taking these agents. Although this approach is expensive in terms of personnel time, preliminary evidence indicates that **directly observed therapy (DOT)** can be effective in reducing the incidence of tuberculosis if it is followed consistently (Frieden et al. 1995).
   d. Some hospitals have developed special negative-pressure rooms in which patients with suspected MDRTB can be tested and treated without risking the spread of drug-resistant infection to other patients (Bellin, Fletcher, and Safyer 1993).
4. Institutional measures
   a. In institutions, methods of primary prevention include the avoidance of overcrowding; the use of improved ventilation; and the introduction of ultraviolet radiation, which kills mycobacteria in the air.

b. Methods of secondary prevention include chest x-rays and tuberculin skin testing.
c. Methods of tertiary prevention include combination therapy, DOT therapy, and use of negative-pressure rooms.

IV. Chemical Substance Abuse
A. Introduction
1. Chemical substances used for nonmedical purposes may cause problems even before all of the criteria for addiction are met; the broader terms "chemical dependency" and "substance abuse" have come into general use.
2. Substance abuse implies both physical dependence (including tolerance) and psychologic dependence on the use of chemicals to modify mood and performance and to escape from anxiety.
3. Abuse of alcohol and abuse of illegal drugs are discussed below; cigarette smoking, which is discussed in other chapters, is also properly considered a form of chemical dependency or substance abuse.

B. Abuse of alcohol
1. The most commonly abused substance is alcohol.
2. Among persons who are heavy drinkers (including alcoholics), the median estimate of the relative risk for a variety of diseases is high: almost 8 for cirrhosis of the liver; about 4 for suicides, accidents, and cancer of the upper digestive and respiratory tracts; and about 1 for stroke. The overall relative risk for mortality is slightly over 2.
3. Primary prevention
    a. The social environment in which one is raised may influence alcohol use. An individual raised without using alcohol may avoid its use for life. The promotion of healthy life-styles in which alcohol use is avoided or controlled is largely a function of families and small social groups such as churches.
    b. Reducing legal accessibility of alcohol to young people tends to reduce alcohol use, but this probably is less effective than are family values.
    c. Limiting the promotion of alcohol in the media may decrease the desire of young people to start using alcohol.
    d. Raising the tax on liquor reduces the amount of alcohol purchased, but it is not clear that it reduces rates of alcoholism.
4. Secondary and tertiary prevention
    a. The long-term treatment of people who are alcohol abusers is difficult.
    b. There are good medications for treating the acute effects of withdrawal, but long-term success often depends on a lifetime commitment to treatment, such as is encouraged in the 12-step program of **Alcoholics Anonymous.**
    c. For some alcoholics, a major change in personal commitment, such as a religious conversion, may be effective.

C. Abuse of illegal drugs
1. Introduction
    a. The effects of **psychoactive drugs** such as cocaine and heroin are due as much to the rate at which the blood level of the drug is increased as they are to the blood level finally achieved (Zahler et al. 1982).
        (1) A route of administration that causes a rapid rise in the blood level of a euphoria-producing drug causes more euphoria than one that delivers the same amount of drug slowly to the bloodstream.
        (2) A drug will have more psychic effect at any blood level while the level is rising than while it is falling.
    b. An upsurge of violence began in the USA in 1986 as a result of the influx of freebase cocaine, called **"crack"** because of the crackling sounds it makes when being manufactured or smoked (Allen and Jekel 1991).
        (1) Crack, the most dangerous form of cocaine, is inhaled while it is heated.
        (2) Inhaled crack is absorbed very rapidly, causing a strong euphoria, and is rapidly addictive.
        (3) In contrast to heroin addicts, crack addicts are at least as dangerous to others when they are high on the drug as when they are seeking money to get high. This is because crack is a stimulant, whereas heroin is a sedative.
        (4) While high on crack, a user may have strong feelings of paranoia and may be in physical danger of myocardial infarction, stroke, or seizures.
    c. Powdered cocaine (the hydrochloride) may be mixed with heroin before intravenous injection; this is an extremely dangerous mixture called a "speedball." To some extent, the stimulant effect of cocaine is counteracted by the depressant effect of heroin, but in high doses, both depress respiration.
2. Primary prevention
    a. Demand reduction involves educating people about the dangers of drugs and also involves the treatment of drug abusers.
    b. Supply reduction implies the reduction of the amount of drug imported, as well as the control of selling of drugs on the street.
    c. It is more difficult to market large amounts of drugs if banking practices and export regulations make it problematic to "launder" the cash obtained from drug deals.
3. Secondary and tertiary prevention
    a. The treatment for illegal drug abuse may be thought of in four stages: assessment, abstinence initiation, relapse prevention, and follow-up.
    b. Abstinence initiation often involves individual psychotherapy as well as medications that suppress craving.
    c. Relapse prevention is best achieved with the help of group therapy.
    d. Long-term success depends on the individual joining support groups and developing new habit patterns.
        (1) For some drug abusers, the 12-step approach to treatment, as embodied in **Narcotics Anonymous,** is helpful.
        (2) A reorientation of the abuser's life, as with religious conversion, increases the likelihood of success.

V. Mental Health
A. In the USA, the most direct evidence of an increase in mental health problems is seen in the rapidly rising rates of major **depression;** 40% of the people born after 1955 have suffered from at least one episode of major depression by the age of 20 years.
B. The reasons for this increase are uncertain but may include increasing urbanization, mobility, and social anomie; changes in family structure (e.g., increases in the number of divorces and single-parent families); changing gender roles in employment and families; and increasing use of drugs and alcohol.
C. Although a discussion of the methods to detect and treat mental illness is beyond the scope of this book, it is important to note that a variety of methods of

primary prevention have been developed with the goal of promoting mental health and providing education and emotional support during times of stress.

VI. Injuries
A. Introduction
1. The impact of injuries is often described in terms of **years of potential life lost (YPLL)**.
2. In the USA, injuries are the leading cause of YPLL before the age of 65 years (see Centers for Disease Control and Prevention 1992).
3. Injuries can be categorized as follows: automobile crashes, home incidents (e.g., falls, burns, poisonings, electrocutions, drownings), occupational incidents, homicides, suicides, and miscellaneous injuries (e.g., plane and train crashes, building collapses).

B. Automobile crashes
1. Risk factors in the preinjury phase
   a. Human factors
      (1) New drivers, young drivers, and drivers suffering from alcohol intoxication, drug intoxication, fatigue, or a combination of these factors are at increased risk for automobile crashes.
      (2) In new drivers, the excess risk of automobile crashes is related to the inability to anticipate and prevent developing hazards, as well as the inability to recognize existing hazards and respond to them quickly and appropriately.
      (3) People who start driving during their teenage years may be at increased risk not only because of their driving inexperience but also because of several "immaturity factors" commonly associated with adolescence: a sense of invulnerability, a refusal to be warned about hazards, and a tendency to let the mind wander and act less cautiously (including taking the eyes off the road) when friends are nearby.
      (4) **Driving while intoxicated (DWI)** with alcohol or drugs interacts with other factors, such as fatigue, and reduces sensory input to increase the risks late at night.
      (5) Dozing and fatigue have been found responsible for many vehicle crashes, including many involving trucks.
   b. Vehicle factors
      (1) The ability of vehicles to brake and other aspects of vehicle construction and maintenance may influence the risk of injuries.
      (2) Research has demonstrated that a tail-light pattern involving two lower red lights at the sides, plus one higher red light in the center of the vehicle, catches the attention of drivers best and reduces rear-end collisions. All new vehicles sold in the USA now have this tail-light pattern.
   c. Environmental factors
      (1) Driving must be slowed during periods of rain, snow, or poor visibility, and this is not always done.
      (2) Poor design and maintenance of roads and highways also increase the risk of vehicle crashes.
2. Risk factors in the injury phase
   a. Human factors
      (1) Seat belt use in automobiles increases resistance to injury.
      (2) Helmet use by motorcycle and bicycle riders reduces the risk of head injury.
   b. Vehicle factors
      (1) Vehicle design has been steadily improving under the influence of federal regulations.
      (2) Vehicle safety features include collapsible steering columns, energy-absorbing construction, in-door side protection, seat belts and air bags, and protected gasoline tanks.
   c. Environmental factors
      (1) The object into which a vehicle crashes helps to determine the seriousness of the crash.
      (2) Energy-absorbing barriers on the shoulder of the road reduce the risk of vehicles going off the road, and median strip barriers reduce injuries from head-on collisions.
3. Risk factors in the postinjury phase
   a. Human factors influencing the fate of crash victims include the ability of individuals at the scene to summon medical help quickly and to help prevent other vehicles from becoming involved in the crash.
   b. Vehicle factors influencing the outcome of a crash include the construction of a vehicle and the extent to which it absorbs energy in a crash while maintaining the integrity of the passenger cage.
   c. Environmental factors
      (1) The extent of injury is influenced by the rapidity and quality of the emergency response. The **advanced life support (ALS)** teams seek to stabilize the condition of injured persons at the crash scene before transport.
      (2) Helicopter ambulance systems appear to improve outcomes, in part because they carry injured persons to trauma centers rather than to the nearest emergency room, which may not be adequately equipped for serious trauma.
4. Surveillance and prevention
   a. One of the most important factors in prevention is improved data on the nature of injuries, the rate at which they occur, and the circumstances under which they occur.
      (1) The **Fatal Accident Reporting System** was developed by the National Highway Traffic Safety Administration and provides valuable epidemiologic data.
      (2) Other injury surveillance systems depend on the use of the E-codes in the **International Classification of Diseases (ICD)** and the use of hospital emergency department and admission diagnoses.
   b. A variety of methods of primary, secondary, and tertiary prevention have been devised to prevent serious injuries from automobile crashes.
      (1) Examples of primary prevention include improvements in driver training, the passage and enforcement of laws concerning driving under the influence of alcohol or drugs, the construction and maintenance of good roads and highways, and the modification of automobiles to make them easier to control under hazardous conditions and to optimize passenger safety in the event of a crash.
      (2) Secondary prevention includes testing of each driver's skill and vision before a license is issued.
      (3) Tertiary prevention includes developing and using effective methods of transporting and caring for victims of automobile crashes so as to limit the degree of impairment they suffer.

C. Common injuries in the home
1. The victims of **poisoning** are usually toddlers and preschool children, who experiment with tasting or swallowing substances that they encounter while exploring. Preventive methods include child-proof caps for containers of medicines and household products; counseling parents to keep cleaning solutions, pesticides, medicines, and other hazardous substances out of the reach of their children; and establishing poison control centers.

2. **Fires** are one of the most common causes of injuries in the home.
   a. Some improvement has been achieved by tightening building codes, particularly the requirement for hard-wired smoke alarms in houses.
   b. The reduction in the prevalence of cigarette smoking has reduced one source of fires, but arson is still common, either for insurance or for revenge.
3. Although people of all ages can be the victims of **falls,** older people are at greater risk of serious injury.
   a. In younger persons, falls are likely to be associated with activities such as climbing ladders, shoveling snow, or walking on a surface covered with ice.
   b. Among older people, falls are frequently due to failing vision, loss of equilibrium or physical strength, or use of medications that decrease stability.
   c. Architectural modifications, such as the provision of handrails in hallways and on stairs, can reduce the incidence of falls in the elderly.
4. **Drowning** occurs most often among school-age children, especially boys. Swimming lessons and water safety instruction at an early age may reduce the number of deaths and injuries associated with activities that occur in and near pools and other bodies of water.

VII. Occupational Health
A. Surveillance of occupational injuries and diseases
1. The surveillance of occupational injuries and diseases is as critical to their prevention as is the surveillance of infectious diseases.
2. Federal efforts in this area were advanced when the **Occupational Safety and Health Administration (OSHA)** and the **National Institute for Occupational Safety and Health (NIOSH)** were established in 1970.
3. The surveillance of occupational diseases is not as adequate or as well understood as the surveillance of infectious diseases (see Centers for Disease Control 1990). There are many reasons for this, including the difficulty of recognizing many occupational diseases and an incomplete understanding by many physicians of their reporting role in occupational illness.
4. Surveillance of occupational diseases serves the same functions as does the surveillance of other diseases, including establishing a background rate, determining significant increases in disease rates, and setting disease control priorities.
5. Currently, the majority of states have mandatory reporting requirements for occupational injuries and diseases, although these requirements, including the types of conditions that must be reported, vary somewhat from state to state.
6. Other sources of reporting certain occupational conditions include laboratories, workers' compensation programs, some industries, and, occasionally, death certificates.
7. A **sentinel health event** is a preventable event (death, disease, or impairment) that can be used to identify a problem in the system of prevention, detection, or treatment of occupational health problems.

B. Work-related injuries
1. Collisions are a common risk to truck drivers, taxi drivers, and other vehicle operators.
2. Falling objects represent a frequent threat to construction workers, longshoremen, and movers.
3. The loss of limbs or digits is a risk of working with agricultural or manufacturing machines.
4. Explosions, fires, and discharge of firearms are threats faced by miners, firemen, and police, respectively.
5. Less striking, but increasingly common, are repetitive motion injuries and back injuries.
   a. Repetitive motion injuries, such as carpal tunnel syndrome, can affect musicians, typists, assembly line workers, and others.
   b. Back injuries are a risk not only to employees whose jobs require lifting and other forms of intense physical labor but even to employees with sedentary office jobs.

C. Work-related diseases
1. Skin diseases
   a. Dermatologic diseases are the most frequent type of occupational disease.
   b. Among the causes of occupational disease are exposure to chemicals, microorganisms, and physical agents (e.g., heat, cold, vibration). Eighty percent of occupational skin diseases are due to exposure to chemicals.
      (1) Either direct irritation or an allergic reaction to a chemical can result in contact dermatitis.
      (2) Oils and greases frequently irritate the skin, causing lesions to become infected.
      (3) Exposure to bacteria or fungi is a frequent cause of occupational skin diseases in persons who work with animals, fish, food, or soil.
      (4) Phototoxic dermatitis and photosensitization are problems that sometimes occur in people who work outdoors without protection from the sun.
      (5) Exposure to severe cold or severe vibration may result in Raynaud's phenomenon (white finger disease), and exposure to temperature extremes may also result in burns, frostbite, trenchfoot, or dermatitis.
      (6) The skin is frequently abraded, penetrated, or cut by physical injury from falls or the use of tools.
   c. A predisposing skin condition makes an occupational skin disease more likely.
2. Lung diseases
   a. Occupational exposure to dusts, gases, fumes, mists, or vapors may cause acute or chronic disease, depending on factors such as the number, duration, and intensity (level) of exposures.
   b. Obstructive lung disease
      (1) **Occupational asthma** is the most prevalent occupational lung disease in developed countries today. Work-aggravated asthma is found in people with preexisting asthma, whereas occupational asthma is due strictly to occupational exposure. When there is no latent period in occupational asthma, it is relatively easy to determine the occupational origin of disease. It is more difficult to determine the causative agent if asthma has its onset months or years after the occupational exposure. Among the many agents that can cause occupational asthma are animal products, plant products, wood dusts, metal dusts, soldering fluxes, drugs, and organic chemicals. Isocyanates, which are used in the manufacture of polyurethane and in many paints, are one of the most common categories of causative agent and are frequently responsible for occupational asthma in automobile painters.

(2) **Byssinosis** is a respiratory disease caused by the dust formed during the processing of cotton. Although it is characterized by bronchoconstriction, with chest tightness and shortness of breath, it is not typical asthma. The symptoms are thought to be due to a toxic effect on the bronchi, rather than to an immunologic reaction; one reason for this supposition is that byssinosis can occur the first time a person is exposed to cotton dust. The symptoms of byssinosis are especially severe after a period of time away from the dust, such as after a vacation or even after a weekend, but they abate somewhat upon continued exposure during the week. Byssinosis may not result in permanent lung damage if the duration and intensity of exposure are limited. It can, however, cause severe disability in people with long-term exposure, particularly those who smoke or have chronic bronchitis. A disease similar to byssinosis sometimes occurs in workers who process flax or hemp.

c. Interstitial lung disease
(1) **Asbestosis, silicosis,** and **coal worker's pneumoconiosis** are examples of interstitial lung diseases associated with the inhalation of dusts.
(2) Chronic inhalation of asbestos or free silica causes the gradual development of fibrosis in the lung.
(3) Inhaled asbestos can also produce two highly fatal cancers, lung cancer and mesothelioma.
(4) Coal worker's pneumoconiosis (black lung disease) can cause a massive, diffuse fibrosis, the major symptom of which is progressive shortness of breath.
(5) The degree of disease and rate of progression are related to the type of dust and to the intensity and duration of exposure.

d. **Hypersensitivity pneumonitis** is an inflammatory response of the lungs to inhaled organic agents (usually bacteria or fungi) found in a variety of work settings. One example is farmer's lung, which is associated with inhaling moldy hay dust.
(1) The names of many types of hypersensitivity pneumonitis indicate the occupation at risk (e.g., cheese handler's lung, grain handler's lung, pigeon-breeder's lung).
(2) Hypersensitivity pneumonitis affects the alveoli and respiratory bronchioles, rather than the bronchi, and usually wheezing is not prominent.
(3) In the early stages, removal of the affected person usually results in complete resolution; chronic exposure, however, may lead to permanent lung damage.

e. The most prominent occupational disease characterized by lung **granuloma** is **berylliosis,** which can be impossible to distinguish from sarcoidosis without measuring tissue levels of beryllium or performing other specialized tests.

3. Liver and kidney diseases
a. Liver disease
(1) The liver is heavily involved in detoxifying absorbed substances, particularly non–water-soluble chemicals absorbed through the gastrointestinal tract.
(2) Occupational exposure to chemicals such as chlorinated hydrocarbons, halogenated aromatics, nitroaromatics, ethanol, vinyl chloride, and epoxy resins can cause acute or subacute toxic hepatitis or fibrosis.
(3) Occupational exposure to hepatitis virus, as sometimes occurs in health care workers, can cause viral hepatitis.

b. Kidney disease
(1) The kidneys are frequently damaged by water-soluble toxins and metals because of their role in the excretion of water-soluble wastes.
(2) Acute tubular necrosis can occur as a result of acute exposure to divalent metals (e.g., mercury, cadmium, and chromium), halogenated hydrocarbons, other hydrocarbons, arsine, and other compounds.
(3) Most cases of work-related chronic renal disease are caused by long-term exposure to metals (especially mercury, lead, and cadmium).

4. Eye damage and hearing loss
a. Most occupational eye damage is due to chemical burns, radiation effects, or mechanical injuries, including lacerations, contusions, and damage to the eye from fractures to surrounding skull bones.
b. Cataracts or corneal damage can result from ionizing radiation (especially to the lid) and from ultraviolet radiation.
c. Workers exposed to the sun a great deal, such as those in the fishing trade, have an increased risk of cataracts.
d. Loud noise on the job may cause loss of high tone hearing, with considerable functional deafness in later life.
e. Chronic exposure to noise at a level of 85 dB frequently results in hearing problems, and many workers are regularly exposed to a level higher than 85 dB at work.
(1) The decibel (dB) scale is a logarithmic scale of sound pressure levels; according to this scale, normal human hearing goes from near 0 to about 120 dB.
(2) The greatest hearing loss tends to occur at a frequency of 3000–4000 Hz, which is a crucial range for hearing the consonants of human speech (for example, f, p, s, and t) and therefore for understanding speech.

D. **Prevention of occupational injuries and diseases**
1. Safety in the workplace demands control over the environment, including providing a drug-free workplace, protective equipment (e.g., safety goggles, hard hats, ear protection devices), tools and equipment in good working order, proper training for workers, and adequate rest periods.
2. As research in the field of **ergonomics** has indicated, prevention of many occupational health problems, including repetitive motion injuries and back injuries, can be accomplished by adjusting the occupational environment to the needs of the workers.
a. In offices, the use of properly designed computer keyboards and the correct placement of screens for each worker can reduce the risk of repetitive motion injuries.
b. In factory warehouses, safe lifting of manufactured goods requires the proper training of individual workers, but well-designed aids to lifting, such as sturdy handles on objects to be lifted, can also help prevent back injuries.
3. The prevention of occupational skin diseases usually can be accomplished by eliminating the causative biologic, physical, or chemical agent or by providing barrier protection (e.g., gloves or protective uniforms) so that the causative agent does not touch the skin.
4. Prevention of a variety of occupational diseases, including those affecting the skin, lung, liver, and kidney, requires preventing or minimizing exposures to toxic agents, including those listed in Table 19–1 and discussed below (see Exposure to Toxins).
5. Appropriate training and equipment for the prevention of contact with infectious agents should be provided to health care workers to prevent infections; the prevention of some infectious diseases, such as

TABLE 19-1. Routes and Effects of Exposure to Toxins That Are Often Found in the Workplace

Toxin	Routes and Effects of Exposure
Metals*	
Arsenic	May enter via the lungs, skin, or gastrointestinal tract. Arsenic compounds are used as insecticides and weed killers. Cause respiratory and gastrointestinal symptoms and, in high doses, can cause death. Can cause lung cancer.
Beryllium	Usually enters via the lungs. Causes granulomas in the lungs; lesions appear similar to those in sarcoidosis.
Cadmium	Usually enters via the lungs. Displaces zinc in enzyme systems, often damaging the renal tubules. Causes metal fume fever.
Lead	Usually enters via the gastrointestinal tract or lungs. Displaces calcium in chemical reactions. Inorganic form of lead causes gastrointestinal and neurologic symptoms. Organic compounds of lead cause diffuse neurologic symptoms.
Mercury	May enter via the lungs, skin, or gastrointestinal tract. Used in the past by hatmakers to make felt for hats. Chronic exposure damages the central nervous system, with the elemental form of mercury tending to cause tremors ("hatter's shakes") and the organic forms tending to cause psychiatric symptoms ("mad as a hatter") and even dementia. Mercury also damages the kidneys.
Zinc	Usually enters via the lungs. Inhaling zinc oxide causes metal fume fever (a disorder that also can be caused by other metal fumes).
Insecticides, herbicides, and fungicides	
Organophosphates	Usually enter via the skin from handling, but can enter via the lungs. Block acetylcholinesterase and produce both central nervous system and peripheral nerve damage.
Pentachlorophenol	Usually enters via the skin. Used as a wood preservative. Interferes with cellular respiration. Causes anorexia and respiratory symptoms and, in high doses, can cause coma and death.
Polychlorinated biphenyls	Enter via the skin or lungs. Are teratogens and possibly also carcinogens.
Hydrocarbon solvents†	
Benzene	Absorbed through the lungs and skin. Is a lipid-soluble aromatic solvent that is used widely in industry. Chronic exposure can result in suppression of the bone marrow, with a possible end result of aplastic anemia.
Carbon tetrachloride	Absorbed readily through the lungs. Is a lipid-soluble chlorinated hydrocarbon that is not used much in industry because it is extremely toxic to kidneys and liver, but it is the prototype of this class of chemical. Damage to either the kidneys or the liver can predominate. Renal tubular necrosis may follow acute exposures, and hepatic centrilobular necrosis tends to predominate in chronic exposure, especially in the presence of ethanol or following hepatic damage by ethanol.
Toluene	Usually inhaled. Is a lipid-soluble aromatic solvent found in products such as glue. Primarily causes central nervous system effects, including hallucinations (which is why glue is sometimes sniffed).
Asphyxiants‡	
Carbon dioxide	Enters the body via the lungs and stimulates the respiratory center. Is a nonreactive asphyxiant. Begins to produce symptoms of rapid breathing when concentration reaches about 3% in the air. In high concentrations, causes coma and death.
Carbon monoxide	Enters the body via the lungs. Is a chemical asphyxiant that is ubiquitous in urban society (a product of automobile exhausts and sometimes of poorly ventilated space heaters). Combines with hemoglobin to form carboxyhemoglobin; when 50% or more of the hemoglobin is in the form of carboxyhemoglobin, fainting and death are likely. Smokers may inhale some carbon monoxide from smoking.
Hydrogen cyanide	Enters the body via the lungs or gastrointestinal tract. Is a chemical asphyxiant. Toxicity is retained in cyanide salts, because it is the reactive cyanide moiety that interferes with cytochrome oxidase. Causes headaches, rapid breathing, and, frequently, death.
Hydrogen sulfide	Absorbed through the lungs. Is a chemical asphyxiant that is as dangerous as hydrogen cyanide, but its smell tends to give warning of its presence before hazardous levels develop. Causes symptoms and effects similar to those of hydrogen cyanide.
Methane	Enters the body via the lungs. Is a nonreactive asphyxiant that is mainly a problem in mines, causing severe respiratory symptoms. The greater danger now is from explosion.
Nitrogen	Enters the body via the lungs. Is a nonreactive asphyxiant that used to be a problem in mines but now presents a hazard primarily to deep sea divers. Divers accumulate nitrogen in the fatty tissues during dives at high pressure. As divers approach the surface, nitrogen reenters their blood, and if the pressure is reduced too fast, it forms small bubbles in the blood, which interfere with circulation, especially to the brain. This process causes the "bends."
Miscellaneous organic compounds	
Resins	Usually enter via the skin, although may enter via the lungs. Produce asthma, irritation and allergic sensitization of the skin, and irritation of the eyes.
Vinyl chloride	Absorbed through the lungs and skin. Is ubiquitous in industry because it is used to make plastics. Can cause sclerodermatous skin lesions, Raynaud's phenomenon, and bone lesions in the hand, in addition to liver damage. The monomer form causes hemangiosarcoma of the liver in a small proportion of persons who are exposed.

* Metals often cause toxicity by interfering with the action of other metals as cofactors in enzyme reactions.
† Lipid-soluble hydrocarbon solvents build up in the body in tissues with high levels of lipids (e.g., the central nervous system), and they all have narcotic effects. Their central nervous system effects are exacerbated by alcohol. Exposure to more than one solvent may result in complex interactions.
‡ Asphyxiants can be divided into two groups: (1) gases that have no direct toxic effect but can reduce the partial pressure of oxygen in the lungs to dangerous levels and (2) gases that interfere with respiration at the cellular level. Nonreactive asphyxiants do not enter into chemical reactions; they reduce the partial pressure of oxygen in the lungs.

hepatitis, can be accomplished by the use of vaccines (see Chapter 16).

E. Health promotion in the workplace
1. Workers in a company tend to be concentrated at certain places and, with the support of the company, industry, or agency, the workers can be reached for education, prevention, and health-promoting services.
2. Activities designed to promote good health in employees often include health education about nutrition, exercise, smoking cessation, and weight reduction.
3. In some cases, companies offer fitness programs and the use of a fitness center; limit the areas in which employees can smoke on the premises; provide immunizations (e.g., influenza and hepatitis B vaccines); and sponsor specific screening programs

(e.g., for hypertension or high cholesterol levels) or the more comprehensive health risk assessments (see Chapter 17).
4. Random screening programs for alcohol or illegal drug abuse are sometimes introduced for certain critical employees, such as pilots, air traffic controllers, or railroad engineers, but it is difficult to justify such activity for all employees.
5. Some companies have developed **employee assistance programs (EAPs)** to provide counseling and treatment services for emotional problems and substance abuse.
   a. These programs are free to the workers and usually have a high degree of confidentiality guaranteed.
   b. Their goal is to increase the attendance and productivity of the workers.

F. Company support for preventive activities
1. Preventive activities are generally of interest to a company when they can be shown to be cost-effective by reducing worker turnover, days missed from work, and medical care costs.
2. The health information and good health habits learned in the workplace may be carried home to improve the health of an entire family.

G. Occupational health regulations
1. The **Occupational Safety and Health Administration (OSHA)** establishes federal standards for exposure to occupational hazards, investigates complaints or reports of problems in the workplace, and enforces the federal standards and regulations in the workplace. OSHA has the right to issue fines or citations when it finds violations of federal laws.
2. Most states have occupational health units in their state health departments or elsewhere in the state government. These have the obligation to receive and investigate complaints and disease reports and, within the limits prescribed by state laws, to enforce workplace changes.

H. Workers' Compensation
1. Workers' Compensation, a mandatory insurance program in each state, provides for medical care and partial replacement of wages for workers with occupational diseases or injuries.
2. Companies, agencies, and businesses are required to purchase a policy or to deposit premiums into a fund; by this mechanism, the companies are liable without fault being assigned to the company or to the worker, and the worker's ability to sue for further damages is markedly limited.
3. The benefits are paid according to the severity and duration of the illness or injury, which is usually defined by law (see Chapter 18 for a discussion of the four major categories of disability). Physicians play an important role in defining the level of impairment due to the occupational injury or disease.

I. Medical ("health") insurance
1. In the USA, most persons under the age of 65 years obtain medical insurance through their place of work.
2. There is increasing demand on the part of companies for a limitation in their costs for such insurance. As a result, more and more companies are requiring managed care plans (see Chapter 21), requiring workers to pay a part of the costs of the insurance, or not offering health insurance at all.
3. The strategy of not offering health insurance is frequently used by businesses that have a rapid turnover of workers (such as restaurants) or operate on a slim profit margin.
4. In the USA, an estimated 40 million people are without medical insurance because of unemployment or the lack of work-related medical insurance plans.

J. Diagnosis and treatment of occupational health problems
1. Traditionally, large companies had in-house physicians who were frequently in the difficult ethical position of depending on the company for their salary but believing that they were ethically responsible to advocate for the welfare of their patients.
2. Today, many companies send their workers to outside occupational medicine practices, whose physicians can be somewhat more objective in their clinical evaluations.

VIII. Exposure to Toxins

A. Forms of toxins
1. Toxins can be solids (such as plastics), particulate matter (such as dusts, fumes, or fibers), gases (including vapors or mists), or liquids.
2. A **dust** consists of very fine solid particles of a larger solid (e.g., a rock or piece of coal or wood) and is created by crushing or sanding mechanisms.
3. A **fume** consists of solid particles that develop by condensation from the gases given off by heating metals or plastics.
4. A **gas** is a chemical that normally exists in the gaseous state at room temperature (such as oxygen or helium).
5. A **vapor** is a gas from the evaporation of a liquid such as gasoline.
6. A **mist** is formed when a liquid is aerosolized, as with an atomizer, forming fine droplets of the liquid.
7. Table 19-1 lists some of the more important toxins.

B. Primary prevention
1. The prevention of toxic effects from a substance found in the environment or the workplace requires knowledge about what levels of the substance are considered safe, both for acute, short-term exposure and for chronic, long-term exposure.
2. Prevention of toxic exposures also requires accurate measurement and surveillance of environmental or workplace levels of the substance.
3. If toxin levels approach the threshold for harm, the options are as follows:
   a. Production or use of the toxin can be reduced (e.g., dry cleaners have tended to switch from using carbon tetrachloride to using less toxic solvents such as tetrachloroethylene).
   b. The substances within the industrial processes can be contained.
   c. The substances can be removed quickly from the working or living environment (e.g., by ventilation) before they can cause injury.
   d. Workers can be protected with special equipment (e.g., ventilatory masks).
4. Increasing knowledge of the risks of environmental exposures, particularly the possibility of fetal damage, has raised new ethical and legal liability questions. It is not clear, for example, whether a company that initially sought to exclude pregnant women from doing certain jobs and was prohibited by law from doing so would later be liable if fetal injury occurred. The company would, however, still be liable for any violations of the Occupational Safety and Health Administration (OSHA) regulations.
5. Recent national laws have required that companies make public the potentially toxic substances used

in the workplace. These laws have made the process of identification and control within a given area somewhat easier.

C. Secondary and tertiary prevention
1. Examples of screening include periodic follow-up tuberculin skin tests for health care workers with negative results in the initial tuberculin test; periodic chest x-rays for health care workers with positive results in the tuberculin test; periodic pure tone audiograms to detect hearing loss in workers exposed to high levels of noise; and screening for lead in the serum of workers with unavoidable exposure to lead.
2. Radiation-sensitive badges for workers in x-ray units or nuclear power stations are more analogous to screening for a risk factor than for a disease.
3. The symptoms of toxic injury and methods of treatment vary, depending on the level of exposure and the type of toxic substance to which the injured person was exposed.

QUESTIONS

DIRECTIONS. (Items 1–10): Each of the numbered items or incomplete statements in this section is followed by answers or by completions of the statement. Select the ONE lettered answer or completion that is BEST in each case. Correct answers and explanations are given at the end of the chapter.

1. Folic acid supplementation is important in the prevention of

   (A) byssinosis
   (B) intrauterine growth retardation
   (C) pellagra
   (D) neural tube defects
   (E) preeclampsia

2. Which one of the following interventions constitutes secondary prevention as a component of prenatal care?

   (A) the provision of food vouchers through the WIC program
   (B) screening for, and treatment of, gestational diabetes
   (C) screening for rubella antibodies
   (D) pelvic ultrasound examination in the second trimester of pregnancy
   (E) a nonstress test prior to parturition

3. Worldwide, the number of people infected with HIV-1 is thought to be approximately

   (A) 20 million
   (B) 200,000
   (C) 2 million
   (D) 200 million
   (E) 5 million

4. By the turn of the twenty-first century, the number of people dying each year of AIDS worldwide is expected to be approximately

   (A) 500,000–700,000
   (B) 20,000–30,000
   (C) 1.5–3 million
   (D) 15 million
   (E) 150,000–200,000

5. In the USA, the leading cause of death among men age 25–44 years is

   (A) homicide
   (B) coronary artery disease
   (C) asthma
   (D) seminoma
   (E) AIDS

6. In central Africa and Southeast Asia, the spread of HIV infection is principally a result of

   (A) heterosexual intercourse
   (B) homosexual intercourse
   (C) intravenous drug use
   (D) a contaminated blood supply
   (E) arthropod vectors

7. The leading cause of death in industrialized nations during the nineteenth century was

   (A) dropsy
   (B) industrial accidents
   (C) tuberculosis
   (D) malaria
   (E) influenza

8. The incidence of tuberculosis in the USA is best described by which one of the following statements?

   (A) The incidence rate has declined steadily since the end of the nineteenth century.
   (B) The incidence rate has risen steadily since the middle of the nineteenth century.
   (C) The incidence rate rose until the 1940s and then began to decline.
   (D) The incidence rate declined until the 1940s, then rose sharply for several decades, and then began to fall in 1985.
   (E) The incidence rate declined from the end of the nineteenth century until 1985, when it rose suddenly, and then it began to decline again in 1993.

9. A Gohn complex on a chest radiograph is evidence of

   (A) active tuberculosis
   (B) latent tuberculosis
   (C) reactivation tuberculosis
   (D) immunodeficiency
   (E) anergy

10. In the presence of immunodeficiency, what may be the result of testing with the tuberculin skin test (PPD)?

(A) It may produce a false-positive result.
(B) It may produce primary infection.
(C) It may cause reactivation of latent disease.
(D) It may produce a false-negative result.
(E) This test may be used reliably in people under age 35 years.

ANSWERS AND EXPLANATIONS

1. The answer is D: neural tube defects. It is now well established that supplemental folic acid (folate) taken at the time of conception and during early embryogenesis substantially reduces the risk of neural tube defects (Daly et al. 1995). The only other choice specifically related to a micronutrient is pellagra, which is a result of niacin deficiency. *Epidemiology*, p. 234.

2. The answer is B: screening for, and treatment of, gestational diabetes. Secondary prevention is the early detection of generally asymptomatic disease and the prevention of adverse sequelae. When gestational diabetes is detected, the disease is already present, so primary prevention is not feasible. Meticulous treatment to maintain a nearly normal serum glucose level at all times prevents adverse consequences of the condition; this is an example of secondary prevention. The provision of food vouchers through the WIC program is a health promotion, or primary prevention, effort. Screening for rubella antibodies, and immunization of patients lacking them, is primary prevention of rubella. Routine pelvic ultrasound studies in the second trimester may be informative but have not been shown to alter the outcome (Ewigman et al. 1993). A nonstress test prior to parturition helps the obstetrician predict and manage the course of labor and delivery. *Epidemiology*, p. 234.

3. The answer is A: 20 million. Currently available data suggest that approximately 20 million people worldwide are infected with HIV-1 (US Public Health Service 1992). The quality of data from certain parts of the world is dubious, however. The incidence rate is relatively stable in the USA but is rising steadily in parts of Southeast Asia and sub-Saharan Africa. *Epidemiology*, p. 235.

4. The answer is C: 1.5–3 million. Projections of HIV-related deaths in the future are based on incidence data available currently. If such data are inaccurate because of underreporting in certain parts of the world (or other reasons), the mortality estimate may be too low. Advances in treatment, such as the recently approved class of drugs known as protease inhibitors, raise the prospect that mortality rates will instead decline in the future. *Epidemiology*, p. 235.

5. The answer is E: AIDS. Since its recognition in 1981, AIDS has become the leading killer of males age 25–44 years in the USA (US Public Health Service 1992). *Epidemiology*, p. 235.

6. The answer is A: heterosexual intercourse. In the USA, AIDS is spread by homosexual and heterosexual intercourse and by use of intravenous drugs. Use of intravenous drugs is beginning to be the predominant manner in which the disease is spread. In parts of the world where the prevalence of HIV infection is similar in men and women, its spread is principally a result of unprotected heterosexual intercourse. *Epidemiology*, p. 235.

7. The answer is C: tuberculosis. *Mycobacterium tuberculosis* has been a human pathogen for centuries but became a scourge in Europe only at the time of the industrial revolution in the nineteenth century. Contributing factors were thought to include population density, poor sanitation, and crowded working conditions. *Epidemiology*, p. 236.

8. The answer is E: The incidence rate declined from the end of the nineteenth century until 1985, when it rose suddenly, and then it began to decline again in 1993. The sudden surge in incidence rates of tuberculosis in the nineteenth century has been ascribed to the industrial revolution and related factors discussed in the chapter. Improvements in sanitation toward the end of the last century as well as some degree of acquired resistance to tuberculosis caused incidence rates to begin falling before the start of the twentieth century. Incidence fell at a variable pace until 1985, when immunodeficiency associated with HIV infection began to facilitate the transmission of tuberculosis. The incidence rate of tuberculosis rose in the USA between 1985 and 1993, and this rate included an alarming rise in the transmission of multiple drug–resistant serotypes (Brudney and Dobkin 1991). The implementation of more aggressive control and surveillance measures may be responsible for a declining incidence rate over the past several years, first detected in 1993. *Epidemiology*, p. 236.

9. The answer is B: latent tuberculosis. The Gohn complex, which is calcified granulomas in the apices and upper lobes of the lungs seen on chest radiograph, develops following containment of primary tuberculosis infection by the cell-mediated immune system. This radiographic finding is evidence of infection in a latent state. Latent disease may lead to symptomatic, communicable tuberculosis, known as reactivation tuberculosis, as a result of immunosuppression due to illness, drugs, environmental stresses, and aging. *Epidemiology*, p. 236.

10. The answer is D: It may produce a false-negative result. The PPD (purified protein derivative) skin test for tuberculosis relies on a cell-mediated immune reaction to the injection of mycobacterial proteins in the dermis. Under conditions of immunosuppression (e.g., AIDS), the immune system is incapable of such a response even when exposure to tuberculosis has occurred. Consequently, those individuals most likely to be infected with tuberculosis are also those most likely to have a false-negative reaction to skin testing. To clarify exposure status in these individuals, an anergy panel, which tests the ability of the immune system to react to any antigen, is applied. Several antigens to which exposure is nearly universal are injected at the same time as the PPD test. A positive reaction to any of the control antigens and a negative reaction to the PPD test may be interpreted as a true-negative test for tuberculosis. When the anergy panel fails to elicit a reaction (induration in the skin), a negative PPD test must be considered unreliable, and a chest x-ray is generally indicated. *Epidemiology*, p. 236.

REFERENCES CITED

Allen, D. F., and J. F. Jekel. Crack: The Broken Promise. London, Macmillan Academic and Professional, Ltd., 1991.

Bellin, E. Y., D. D. Fletcher, and S. M. Safyer. Association of tuberculosis infection with increased time in or admission to the New York City Jail System. Journal of the American Medical Association 269:2228–2231, 1993.

Brudney, K., and J. Dobkin. Resurgent tuberculosis in New York City: human immunodeficiency virus, homelessness, and the decline of tuberculosis programs. American Review of Respiratory Diseases 144:745–749, 1991.

Centers for Disease Control. A Child-Centered Program to Prevent Tuberculosis. Publication No. (PHS) 1280. Washington, D. C., Government Printing Office, March 1965.

Centers for Disease Control. Mandatory reporting of occupational diseases by clinicians. Morbidity and Mortality Weekly Report 39:19–28, 1990.

Centers for Disease Control and Prevention. Years of potential life lost before age 65, by race, Hispanic origin, and sex: United States, 1986–1988. Morbidity and Mortality Weekly Report 41:13–23, 1992.

Clemens, J. D., J. J. Chuong, and A. R. Feinstein. The BCG controversy: a methodological and statistical reappraisal. Journal of the American Medical Association 249:2362–2368, 1983.

Daly, L. E., et al. Folate levels and neural tube defects: implications for prevention. Journal of the American Medical Association 274:1698–1702, 1995.

Ewigman, B. G., et al. Effect of prenatal ultrasound screening on perinatal outcome. New England Journal of Medicine 329:821–827, 1993.

Frieden, T. R., et al. Tuberculosis in New York City: turning the tide. New England Journal of Medicine 333:229–233, 1995.

Iseman, M. D. Treatment of multidrug-resistant tuberculosis. New England Journal of Medicine 329:784–791, 1993.

Jekel, J. F. Epidemiology, Biostatistics, and Preventive Medicine. Philadelphia, W. B. Saunders Company, 1996.

Kaplan, E. Evaluating needle-exchange programs via syringe tracking and testing (STT). AIDS and Public Policy Journal 6:109–115, 1992.

Mount, F. W., and S. H. Ferebee. Preventive effects of isoniazid in the treatment of primary tuberculosis in children. New England Journal of Medicine 265:713, 1961.

Selwyn, P. A., et al. High risk of active tuberculosis in HIV-infected drug users with cutaneous anergy. Journal of the American Medical Association 268:504–509, 1992.

US Public Health Service, Department of Health and Human Services. Strategic plan to combat HIV and AIDS in the United States. Washington, D. C., Government Printing Office, 1992.

Zahler, P., et al. Kinetics of drug effect by distributed lags analysis: an application to cocaine. Clinical Pharmacology and Therapeutics 31:775–782, 1982.

CHAPTER TWENTY

Public Health Responsibilities and Goals

SYNOPSIS

OBJECTIVES
- To define public health.
- To discuss the administrative structure of public health in the USA, and to characterize the roles and responsibilities of different branches of government.
- To discuss the goals of public health in the USA and the health goals for the year 2000, in particular.
- To characterize health objectives as health promotion, health protection, or disease prevention.

I. Definition of Public Health
  A. Public health refers to the health status of the public—that is, of a defined population (see Chapters 2 and 14).
  B. The second meaning, which is the focus of Chapters 20 and 21, refers to the organized social efforts made to preserve and improve the health of a defined population.
  1. The best-known definition of public health in terms of this second meaning was written in 1920 by C.-E. A. Winslow: "Public health is the science and art of preventing disease, prolonging life, and promoting physical health and efficiency through organized community efforts for the sanitation of the environment, the control of community infections, the education of the individual in principles of personal hygiene, the organization of medical and nursing service for the early diagnosis and preventive treatment of disease, and the development of the social machinery which will ensure to every individual in the community a standard of living adequate for the maintenance of health."
  2. Winslow's definition states the central emphasis of all public health work—namely, promoting health and preventing disease.
  3. Winslow also emphasized the diverse strategies that are required to bring this about, including environmental sanitation, specific disease control efforts, health education, medical care, and an adequate standard of living.
  4. Winslow also made clear that for these goals to be achieved, organized social action is required. This action is largely expressed in the policies of the federal, state, and local government bodies.

II. Administration of Public Health
  A. Responsibilities of the federal government
  1. **Regulation of commerce**
     a. The regulation of commerce involves controlling the entry of people and products into the USA, as well as regulating commercial relationships among the states.
     b. People who may be excluded from entry to the USA include those with particular health problems such as active tuberculosis or human immunodeficiency virus (HIV) infection.
     c. Products excluded from entry include fruits and vegetables that are infested with certain organisms (e.g., the Mediterranean fruit fly) or have been treated with prohibited insecticides or fungicides.
     d. The regulation of commercial relationships between states has caused much dissent in the USA in recent years.
        (1) Contaminated food products that cross state lines are considered to be "interstate commerce" in harmful microorganisms. Therefore, the federal government takes the responsibility for inspection of all milk, meat, and other food products at their site of production and processing.
        (2) Likewise, polluted air and polluted water that may flow from state to state are deemed to be "interstate commerce" in polluting agents and therefore come under federal regulation.
  2. **Taxation for the general welfare** is the constitutional basis for the federal government's development of most of its public health programs and agencies, including the Centers for Disease Control and Prevention (CDC) and the Occupational Safety and Health Administration (OSHA); for research programs, such as those of the National Institutes of Health (NIH); and for the payment for medical care, such as Medicare and Medicaid (see Chapter 21).
  3. **Provision of care for special groups** is a responsibility of the federal government. Such groups include

active military personnel and their families, provided for through Medicare; veterans, provided for through the Veterans' Administration hospital system; and Native Americans and Alaska natives, provided for through the Indian Health Service of the US Public Health Service.

4. **Coordination of federal agencies**
   a. In the USA, the major federal department concerned with health is the **Department of Health and Human Services** (DHHS) (see Table 20-1).
   b. The **Administration on Aging** provides advice to the Secretary of the DHHS on issues and policies regarding aging persons in the USA. It also administers certain grant programs for the benefit of the aging population.

   c. The **Administration for Children and Families** is responsible for administering child welfare programs through the states, Headstart programs, child abuse prevention and treatment programs, developmental disabilities programs, and child support enforcement.
   d. The **Health Care Financing Administration** is responsible for setting standards for programs and institutions that provide medical care, developing payment policies, contracting for third-party payers to pay the bills, and monitoring the quality of care provided. It is also responsible for administering two major programs from the Social Security Act: Medicare and Medicaid.
      (1) **Medicare,** which is covered under Title 18, pays for medical care for the elderly.
      (2) **Medicaid,** which is covered under Title 19, pays for medical and nursing home care for the poor (in cooperation with the states) (see Chapter 21).
   e. The **Public Health Service** (PHS) has seven constituent agencies:
      (1) The **Agency for Health Care Policy and Research** is the main federal agency for research and policy development in the area of medical care organization, financing, and quality assessment.
      (2) The **Agency for Toxic Substances and Disease Registry** provides leadership and direction to programs designed to protect workers and the public from exposure to and adverse health effects of hazardous substances that are kept in storage sites or are released by fire, explosion, or accident.
      (3) The **Centers for Disease Control and Prevention** (CDC) has the responsibility for directing and enforcing federal quarantine activities, working with states in disease surveillance and control activities, developing immunization and other preventive programs, conducting research and training, working to promote environmental and occupational health and safety, providing consultation to other nations in the control of preventable diseases, and participating with international agencies in the eradication and control of diseases around the world. The CDC has 11 major operating components, as shown in Table 20-1.
      (4) The **Food and Drug Administration** (FDA) is the primary agency for regulating the safety and effectiveness of drugs for use in humans and animals; vaccines and other biologic products; diagnostic tests; and medical devices, including ionizing and nonionizing radiation–emitting electronic products. The FDA is also responsible for the safety, quality, and labeling of cosmetics, foods, and food additives and colorings.
      (5) The **Health Resources and Services Administration** is responsible for improving access to health services and the equity and quality of health care.
      (6) The **Indian Health Service** provides medical care and health promotion for American Indians and Alaska natives.
      (7) The **National Institutes of Health** conduct research on particular diseases or organ systems, sponsor extramural research through competitive grant programs, and are responsible for some disease control programs and public and professional education (see Table 20-1).
   f. The **Social Security Administration** administers the national program of retirement payments for social security program recipients and disability payments for the disabled.
   g. The **Substance Abuse and Mental Health Services Administration** provides national leadership in the prevention and treatment of addictive

TABLE 20-1. Major Operating Units and Subunits of the US Department of Health and Human Services

Administration on Aging
Administration for Children and Families
Health Care Financing Administration
    Medicaid
    Medicare
    Quality Assurance
Public Health Service
    Agency for Health Care Policy and Research
    Agency for Toxic Substances and Disease Registry
    Centers for Disease Control and Prevention
        Epidemiology Program Office
        International Health Program Office
        National Center for Chronic Disease Prevention and Health Promotion
        National Center for Environmental Health
        National Center for Health Statistics
        National Center for Infectious Diseases
        National Center for Injury Prevention and Control
        National Center for Prevention Services
        National Immunization Program Office
        National Institute for Occupational Safety and Health
        Public Health Practice Program Office
    Food and Drug Administration
    Health Resources and Services Administration
    Indian Health Service
    National Institutes of Health
        Fogarty International Center
        National Cancer Institute
        National Center for Human Genome Research
        National Eye Institute
        National Heart, Lung, and Blood Institute
        National Institute on Aging
        National Institute of Alcohol Abuse and Alcoholism
        National Institute of Allergy and Infectious Diseases
        National Institute of Arthritis and Musculoskeletal and Skin Diseases
        National Institute of Child Health and Human Development
        National Institute on Deafness and Other Communication Disorders
        National Institute of Dental Research
        National Institute of Diabetes and Digestive and Kidney Diseases
        National Institute on Drug Abuse
        National Institute of Environmental Health Sciences
        National Institute of General Medical Sciences
        National Institute of Mental Health
        National Institute of Neurological Disorders and Stroke
        National Institute of Nursing Research
        National Library of Medicine
Social Security Administration
Substance Abuse and Mental Health Services Administration
    Center for Mental Health Services
    Center for Substance Abuse Prevention
    Center for Substance Abuse Treatment

Source of data: Office of the Federal Register, National Archives and Records Administration. United States Government Manual 1994/1995. Washington, D. C., Government Printing Office, 1994.

and mental disorders. Its three major operating divisions are the Center for Mental Health Services, the Center for Substance Abuse Prevention, and the Center for Substance Abuse Treatment.
  B. **Responsibilities of states**
    1. In the USA, the fundamental responsibility for the health of the public lies with the states. This authority is clarified in the Tenth Amendment to the US Constitution.
    2. In each state, there is a state health department to oversee the implementation of the **public health code,** a compilation of the state laws and regulations regarding public health and safety.
    3. As a part of its duty to ensure the health of the public, every state licenses medical and other health-related practitioners, as well as medical care institutions such as hospitals, nursing homes, and home care programs.
  C. **Responsibilities of municipalities and counties**
    1. Municipalities accept public health responsibilities in return for a considerable degree of independence from the state in running their affairs, including matters concerning property ownership and tax levies.
    2. Counties are bureaucratic subdivisions of the state created for the purpose of administering (with varying degrees of local control) state responsibilities such as courts of law, educational programs, highway construction and maintenance, police and fire protection, and health services.
    3. Local public health departments usually are administrative divisions of municipalities or counties, and their policy is established by a city or county board of health.
      a. Boards of health have the right to establish public health laws and regulations, provided that they are at least as strict as similar laws and regulations in the state public health code and that they are "reasonable." Anything that is too strict can be challenged in the courts as being "unreasonable."
      b. The courts have generally upheld local and state health department laws and regulations when they have to do with the control of communicable diseases.
      c. Neither legislatures nor courts, however, have been as supportive of laws and regulations that control human behavior when communicable diseases are not involved.
        (1) Bills requiring motorcyclists and bicyclists to wear helmets often fail to be enacted into law or are repealed following passage, despite abundant evidence of their benefits.
        (2) If an individual risk factor for disease can be shown to have a negative public impact, such as passive smoke inhalation, legislatures are supporting increasingly strict controls.
  D. **Responsibilities of local public health departments**
    1. The six primary responsibilities of local public health departments are vital statistics, communicable disease control, maternal and child health, environmental health, health education, and public health laboratories.
    2. Because the "basic six" functions are not adequate to deal with more recent public health problems such as environmental pollution, occupational toxins and safety hazards, and the increased incidence of chronic degenerative diseases, public health leaders have continued a lively debate concerning the proper functions and responsibilities of health departments at both the local and state level (see Table 20–2).
  E. **The mission of public health**
    1. The **assessment** role requires that "every public health agency regularly and systematically collect, assemble, analyze, and make available information on the health of the community, including statistics on health status, community health needs, and epidemiologic and other studies of health problems." (Institute of Medicine 1988)
    2. The **policy development** role requires that "every public health agency exercise its responsibility to serve the public interest in the development of comprehensive public health policies by promoting the use of the scientific knowledge base in decision-making about public health, . . . by leading in developing public health policy, [and by taking] a strategic approach, developed on the basis of a positive appreciation for the democratic political process." (Institute of Medicine 1988)
    3. The **assurance** role requires that "public health agencies assure their constituents that services necessary to achieve agreed upon goals are provided, either by encouraging action by other entities (private or public sector), by requiring such action through regulation, or by providing services directly" (Institute of Medicine 1988).
    4. It should be noted that the current view of "public health policy" in the USA is much narrower than that in the world public health scene. According to the Ottawa Charter for Health Promotion (1986), which guides much of the international work in this area, health promotion requires that all policies be reviewed for their health impact and adjusted to strengthen the effort to achieve good health.
  F. **An intersectoral approach to public health**
    1. Many duties with public health implications are carried out by government agencies that are not usually considered health agencies (as emphasized by the Ottawa Charter).
      a. Departments of agriculture may be responsible for monitoring the safety of milk, meat, and other agricultural products, as well as controlling zoonoses (animal diseases that can be spread to human beings).
      b. Departments of parks and recreation are responsible for the safety of water and sewage disposal in their facilities.
      c. Highway departments are responsible for the safe design and maintenance of roads and highways.
      d. Education departments have the responsibility for health education, as well as for providing a safe and healthful environment in which to learn.
      e. Government departments that have to do with promoting a healthy economy are critical, because when an economy is failing, the health of the people will falter as well.
    2. The USA is home to many voluntary health agencies, whose focus is to prevent or control certain diseases, either those of one organ system (e.g., the American Heart Association and the American Lung Association) or a related group of diseases (e.g., the American Cancer Society).
III. **Goals of Public Health**
  A. **US health goals for the year 2000**
    1. *Healthy People 2000* is a national consensus strategy of the government, public health organizations, and public-spirited citizens (see US Department of

TABLE 20-2. Types of Services Often Provided by Local Health Departments in the 1990s

Vital statistics
- Recording of birth and death certificates
- Transmittal of certificates to the state and others
- Analysis and monitoring of rates

Communicable disease control
- Immunization
- Tuberculosis control
  - Clinics for diagnosis and treatment
  - Epidemiologic follow-up, contact testing, and prophylaxis
  - Home supervision of patients and contacts (nursing)
  - Tuberculin screening programs (high-risk areas only)
- Sexually transmitted disease control
  - Clinics for diagnosis, treatment, interviewing, and prophylaxis
  - Contact tracing and cluster testing
  - Screening programs
  - Education programs
- Surveillance of disease
  - Receipt and transmittal of reports
  - Analysis of data and provision of feedback to physicians and agencies
  - Dissemination of information to physicians and clinics
- Epidemic investigation and control
  - Work with other local agencies (e.g., hospitals)
  - Work with the state health department
- Other services
  - Rabies control
  - Screening for parasites
  - Screening of food handlers and persons having extensive contact with children

Maternal and child health
- Genetic counseling
- Genetic screening programs for phenylketonuria, thyroid disorders, etc. (usually a service provided by the state health department)
- Family planning programs
- Prenatal care
- Postpartum care, usually with home nursing
- Well child care (and, in some areas, care for ill children)
- School health services
- Crippled children's services (usually provided in conjunction with the state health department)
- Medical social work

Environmental health
- Promulgation of laws, standards, and regulations
- Water and sewerage engineering
- Monitoring and control of the drilling of wells
- Testing of water safety (e.g., pools, beaches, and public supplies)
- Monitoring of sewage treatment plants and subsurface sewage disposal
- Monitoring of solid waste storage and disposal
- Insect control
- Rodent control
- Air pollution control
- Radiologic health services
- Restaurant inspection and licensing
- Facilities licensure (e.g., day-care centers, schools, hospitals, and other public buildings)
- Nuisance abatement and control
- Investigation and control of toxic waste dumping
- Housing inspection and control of health hazards in homes
- Consumer protection

Health education
- Dissemination of information (e.g., to media, schools, and organizations)
- Fostering of community action around issues
- Developing a community constituency around health department programs
- Assisting in community health planning

Laboratory services
- Support for environmental programs
- Support for disease control efforts

Chronic disease control
- Screening programs (e.g., for hypertension, cholesterol, or cervical and breast cancer)
- Home nursing programs
- Educational efforts (including behavior change programs, such as "Stop Smoking")
- Medical social work
- Lead poisoning detection and treatment programs

Occupational health
- Environmental inspections
- Health education
- Epidemic investigation
- Immunization programs
- Screening efforts (e.g., for hypertension or tuberculosis)

Dental health
- Screening and referral, especially in schools
- Primary prevention by cleaning and use of topical fluoride
- Water fluoridation advocacy
- Treatment clinics for children
- Education in nutrition and dental hygiene
- Control of vending machines in schools

Mental health
- Education
- Community mental health clinics
- Support services for deinstitutionalized patients
- Home nursing services
- Alcohol and addiction services (e.g., educational programs and halfway houses)
- Promotion of self-help groups

Emergency medical services
- Planning and development
- Coordination of community efforts
- Operation of emergency services

Nutrition services
- Women, Infants, and Children (WIC) Program
- Education and counseling programs, often in conjunction with nursing

Other programs and services
- Health planning and coordination
- Operation of medical institutions and provision of medical care
- Injury prevention
- Blindness prevention
- Substance abuse programs

Health and Human Services 1990). It has had a major impact on the way government and other institutions in the USA are directing their resources in public health.

2. The first goal recommended in *Healthy People 2000* is to "increase the span of healthy life for Americans." Because a portion of the lives of many people will be spent with illness and disability, the goal is to improve the quality of remaining life and not just the length of life (see Chapter 14).

3. The second goal is to "reduce health disparities among Americans," including disparities in life expectancy, infant mortality, and premature death due to AIDS, violence, and preventable diseases.

4. The third goal is to "achieve access to preventive services for all Americans," especially services such

as prenatal care, immunizations, primary medical care, and health insurance coverage.

B. Priority objectives
1. Health promotion
    a. Physical fitness and exercise
        (1) The *objective* is to increase the proportion of the population who engage in regular exercise to at least 30% of those who are 6 years of age or older.
        (2) Regular exercise is expected to reduce the incidence of obesity, hypertension and other cardiovascular diseases, non–insulin-dependent diabetes mellitus, and osteoporosis, among other problems.
        (3) Suggested *methods* for achieving the objective are as follows:
            (a) Emphasize regular exercise programs in schools.
            (b) Increase the availability of employer-sponsored and community-sponsored physical fitness opportunities.
            (c) Increase the proportion of primary care providers who counsel their patients on physical activity.
    b. Nutrition
        (1) The *objectives* include the following:
            (a) Reduce dietary fat intake to an average of 30% or less of total calories and saturated fat intake to no more than 10% of calories.
            (b) Increase the intake of complex carbohydrates, fiber, calcium, and iron.
            (c) Decrease the intake of salt and other forms of sodium.
        (2) Suggested *methods* for meeting these objectives include the following:
            (a) Improve the labeling of food (already partly accomplished).
            (b) Increase the availability of food products with reduced total and saturated fat.
            (c) Increase the proportion of restaurants and school lunchrooms that offer low-fat and low-calorie alternative meals.
            (d) Devote more attention to nutrition education in the schools.
            (e) Increase the proportion of primary care practitioners who regularly perform nutritional evaluation and counseling as a part of the care they provide.
    c. Tobacco use
        (1) The *objective* is to reduce tobacco use, particularly among teenagers.
        (2) Suggested *methods* for control of the problem include the following:
            (a) Establish tobacco-free environments (particularly in schools and worksites).
            (b) Enact laws to control indoor air pollution.
            (c) Prohibit the sale and distribution of cigarettes to minors.
            (d) Restrict tobacco product advertising.
            (e) Increase the proportion of primary care and oral health providers who advise and assist with smoking cessation.
    d. Alcohol and drug use
        (1) The *objective* is to reduce the incidence of substance abuse.
        (2) Suggested *methods* for control of substance abuse include the following:
            (a) Develop state plans for dealing with the problem.
            (b) Educate primary school children about substance abuse.
            (c) Adopt alcohol and drug policies in worksites.
            (d) Suspend or revoke the driver's license of anyone caught driving under the influence of drugs or alcohol.
            (e) Control the advertisement of alcohol.
            (f) Increase the proportion of primary care providers who look for and treat substance abuse.
    e. Family planning
        (1) The *objectives* include the following:
            (a) Decrease the number of cases of infertility, unintended pregnancy, and early initiation of sexual intercourse.
            (b) Increase the effectiveness of family planning methods used.
            (c) Increase the use of contraceptive methods by unmarried people who are sexually active.
        (2) The suggested *methods* for achieving these objectives include the following:
            (a) Improve education about sex and family life.
            (b) Improve counseling about pregnancy and adoption.
            (c) Increase the percentage of primary care providers who offer appropriate preconception care and counseling.
    f. Mental health
        (1) The mental health *objectives* include the following:
            (a) Reduce the rate of suicide and attempted suicide.
            (b) Reduce the prevalence of mental disorders among children and adults.
            (c) Reduce the proportion of persons 18 and older who experience adverse health effects from stress.
        (2) Suggested *methods* for achieving these objectives include the following:
            (a) Increase the number of states with official plans to reduce the number of suicides by prison inmates.
            (b) Increase the number of worksites that provide programs to reduce employee stress.
            (c) Establish mutual help clearinghouses in at least 25 states.
            (d) Increase the percentage of primary care providers who routinely review the cognitive, emotional, and behavioral functioning of their patients, especially their patients who are children and adolescents.
    g. Violent and abusive behavior
        (1) The health status *objectives* include reducing the incidence of homicide, suicide, child abuse and neglect, spouse or partner abuse, assault injury, rape, physical fighting and weapon-carrying by adolescents, and inappropriate weapon storage.
        (2) Suggested *methods* for achieving these objectives include the following:
            (a) Extend the surveillance and data systems to cover violent behavior.
            (b) Increase the number of states that use unexplained child death review systems.
            (c) Provide mental health evaluations for children who have suffered from physical or sexual abuse.
            (d) Increase the capacity of shelters to accept battered women and their children.
            (e) Teach nonviolent conflict resolution skills in schools.
            (f) Increase the number and coverage of comprehensive violence prevention programs.
            (g) Increase the number of programs seeking to prevent suicide attempts by prison inmates.
    h. Educational and community-based health programs
        (1) The *objectives* are as follows:
            (a) Increase the expected years of healthy life (also referred to as quality-adjusted life years) to at least 65 years.
            (b) Increase the high school graduation rate to at least 90%.
        (2) Suggested *methods* for achieving these objectives include the following:
            (a) Provide adequate preschool programs for disadvantaged children, including children with disabilities.

(b) Increase the percentage of primary schools that provide kindergartens.
(c) Increase the proportion of postsecondary educational institutions with health promotion programs.
(d) Increase the proportion of workplaces that offer health promotion programs and the proportion of workers who use them.
(e) Increase the number of health promotion programs for all citizens, including specific programs for senior citizens.
(f) Increase the proportion of medical care organizations (hospitals, health maintenance organizations, group practices, and so forth) that provide patient education programs.
(g) Increase the proportion of people who are served by a full-time local health department.

2. Health protection
   a. Unintentional injuries
      (1) Health status *objectives* include the following:
         (a) Reduce the number of deaths and nonfatal injuries due to motor vehicle crashes, falls, drownings, fires, poisonings, and similar events.
         (b) Areas of special concern include hip fractures, head injuries, and spinal cord injuries.
      (2) Suggested *methods* for achieving these objectives include the following:
         (a) Increase the use of automobile occupant protection systems (e.g., seat belts and air bags).
         (b) Increase the use of helmets by motorcyclists and bicyclists.
         (c) Enact laws to require the use of devices such as seat belts and helmets.
         (d) Develop hand gun safety methods to minimize the use of these weapons by children.
         (e) Increase the use of smoke detectors and sprinkler systems.
         (f) Provide adequate protection for sports participants.
         (g) Improve highway safety signs and markers.
         (h) Improve trauma care networks and systems.
         (i) Increase the proportion of primary care providers who counsel about safety precautions.
   b. Occupational safety and health
      (1) The *objectives* are as follows:
         (a) Reduce work-related deaths and injuries.
         (b) Injuries cited for special attention include cumulative trauma (e.g., repetitive motion) disorders, occupational skin disorders, and hepatitis B infections.
      (2) Suggested *methods* for achieving these objectives include the following:
         (a) Increase employee use of occupant protection systems such as seat belts during vehicle travel.
         (b) Reduce the proportion of workers exposed to noise levels exceeding 85 decibels.
         (c) Reduce exposures to lead.
         (d) Increase hepatitis B immunization levels among health care workers.
         (e) Implement occupational safety and health plans in all 50 states.
         (f) Establish standards in every state to limit exposures to safe levels of asbestos, coal dust, cotton dust, and silica.
         (g) Increase the worksites with programs on worker health and safety, including back injury prevention and rehabilitation.
         (h) Increase the proportion of primary care providers who routinely elicit occupational health and safety exposures as a part of the medical history.
   c. Environmental health
      (1) The *objectives* include the following:
         (a) Reduce the incidence of asthma, serious mental retardation, chemical poisoning, and waterborne infectious disease.
         (b) Reduce blood lead levels that exceed 15 μg/dL in children 6 months to 5 years of age.
         (c) Reduce the proportion of people who live in areas exceeding the Environmental Protection Agency (EPA) standards for air pollution.
         (d) Reduce the number of homes with high concentrations of radon.
         (e) Reduce the level of toxic agents in the air, water, and soil.
         (f) Increase the proportion of people who receive drinking water that meets the EPA standards for safety.
      (2) Suggested *methods* for achieving these objectives include the following:
         (a) Increase regulations on new home construction to eliminate unsafe radon levels.
         (b) Increase the number of states that require prospective buyers to be informed of radon levels.
         (c) Increase testing for lead-based paint in homes built before 1950.
         (d) Increase the pace of cleanup at hazardous waste sites.
         (e) Increase programs for recycling materials and proper disposal of hazardous wastes.
         (f) Establish plans in at least 35 states to define sentinel environmental diseases and institute surveillance regarding these diseases.
   d. Food and drug safety
      (1) The *objectives* include the following:
         (a) Reduce the number of infections caused by key food-borne pathogens, particularly *Salmonella enteritidis*.
         (b) Improve food-handling practices in homes.
         (c) Reduce the number of adverse drug reactions.
      (2) Suggested *methods* for achieving these objectives include the following:
         (a) Increase the proportion of states that have implemented model food codes for institutions.
         (b) Increase the proportion of pharmacies that use linked systems to provide alerts of possible drug interactions among medications prescribed for individual patients.
         (c) Increase the proportion of primary care providers who routinely discuss the use of all prescribed and over-the-counter medications with their patients who are 65 years or older.
   e. Oral health
      (1) Health status *objectives* include the following:
         (a) Reduce the incidence of dental caries (cavities) in children.
         (b) Reduce the incidence of the loss of teeth in adults (especially the total loss of teeth in older adults).
         (c) Reduce the incidence of gingivitis and periodontal diseases in adults.
         (d) Reduce the incidence of death due to cancer of the oral cavity and pharynx.
         (e) Increase the proportion of children who receive protective sealants on permanent teeth.
         (f) Increase the proportion of people who drink fluoridated water or are provided with systemic and topical fluoride treatment.
         (g) Increase the use of feeding practices that prevent tooth decay in bottle-fed infants.
      (2) Suggested *methods* for achieving these objectives include the following:
         (a) Increase the oral health screening of children in primary school and in long-term institutional facilities.
         (b) Increase the proportion of people using oral health care providers.
         (c) Increase the requirement for effective head, face, and mouth protection in sports and recreational events.

3. Preventive health services
   a. Maternal and infant health
      (1) The primary health status *objectives* include the following:
         (a) Reduce the infant mortality rate to no more than 7 per 1000 live births.
         (b) Reduce the fetal death rate (death at 20 or more weeks of gestation) to no more than 5 per 1000 live births.
         (c) Reduce the maternal mortality rate to no more than 3.3 per 100,000 live births.
         (d) Reduce the incidence of fetal alcohol syndrome to no more than 0.12 per 1000 live births.
         (e) Reduce the incidence of low birth weight (less than 2500 g) to no more than 5% of live births and the incidence of very low birth weight (less than 1500 g) to no more than 1% of live births.
         (f) Reduce the cesarean section rate to no more than 15 per 1000 deliveries.
         (g) Increase to 85% the proportion of mothers who achieve the minimum recommended weight gain in pregnancy.
         (h) Increase to 75% the proportion of mothers who breast-feed their infants.
         (i) Increase the proportion of women who abstain from alcohol, tobacco, cocaine, and other illegal drugs during pregnancy.
      (2) Suggested *methods* for achieving these objectives include the following:
         (a) Increase to 90% the proportion of pregnant women who receive prenatal care during the first trimester of pregnancy.
         (b) Increase the proportion of infants who receive regular care during the first 18 months of life.
   b. Heart disease and stroke
      (1) Health status *objectives* include the following:
         (a) Reduce the age-adjusted rate of deaths caused by coronary artery disease to no more than 100 per 100,000 people each year (in contrast to 135 per 100,000 in 1987) and of deaths caused by stroke to no more than 20 per 100,000 people each year (in contrast to 30.3 per 100,000 in 1987).
         (b) Reverse the increase in end-stage renal disease requiring dialysis or transplantation.
         (c) Increase to at least 50% the proportion of hypertensive people whose blood pressure is under control (in contrast to 24% in 1984).
         (d) Increase to 90% the proportion of hypertensive people who are taking steps to control their blood pressure (in contrast to 79% in 1985).
         (e) Reduce the mean serum cholesterol level in adults to no more than 200 mg/dL.
         (f) Reduce the prevalence of high blood cholesterol levels (greater than or equal to 240 mg/dL) to 20% of adults or less.
         (g) Reduce dietary fat intake to an average of 30% of calories or less, with 10% or less from saturated fat.
         (h) Reduce the prevalence of overweight adults to no more than 20%.
         (i) Increase to at least 30% the proportion of people age 6 years or older who engage regularly in light to moderate physical exercise.
      (2) Suggested *methods* for achieving these objectives include the following:
         (a) Increase to 90% the proportion of adults who have had a blood pressure reading within the past 2 years.
         (b) Increase to 75% the proportion of adults who have had a cholesterol check within the past 5 years.
         (c) Increase the proportion of primary care providers who initiate diet and other needed interventions for elevated cholesterol levels.
         (d) Increase the proportion of worksites that offer education concerning high blood pressure or high cholesterol levels.
         (e) Increase to at least 90% the proportion of laboratories that meet the accuracy standard for cholesterol measurement.
   c. Cancer
      (1) The *objectives* include the following:
         (a) Reverse the recent rise in the overall cancer death rate.
         (b) Reduce the rate of deaths due to lung cancer (a rate that has been rising).
         (c) Reduce the rate of deaths due to breast cancer (a rate that has been steady).
         (d) Speed up the declines in the rates of death due to cervical cancer and colorectal cancer.
      (2) Suggested *methods* for achieving these objectives include the following:
         (a) Reduce cigarette smoking to no more than 15% in people age 20 years or older.
         (b) Reduce dietary fat intake.
         (c) Increase the intake of complex carbohydrates and foods containing fiber and carotenoids (fruits and vegetables).
         (d) Limit exposure to direct sunlight and increase use of sunscreens when exposure to sun is not avoidable.
         (e) Increase the activity of primary care providers in counseling about cessation of tobacco use, diet modification, and the need for mammography or other measures to detect the presence of cancer.
         (f) Increase the proportion of women who receive a regular Papanicolaou test and the proportion of all adults who have fecal occult blood testing and skin examination.
         (g) Increase the quality standards for cytology laboratories and mammography facilities.
   d. Diabetes and chronic disabling conditions
      (1) The primary health status *objective* is to increase the healthy life expectancy (i.e., the years free of chronic disabling conditions) by reducing the prevalence of conditions causing activity limitation, such as asthma, chronic back conditions, hearing impairment, visual impairment, obesity, osteoporosis, and diabetes.
      (2) Suggested *methods* for achieving these objectives include the following:
         (a) Improve screening of vision and hearing in adults as well as children.
         (b) Improve evaluation and management of conditions such as osteoporosis.
   e. Human immunodeficiency virus (HIV) infection
      (1) The *objectives* include the following:
         (a) Limit the incidence of acquired immunodeficiency syndrome (AIDS) to no more than 98,000 cases each year.
         (b) Limit the prevalence of HIV infection to no more than 800 per 100,000 people each year.
         (c) Reduce the proportion of adolescents who engage in sexual intercourse.
         (d) Reduce the proportion of sexually active unmarried adults who do not use condoms.
         (e) Increase the proportion of intravenous drug users who are in treatment programs and who use only uncontaminated drug paraphernalia.
         (f) Reduce the risk of HIV transmission from blood products to no more than 1 per 250,000 units.
         (g) Increase the proportion of HIV-infected persons who know their infection status.
         (h) Increase the proportion of schools that have age-appropriate HIV education in the curriculum.
         (i) Increase both the number of training programs and the availability of protective equipment for people whose work places them at increased risk of HIV infection.

(2) *Methods* for achieving these objectives include increased education about HIV infection and AIDS (discussed in Chapter 19).
  f. Sexually transmitted diseases
    (1) The *objectives* include reducing the incidence rates of gonorrhea, genital herpes, genital warts, pelvic inflammatory disease, and sexually transmitted hepatitis B infection, as well as primary, secondary, and congenital syphilis.
    (2) Suggested *methods* for achieving these objectives include the following:
      (a) Increase the number of clinics and practitioners who correctly diagnose and treat sexually transmitted diseases.
      (b) Provide age-appropriate education concerning how to avoid acquiring and spreading these diseases.
  g. Vaccine-preventable and other infectious diseases
    (1) The *objectives* include the following:
      (a) Reduce the number of cases of vaccine-preventable and other infectious diseases in hospitals and in travelers.
      (b) Increase national immunization levels (discussed in Chapter 16).
      (c) Reduce the need for postexposure rabies treatments.
    (2) Suggested *methods* for achieving these objectives include the following:
      (a) Expand immunization laws to cover children entering schools, preschools, and day-care settings.
      (b) Improve the financing and delivery of immunizations, so that there is no financial barrier.
      (c) Increase the proportion of primary care providers and health departments that offer immunizations.
      (d) Improve the compliance of patients who are undergoing treatment for tuberculosis.
  h. Clinical preventive services
    (1) The health status *objective* is to increase the healthy life expectancy (as defined above) of the population by ensuring that everyone has access to preventive health services.
    (2) The key *method* for achieving this objective is to remove the financial and other barriers that keep children and adults from having a primary care provider who will make sure that they receive adequate immunization, counseling, and screening services.
4. Surveillance and data systems
  a. To measure progress toward achieving the goals outlined in *Healthy People 2000*, one of the objectives is to develop improved health status indicators that are appropriate for the use of local, state, and federal health agencies.
  b. Another objective is to identify—or, if necessary, create—national data sources that are not only capable of monitoring the progress in health promotion, health protection, and preventive health services but are also designed to facilitate the rapid exchange of data among the various levels of government.

QUESTIONS

DIRECTIONS. (Items 1–10): Each of the numbered items or incomplete statements in this section is followed by answers or by completions of the statement. Select the ONE lettered answer or completion that is BEST in each case. Correct answers and explanations are given at the end of the chapter.

1. The authority of the US federal government to inspect food and regulate air pollutants derives from

  (A) the Environmental Protection Agency
  (B) the US Department of Agriculture
  (C) the Food and Drug Administration
  (D) the interstate commerce clause
  (E) tax law

2. The Food and Drug Administration and the Centers for Disease Control and Prevention are appropriately categorized as

  (A) major operating units of the Environmental Protection Agency
  (B) major operating units of the Department of Health and Human Services
  (C) constituent agencies of the Public Health Service
  (D) independent of the federal government
  (E) subject to state law

3. A requirement of public health laws established by city or county boards is that they

  (A) be approved by the federal government
  (B) be approved by referendum
  (C) be revised annually
  (D) pertain to communicable disease
  (E) be at least as strict as regulations in the state public health code

4. The basic responsibilities of local health departments include all of the following EXCEPT

  (A) communicable disease control
  (B) setting radiation exposure standards
  (C) health education
  (D) collating vital statistics
  (E) ensuring maternal and child health

5. All of the following are methods recommended in *Healthy People 2000* to increase the percentage of physically active individuals in the USA EXCEPT

  (A) screening for cardiovascular fitness with exercise stress testing
  (B) emphasizing school-based exercise programs
  (C) increasing the availability of employer-sponsored fitness facilities
  (D) increasing the proportion of primary care providers who provide exercise counseling
  (E) increasing the availability of community-sponsored fitness programs

6. Recently, the rate of cigarette smoking has been rising among

  (A) elderly women
  (B) elderly men
  (C) teenage women
  (D) minority men
  (E) college students

7. To comply with the recommended methods in *Healthy People 2000,* primary care providers should

   (A) interview and counsel patients about a variety of life-style issues
   (B) perform a comprehensive physical examination annually
   (C) consider the cost-effectiveness of any screening test
   (D) make public health a priority over individual patient care
   (E) restrict the use of high-cost diagnostic techniques

8. Which one of the following is a true statement regarding radon?

   (A) Radon levels are regulated by OSHA.
   (B) Radon is a component of cigarette smoke.
   (C) Radon is a cause of occupational asthma.
   (D) Radon exposure contributes to cardiovascular disease.
   (E) Reducing exposure to radon is a goal expressed in *Healthy People 2000.*

9. Carotenoids are abundant in which of the following foods?

   (A) shellfish
   (B) fish
   (C) legumes
   (D) fruits and vegetables
   (E) seeds and nuts

10. The targeted incidence of HIV infection in the USA in the goals for the year 2000 is approximately

    (A) 1000 cases per year
    (B) 800 cases per 100,000 population
    (C) 1 million cases per year
    (D) 100,000 cases per year
    (E) 9.8 cases per 1000 population

ANSWERS AND EXPLANATIONS

1. The answer is D: the interstate commerce clause. Most of the responsibility for protecting the public health resides with the states. Factors that influence health and cross state lines often fall under federal jurisdiction. Contaminated food products and air pollutants are apt to cross state lines and are therefore considered a form of (undesirable) interstate commerce. As such, their regulation is a federal responsibility under the interstate commerce clause of the US Constitution. *Epidemiology,* p. 250.

2. The answer is C: constituent agencies of the Public Health Service. The CDC is responsible for the investigation and control of communicable diseases and other public health threats. The FDA is responsible for regulating the safety and effectiveness of drugs and additives to the food supply. The FDA and the CDC, along with five other agencies, are incorporated within the US Public Health Service (see Table 20–1). *Epidemiology,* p. 251.

3. The answer is E: be at least as strict as regulations in the state public health code. City and county boards of health are delegated their authority by the state government, which has the fundamental responsibility for protecting the public health. These boards may generate local policy regulations that comply with statewide standards. Local regulations that are stricter than state laws must be deemed "reasonable" or they may be challenged in court. *Epidemiology,* p. 252.

4. The answer is B: setting radiation exposure standards. Controlling communicable diseases, providing health education, collating vital statistics, and ensuring maternal and child health are all included in the "basic six" functions of local health departments, first espoused in 1940 (Jekel 1991). The list of local health department responsibilities has expanded but does not include setting standards for occupational and environmental exposures. Depending on the situation, this is either a state or a federal responsibility. *Epidemiology,* p. 252.

5. The answer is A: screening for cardiovascular fitness with exercise stress testing. Exercise stress testing is not used to screen for fitness but rather, as are most tests used for case finding, to look for evidence of disease (specifically, ischemic heart disease). The use of stress testing is not recommended in *Healthy People 2000* as a means of increasing the fitness of the population, and there is no evidence to support such an effort. Detection of disease often provides an opportunity for health promotion, however, because a patient with symptomatic disease may be frightened and therefore highly motivated to change. An abnormal stress test in a patient with recurrent chest pain might cause the patient to make life-style changes that would enhance fitness. Routine stress testing in the general population would be costly and inefficient because the low prevalence of ischemic heart disease in an unselected population would lead to many false-positive results (see Chapters 7 and 8). *Epidemiology,* p. 255.

6. The answer is C: teenage women. While smoking rates in the USA have been declining in general, there has been a recent increase in the rates for adolescent women. Much of the federal effort to control tobacco use is currently directed at regulating tobacco company advertising intended to attract adolescents. *Epidemiology,* p. 256.

7. The answer is A: interview and counsel patients about a variety of life-style issues. The goals of *Healthy People 2000* represent disease prevention and health promotion. They are not specifically directed at cost containment. Cost-effectiveness will be one of many factors that influence the feasibility of the goals in *Healthy People 2000.* A physician should consider public health but is professionally bound to make the care of the individual patient the first priority. A comprehensive physical examination is often a low-yield procedure and should be modified to match the characteristics of the patient and the occasion. To comply with the recommendations in *Healthy People 2000,* the primary care provider should routinely discuss a variety of life-style issues with patients, such as dietary pattern, activity level, sexual behavior, and exposure to domestic violence, and should offer counseling when it is appropriate. *Epidemiology,* p. 255.

8. The answer is E: Reducing exposure to radon is a goal expressed in *Healthy People 2000.* This is an environmen-

tal health goal listed in *Healthy People 2000*. Radon represents an environmental rather than an occupational exposure, and levels are therefore not regulated by OSHA. Radon enters houses from radiation sources in the ground. It is not a component of cigarette smoke. Radon increases the risk of lung cancer. It is not a cause of occupational asthma nor is it known to contribute to the risk of cardiovascular disease. *Epidemiology*, p. 257.

9. The answer is D: fruits and vegetables. Carotenoids, such as beta carotene, are abundant in dark green and brightly colored fruits and vegetables, such as broccoli, carrots, pumpkins and other squashes, apricots, and melons. These micronutrients are thought to promote health, possibly through an antioxidant mechanism. *Epidemiology*, p. 258.

10. The answer is D: 100,000 cases per year. To be more exact, the year 2000 objective is to reduce the annual incidence of AIDS to 98,000 cases. The prevalence goal is no more than 800 cases per 100,000 people. *Epidemiology*, p. 258.

REFERENCES CITED

Institute of Medicine. The Future of Public Health. Washington, D. C., National Academy Press, 1988.

Jekel, J. F. Epidemiology, Biostatistics, and Preventive Medicine. Philadelphia, W. B. Saunders Company, 1996.

Jekel, J. F. Health departments in the US, 1920–1988: statements of mission with special reference to the role of C.-E. A. Winslow. Yale Journal of Biology and Medicine 64:467–479, 1991.

Levin, L. [Yale University School of Medicine, Department of Epidemiology and Public Health, International Health Division.] Personal communication, 1995.

Ottawa Charter for Health Promotion. Report of an International Conference on Health Promotion, Sponsored by the World Health Organization, Health and Welfare Canada, and the Canadian Public Health Association, Ottawa, Ontario, Canada, November 17–21, 1986.

US Department of Health and Human Services, Public Health Service. Healthy People 2000: National Health Promotion and Disease Prevention Objectives. DHHS Publication No. (PHS)91-50212. Washington, D. C., Government Printing Office, 1990.

Winslow, C.-E. A. The untilled fields of public health. Science 51:22–23, 1920.

CHAPTER TWENTY-ONE

MEDICAL CARE POLICY AND FINANCING

SYNOPSIS

OBJECTIVES
- To describe the structure of medical care systems.
- To discuss the goals and obligations of medical care.
- To discuss the historical and current organization of health care in the USA.
- To summarize the various means by which medical care is financed.
- To provide a historical perspective on health care in the USA.
- To explain the rapid rise in medical care costs over recent years, and to describe cost-containment strategies, both traditional and modern.
- To discuss the history and prospects for further health care reform and national health insurance.
- To note the difficulty in determining the optimal system for apportioning health care resources.

I. A Framework for Understanding Medical Care Systems
A. Unresolved tensions
1. Physician's role
   a. There is debate regarding the appropriate level of involvement of physicians in the lives of their patients.
      (1) Traditionally, physicians knew whole families and played a nurturing role.
      (2) Modern times have created physicians who may at times be masters of technology with little direct knowledge of their patients' lives.
      (3) Public health and preventive medicine priorities support the physician as nurturer, involved in the lives and environments, not merely the diseases, of patients.
   b. There is debate regarding the appropriate role of physicians in the public health.
      (1) The principal obligation of the physician is to provide care and advocacy for the individual patient.
      (2) There are at times tensions between individual patient care and the delivery of cost-effective, equitable care to the public.
2. In the USA, there is debate concerning the ways in which medical care is or should be organized and financed. The following questions are unresolved:
   a. Should the emphasis be on **prevention or cure?**
   b. Within prevention, should the emphasis be on **health promotion or disease prevention?**
   c. What type of practice should be emphasized in medical education: **primary care or specialty practice?**
   d. Should hospitals, health maintenance organizations, and group practices focus exclusively on the people who enter their institutions, or should they be actively involved in promoting the health of the entire community?
   e. To what extent should (and, indeed, to what extent can) physicians share medical decision-making with their patients?
   f. To what extent is there a tension between low cost and high quality in the care process?

B. Terminology in health care
1. A **disease** is a medically definable process, in terms of pathophysiology and pathology, whereas **illness** is what the patient experiences.
   a. Several different diseases might produce similar illness experiences.
   b. The same basic disease process, such as diabetes mellitus, can produce different illness experiences in different patients.
2. **Impairment** is defined as a limitation of capacity or functional ability, usually as determined by a licensed physician.
3. **Disability** is a social definition of limitation, based on the degree of impairment.
   a. Disability is defined by society in laws dealing with social benefits (such as Social Security benefits) and rights for the handicapped.
   b. For purposes of social benefits, there usually are four categories of disability:
      (1) **Temporary partial disability** (e.g., a fractured arm).
      (2) **Temporary total disability** (e.g., a broken back without paralysis).
      (3) **Permanent partial disability** (e.g., permanent loss of one eye or one limb).
      (4) **Permanent total disability** (e.g., permanent loss of two eyes, two arms, or two legs).

4. **Need for medical care**
   a. The need for care is usually considered a professional judgment.
   b. A patient's judgment about the need for care is referred to as **felt need** or **demand for medical care.**
      (1) Demand has both a medical and an economic definition.
      (2) The medical definition of demand is the amount of care people would use if there were no barriers to care.
      (3) The economic definition of demand is the quantity of care that is purchased at a given price.
         (a) For the economic definition to pertain, there must be price elasticity (i.e., an assumption that as demand falls, the price for a given amount of care will also fall).
         (b) In the medical market, the assumption of price elasticity sometimes proves false.
   c. Because of difficulties in measuring demand, the effective (realized) demand, or **utilization,** is often measured instead.
   d. Utilization is usually less than need; the difference is **unmet need,** which is shown by the following equation:

   Unmet need = Need − Utilization

C. **Factors influencing need and demand**
   1. **Demographic factors**
      a. The **age of the population** is defined either as the median age of persons in the population or as the percentage of persons over a certain age, such as 65 years.
         (1) The median age and the percentage over 65 are both determined more by **fertility patterns** than by **mortality rates.**
         (2) As the birth rate falls, the population ages because fewer young people are born to counteract the continual aging of those already in the population.
         (3) The US crude birth rate (the number of births per 1000 population) and the US general fertility rate (the number of children born to women age 15–44 years divided by the population of women age 15–44 years) have been declining during most of this century.
         (4) The exception to this was the period immediately after World War II (the postwar "baby boom" period), when the birth rates were high.
         (5) A rather sudden decline in birth rates occurred around 1970, when induced abortion became legally available.
         (6) The process of aging of the population, which had been occurring somewhat gradually, accelerated after 1970.
         (7) While the reduction in the rate of children born has reduced the demand for pediatricians and, to a lesser extent, has reduced the demand for obstetrician-gynecologists, the aging of the population has increased the overall utilization of hospitals and other sources of medical care.
         (8) Currently, more than 50% of the patients hospitalized at any point in time are likely to be 65 years of age or older, whereas only 11–12% of the population is likely to be in this age group.
      b. **Employment patterns**
         (1) The long-term result of lowered fertility will be an extended period when the number of workers will be comparatively small; this period is expected to begin around the year 2015, when large numbers of "baby boom" children will be of retirement age.
         (2) A major concern is whether the small number of workers will be able to support the large older population during its retirement with such benefits as Social Security retirement funds and medical care.
         (3) The probable shortage of workers is expected to drive up wages, making medical and nursing home care more expensive than it already is.
         (4) Chronic disease of the elderly, with its high demand for medical and nursing home care, will be common.
   2. **Advances in medical technology** produce new methods of prevention, diagnosis, and treatment that come to be viewed, by both physicians and patients, as desirable and necessary.
   3. **Poverty**
      a. People below some percentage of the **poverty line** (often 125%) are eligible for Medicaid under Title 19 of the Social Security Act.
      b. Medicaid provides coverage for medical care, as well as nursing home stays for people who are poor.
      c. People whose incomes are too high to be eligible for Medicaid, who do not receive medical insurance in their jobs, and who are not able to pay for individual medical care insurance policies are known as **medically indigent.**
         (1) Many of the **medically uninsured** (i.e., those who have no health insurance) and **medically underinsured** (those who have some but inadequate health insurance) are medically indigent.
         (2) The medically indigent are not on welfare, but they cannot financially tolerate major medical bills.
         (3) In the USA, about 37 million people (15% of the US population) were uninsured in 1993.

II. **The Medical Care System**
   A. The **goals of medical care** are as follows:
      1. Increase the length of life.
      2. Improve the function of individuals.
      3. Increase the comfort of ill persons and their families.
      4. Explain medical problems to patients and their families.
      5. Provide a prognosis for the patient.
      6. Provide support and care for patients and their families.
      7. These goals are not always fully compatible; extending biologic life to the maximum, for example, may increase pain and anxiety for the patient and family.
   B. The **basic requirements for good medical care** are known as the 7 A's and the 3 C's.
      1. **Availability** of medical care means that care can be obtained during the hours and days when people need it.
      2. **Adequacy** is a sufficient volume of care to meet the need and demand of a community.
      3. **Accessibility** refers to both geographic and financial accessibility; facilities that cannot be reached easily by public transportation and facilities that deny care to those without adequate insurance or financial resources are inaccessible.
      4. **Acceptability** of care depends on a variety of factors, including whether the providers can communicate well with their patients (e.g., can speak the patients' languages), whether the care is seen as warm and humane and concerned with the whole person, and whether the patients believe in the confidentiality and privacy of information shared with their providers.
      5. **Appropriateness** of care means that the procedures being performed are properly selected and

carried out by trained personnel in the proper setting.
6. **Assessability** means that the medical care can be readily evaluated.
7. **Accountability** incorporates public representation on the board of directors of the health care facility, regular review of the financial records by certified public accountants, and appropriate public disclosure of financial records and of quality of care studies.
8. **Completeness** of care requires adequate attention to all aspects of a medical problem, including prevention, early detection, diagnosis, treatment, follow-up measures, and rehabilitation.
9. **Comprehensiveness** of care means that care is provided for all types of health problems, including dental and mental health problems.
10. **Continuity** of care requires that the management of a patient's care over time be coordinated among providers.

III. The Organization of Medical Care
  A. Historical overview
    1. In the late **1800s,** most medical care was ambulatory care or care in the home, with local practitioners paid on a fee-for-service basis; the hospital tended to be viewed as a death house and a place for the sick poor.
    2. In the early **1900s,** as medicine became more scientific and efficacious, the hospital came to be seen as the doctors' workshop; technology and ancillary personnel and services were usually provided at no charge to the doctors and helped them to perform their craft.
    3. Acute, general hospitals, however, did not usually offer care to the mentally ill; **mental hospitals** were dismal facilities, usually built in sparsely populated areas.
      a. Eventually, the development of psychoactive medications enabled many patients with mental disorders to be treated in community settings, and state mental hospitals were progressively closed in the late 1960s and early 1970s.
      b. **Deinstitutionalization** occurred without adequate community resources to receive the large numbers of mentally ill patients being discharged.
      c. Some of these patients benefited from treatment in an outpatient setting; others, however, could not function well in society. As a result, many discharged patients became poverty-stricken and homeless, particularly if they forgot or refused to take their medications.
    4. General hospitals in the past usually did not offer care to patients with **tuberculosis.** These patients were treated in special, state-run tuberculosis hospitals until after World War II, when the development of antibiotics permitted them to be treated in general hospitals or at home.
    5. With the founding of the **National Institutes of Health** in 1948 came a push for improved biomedical technology.
      a. The research done since that time has made the practice of medicine much more effective but also far more complex and costly.
      b. The increased complexity has not only resulted in increasing specialization of physicians and other health care workers but has also required an increasing rationalization of the levels of care to make the complexity of care and the resulting costs appropriate to each patient's needs.
  B. Levels of medical care
    1. A patient is initially assigned to an appropriate level of care and is reassigned to another level whenever there is an improvement or setback in the patient's condition.
    2. Although the movement from one level to another should be easy, rapid, and smooth, the lack of available hospital beds or lack of planning sometimes causes difficulties.
    3. At the top of the scale of complexity are three types of **acute, general hospital facilities.**
      a. The **tertiary medical center** has most or all of the latest technology and usually participates actively in medical education and even in clinical research.
      b. The **intermediate hospital** is a medium to large community hospital that has a considerable amount of the latest technology but less research and investigational activity.
      c. The **local community hospital** provides services such as routine diagnosis, treatment, and surgery but lacks the personnel and facilities for complex procedures.
    4. Next on the scale of complexity are two types of **rehabilitation or convalescent care facilities.**
      a. The first type is a **special unit in a regular hospital.**
      b. The second type is a **rehabilitation hospital.**
    5. If patients are not discharged from the hospital directly to their homes, they are most likely to be discharged to one of three different types of **extended care facilities** (ECFs).
      a. The **skilled nursing facility** (SNF) is commonly called a nursing home and usually provides special kinds of care, such as intravenous fluids and medicines.
      b. The **intermediate care facility** (ICF) is suitable if the patient's primary need is for help with the activities of daily living (eating, bathing, grooming, transferring, toileting, and so forth). Unlike an SNF, an ICF is not required to have a registered (skilled) nurse on duty at all times.
      c. The **hospice** is a nursing home that specializes in providing terminal care, especially for patients with cancer or acquired immunodeficiency syndrome (AIDS).
    6. **Organized home care** is necessary for patients who are discharged from the hospital to the home, where they continue to receive treatment or follow-up procedures that require specialized skills. Examples include the placement and monitoring of intravenous lines for therapy and the drawing of blood for tests.
    7. The least complex level of medical care is **self-care** in the home.
      a. The majority of medical care decisions are not made by professionals but instead are made by people for themselves, for friends, or for members of their families.
      b. Many home diagnostic tools, such as blood pressure cuffs and blood glucose testing equipment, have given patients greater power to monitor their own health status.

C. Medical care institutions

1. **Hospitals**
   a. Some hospitals focus on a special group of patients (e.g., a children's hospital).
   b. Some hospitals focus on a special type of medical problem (e.g., a psychiatric hospital) or a particular type of service (e.g., a rehabilitation hospital).
   c. Hospitals may be for-profit or not-for-profit.
      (1) A for-profit hospital may be independent or part of a for-profit chain of hospitals.
      (2) Not-for-profit hospitals may be sponsored by the community in which they are located; a church or other religious group; a charitable organization (e.g., Shriners' hospitals for children); a city, county, or state government; or the federal government (e.g., Veterans Administration Hospitals).

2. **Ambulatory primary care systems**
   a. Historically, most US physicians were in **solo medical practices,** although they might share night and weekend coverage with other solo practitioners. This type of practice could be quite rewarding, but it frequently was exhausting.
   b. Gradually, US physicians began to develop practice **partnerships,** partly to solve the problem of sharing weekend and nighttime coverage and partly to achieve efficiencies and economies by sharing the cost of office space, equipment, and staff.
   c. A logical extension of partnerships was the formation of **group practices** consisting of three or more (often many more) physicians.
      (1) This increased the efficiencies of sharing office space and staff and increased the free time available to physicians.
      (2) Group practices provide the advantage of ready access to other physicians for consultation concerning complex cases.
      (3) Group practices could be of either the single-specialty or the multiple-specialty type.
      (4) Although most group practices initially operated on a fee-for-service basis, some began to develop the concept of prepaid group practice—a concept that goes at least as far back as the final report of the Committee on the Costs of Medical Care (1932).
         (a) On the West Coast, the Kaiser Corporation set up its own multispecialty group practice before World War II to care for its own workers, but membership has since been opened to the general public, and it is now known as **Kaiser Permanente.**
         (b) This was the first example of a large prepaid group practice in the USA.
   d. Prepaid group practices that met certain standards and contractual arrangements were named **health maintenance organizations** (HMOs) by the Nixon administration.
      (1) The national HMO law was passed in 1973 to encourage the large-scale development of HMOs.
      (2) People who enrolled in an HMO were usually part of some economic group, such as workers in a company or industry, but their enrollment had to be voluntary.
      (3) In return for a fixed monthly fee from members, the HMO had the contractual obligation to provide the types of medical care specified in the contract (rather than to provide financial reimbursement, as in the case of an insurance company) or at least to ensure that the stipulated care was provided.
      (4) The HMO assumed some of the risk when income was less than expenses and made a profit when income was greater than expenses.
      (5) There are three main functional parts to an HMO:
         (a) There is a legal and fiscal entity that does the contracting and financial transactions.
         (b) There is a group of physicians that provides the outpatient and inpatient medical care.
         (c) There is an associated hospital or hospitals.
      (6) Usually, HMOs are divided into two fundamental types:
         (a) In the **staff model HMO,** most of the physicians are salaried, full-time employees who work exclusively in the health plan or belong to a physician group that contracts to provide all of the medical services to the plan (as is typical in Kaiser Permanente). Some specialists may be retained on part-time contracts. The HMO may have its own hospitals (as does Kaiser Permanente) or may hospitalize its patients in one or more local hospitals, in which case the local hospitals are not usually a formal part of the HMO.
         (b) In the **independent (or individual) practice association** (IPA), the patients enrolled in the program can choose a primary care physician from a list of physicians (known as a **panel**) who have contracted to provide services for the IPA. The IPA pays each physician on a fee-for-service basis whenever an HMO member uses that physician's services. The IPA physicians limit their fees to the rates specified in the contract, and they agree to certain kinds of quality review and practice controls that are often similar to those of managed care (see below).
         (c) A **preferred provider organization** (PPO) is a variation on the IPA theme. The PPO is formed when a third-party payer (e.g., an insurance plan or a company) establishes a network of contracts with independent practitioners. The patients in a PPO can see physicians who are not on the panel, although they will have to pay a surcharge for their services.
   e. Traditionally, many **hospital outpatient clinics** have served people who are poor and may not be well insured. Patients receive treatment (although often following a long wait) in return for being cared for by physicians in training, with proper supervision.
   f. **Hospital emergency departments** are becoming increasingly complex, and because of the heavy patient loads, their staff may triage nonurgent patients to a satellite convenience clinic.
   g. **Community or neighborhood health centers** were promoted by federal health programs in the 1960s and 1970s; most were placed in underserved areas in big cities or rural locations.
   h. Freestanding outpatient surgical centers (**"surgicenters"**) have become increasingly popular and may be owned by hospitals or group practice associations.
   i. **Urgent care centers** are freestanding clinics that are conveniently located (e.g., often in shopping centers) and allow ambulatory patients to be seen on a drop-in basis not only during the usual weekday hours but also during evenings and on weekends. Most of the urgent care centers are proprietary, and many specialize in evaluating and treating injured workers promptly in order to get them back to work.

IV. Payment for Medical Care
A. Physician payments
1. In the **fee-for-service method,** physicians are paid for each major item of service provided.
   a. Charges are established on the basis of the type and complexity of service (complete workup, follow-up visit, hospital visit, major surgical procedure, etc.).
   b. The amount charged by a physician may exceed the amount that a third-party payer is willing to reimburse, in which case the patient is expected to pay the difference between the charges and the third-party payment.
2. Sometimes primary care physicians are paid on a **capitation basis,** meaning on a "per head" basis.
   a. Regardless of the number of services needed by a patient, the physician receives the same amount of money per year.
   b. This method of payment has much lower administrative costs, and it is thought to promote physicians' efforts in preventive care.
   c. It may lead to poor gatekeeping, however, because it may be easier to refer a patient than to provide a service.
   d. This payment method is sometimes used in the USA to pay practitioners working in HMOs and is commonly used in Britain to pay general practitioners.
3. Physicians who work full-time for HMOs, hospitals, universities, or companies may be paid by **salary.**

B. Insurance and third-party payers
1. A century ago, physicians were paid directly by patients for their services; as medicine became more scientific and technical, often requiring long hospital stays, the out-of-pocket payment method became inadequate.
2. Today, it is not unusual for a very sick patient to be charged from $1500 to $2000 per day for a hospital stay, even if no surgery or major procedures are done.
3. One solution to the cost problem was to have a **third-party payer,** such as an insurance company. The third-party payer collected money regularly from a large population in the form of medical insurance premiums and paid the hospitals and physicians when care was required.
4. **Modern US hospital insurance** had its foundation in Dallas, Texas, where a group of schoolteachers entered into a contract with Baylor University Hospital in 1929.
   a. The teachers paid the hospital 50 cents per person per month and, in turn, the hospital promised to cover 21 consecutive days in the hospital in a semiprivate room, along with medications, laboratory studies, the use of operating rooms, and so forth.
   b. This led to the development of **Blue Cross,** which is a form of insurance that covers only hospital care.
5. If an insurance policy covers **indemnity benefits,** this means that the insurance company (carrier) will reimburse the insured patient a fixed number of dollars per service, regardless of the actual charges incurred for the service.
6. If an insurance policy covers **service benefits,** this means that the carrier will be responsible for full payment for the needed services, regardless of their costs.
7. Actuaries, the statisticians who estimate risks and establish premiums for insurance companies, have a standard set of **actuarial principles** that guide the process of underwriting (insuring) medical and other risks.
   a. Originally, insurance was designed to pool the risk from large groups to protect individuals from rare but devastating losses, such as those resulting from fires.
   b. The actuarial principles developed to accomplish this objective do not adapt well to medical care for two reasons:
      (1) Medical care involves frequent and fairly predictable costs as well as rare and catastrophic costs.
      (2) Those at greatest risk of ill health and hospitalization can least afford the cost of insurance, although according to actuarial principles they should be charged the most.
   c. So far, the solution applied to this dilemma in health care has been the concept of **pooling risk.**
      (1) If all of the people in a large, natural community (i.e., a community consisting of people of various ages and degrees of health) were to be insured by the same carrier and were to pay the same monthly premium rate, then the law of averages would work so as to protect the carrier from excessive loss except, perhaps, in times of disaster.
      (2) In effect, the low-risk people in the population would be helping to pay the premiums for the high-risk people, because the risk would be averaged according to the "experience rating" of the entire group.
      (3) This is not a complete solution, because the poor still may not be able to pay the established premium, and the plan may not be offered where they work.
8. Insurance companies prefer to insure groups, particularly working groups.
   a. **Group insurance** requires less sales effort and paperwork.
   b. By insuring a large group of people, a carrier is less likely to insure a high percentage of bad risks (i.e., people who will probably require a lot of medical care).
   c. By insuring a working group, the carrier benefits from the **healthy worker effect.**
      (1) Because people with jobs must be in reasonably good health to work in the first place, they have a lower risk of death and illness than the population as a whole.
      (2) Unfortunately, many of the individuals who need the most medical care are unemployed, and this leaves them with multiple problems:
         (a) They are unable to purchase insurance because they represent undesirable risks to an insurance company.
         (b) They are unable to pay for the health care they need because they have no earning power.
         (c) They remain in a poor state of health that keeps them from finding employment.
9. Many insurance carriers sought to attract the business of low-risk individuals and low-risk companies by offering lower premiums. As the people with low risks were removed from the community pool (a process called **"cream skimming"**), the people remaining in the pool were, on the average, at higher risk, so they had to be charged a higher premium, making the community pool still less attractive.
10. Some insurance companies have sought to use epidemiologic information to undersell the Blue Plans (Blue Cross and Blue Shield). These companies of-

fer low-cost medical insurance to individuals with low epidemiologic risk profiles, usually those who do not drink alcohol or smoke.
11. Most Blue Cross plans cover the first several thousand dollars of their liability, but they insure themselves against disastrous medical care liability by **reinsurance**. They pay a fee to another (reinsurance) company, which then assumes the liability for Blue Cross commitments above a certain agreed-upon amount.
12. In order for patients to be covered for huge medical costs, many plans also include a premium for **major medical insurance**.
    a. This insurance usually pays 80% of the amount by which a hospital bill exceeds the Blue Cross maximum, and the patient is obligated to pay the other 20% of the bill for major expenses.
    b. The patient's portion is called a **copayment**, because the patient and the insurance company are sharing the total cost.
    c. The copayment is also seen as a means to enlist the patient or the family in the effort to get the patient out of the hospital as soon as possible.

C. Social insurance
1. Compulsory insurance for a population group is often called social or public insurance.
2. Most people employed in the USA must make payments into the Social Security Trust Fund for two national social insurance programs: Medicare and retirement benefits.
3. **Medicare** is authorized under Title 18 of the Social Security Act and is administered by the federal government, although it uses insurance carriers as fiscal intermediaries.
    a. The people eligible for Medicare include most individuals who are 65 years of age or older and those individuals who receive Social Security benefits due to disability.
    b. Part A and Part B of Medicare provide partial coverage for hospital expenses and physician expenses, respectively.
    c. Social Security beneficiaries do not pay premiums for Part A coverage, but they do pay premiums on a regular basis if they elect to have Part B coverage.
    d. Medicare also will pay for a certain amount of home care or nursing home care for a medical problem that follows directly from a Medicare-covered hospitalization.
    e. Since Medicare does not cover all hospital expenses, patients are billed for the portion of charges not covered by Medicare. Therefore, insurance plans have developed **Medicare supplemental insurance**, so that patients have few, if any, bills after a hospital admission.
4. By 1948, all states in the USA had some sort of **workers' compensation program**.
    a. Workers' compensation laws stipulate that people with a job-related injury or illness have their medical and rehabilitation expenses paid and also receive a certain amount of cash payments in lieu of wages while they are recuperating.
    b. The size of a company's premiums depends on that company's **experience rating**, which is the amount of claims paid in recent years. This serves as a stimulus for companies to invest in safety on the job.

D. Social welfare
1. **Medicaid** is authorized under Title 19 of the Social Security Act.
2. Unlike Medicare recipients, Medicaid recipients have not previously paid money into a trust fund.
    a. Medicaid is paid from general tax revenues of the federal and state governments.
    b. The benefits of Medicaid are considered to be social welfare, instead of social insurance.
3. The people covered by Medicaid are poor and usually receive additional types of welfare assistance, such as Aid to Families with Dependent Children (AFDC).
4. In contrast to Medicare, which is entirely federally administered, Medicaid is administered by the states, which share the costs of the program with the federal government.
    a. Although the federal government usually reimburses a state for approximately half of its Medicaid costs for a given year, poorer states get slightly more.
    b. The eligibility criteria for Medicaid, as well as the size of the benefits, vary from state to state.
5. Medicaid basically covers two things:
    a. Medicaid covers medical care expenses, including both hospital and physician bills. The amount of reimbursement is usually far below the customary charges of physicians, making the program unpopular with many physicians.
    b. Medicaid also covers long-term nursing home care, but only after people have largely exhausted their resources, a process called spend-down (see Long-Term Care, below).

V. The Current Situation in the USA
A. Historical overview
1. Organized medicine has spent vast amounts of money fighting government-provided medicine, national health insurance, and any increase in government control or regulation of the practice of medicine, but did not defend against reform by entrepreneurial forces within health care.
2. During the past two decades, for-profit hospitals, hospital chains, nursing homes, and insurance companies have rapidly gained control of the organization and financing of medical practice and have instituted **managed care** to achieve cost control.

B. Reasons for the rapid increase in the cost of medical care
1. The costs of medical care were increasing much faster than the general inflation rate and in 1994 represented about 14% of the gross domestic product.
2. Reasons for this rapid increase in costs include the following:
    a. The increasing effectiveness of medical care has led to an **increased demand for care**, thereby increasing the total costs.
    b. There have been increases in the **wages of health care workers**.
    c. Medical costs have been and continue to be increased by the use of complex but only **partially effective technology** for the diagnosis and treatment of disease.
       (1) Before polio vaccines were developed, for example, **iron lungs** were used to extend the lives of paralytic poliomyelitis victims.
       (2) In contrast to the polio immunization program, which has proved to be highly cost-effective, the

iron lung was an expensive and ineffective ("halfway") technology.
            (3) Two current examples of "halfway" technologies are renal dialysis and coronary artery bypass grafts, which are more costly and less effective than implementing programs to prevent renal damage and coronary atherosclerosis.
    d. As medical care increasingly is dependent on technology, the **cost of medical education** increases.
        (1) Medical education is now so expensive that many new physicians begin practice with more than $100,000 of educational debt.
        (2) The median debt of graduating US physicians exceeds $60,000.
    e. **Underutilization of hospitals** hurts the financial stability of the institutions.
        (1) Hospitals need to have a steady bed occupancy rate of greater than 80% to remain solvent over the long run.
        (2) Many hospitals, especially in rural areas, have closed in recent years because of low occupancy rates.
    f. **Lack of insurance** leads to inappropriate utilization of emergency departments and to delayed care, with resulting increased expense because disease is found at a later and less treatable stage and in a more costly setting.
    g. **Planning failures** have contributed to the problem of increasing costs of medical care.
        (1) The Hill-Burton Act of 1946 encouraged hospital construction.
        (2) Beginning in the mid-1960s and continuing for almost 20 years, the federal government supported official health planning strategies, largely in an effort to control costs.
        (3) Among the primary strategies it supported were the appointment of rate-setting authorities within states and the issuance of certificates of need (CON) for the construction of new hospitals or purchase of expensive equipment in particular locales.
        (4) Because many of the planning efforts were underfunded or difficult to enforce, they were often ineffective in preventing the duplication of facilities and expensive equipment.
        (5) To ensure that cost-cutting efforts did not decrease the quality of care, regulations were instituted to control the medical care institutions and the process of care. The effects of these regulations on quality is uncertain, but they add considerably to the costs of care in the following ways:
            (a) The amount of paperwork to be completed is increased.
            (b) The complexity of the administrative process is increased.
            (c) The number of managers and administrators required is increased.
        (6) Sweet (1993) predicted that if the present trends in health care management and utilization continue, by the year 2026 there will be over two million administrators and no patients!

C. Traditional cost-containment strategies
1. The first committee focusing on the cost of medical care in this country was established in 1929 and was called the Committee on the Costs of Medical Care; it published its landmark report in 1932.
2. The first and most basic method of discouraging the overuse of medical care has been to create **deductibles,** which are out-of-pocket payments made by the patient, often at the beginning of the care process; physicians have worried that deductibles might discourage patients from coming in for early symptoms of serious disease.
3. The second basic cost-control method has been **copayments,** as discussed in the section entitled Insurance and Third-Party Payers (see above). Copayments were thought to "encourage" patients not to stay in hospitals longer than necessary.
4. The third common method has been **exclusions** in the insurance.
    a. Some insurance policies totally excluded psychiatric care and dental care from coverage, while others severely restricted the reimbursement for these types of care.
    b. Psychiatric care, in particular, was perceived by third-party payers as a potentially bottomless pit that could consume large amounts of money in endless visits.
5. Another cost-control method used in the 1960s, that of requiring hospitalization for most diagnostic procedures, seemed counterproductive, even at the time.
    a. Most insurance plans required all major diagnostic workups to be done in hospitals, even though there often was no medical reason that they could not be done on an outpatient basis.
    b. Insurance companies worried that there might be no end to ambulatory diagnostic testing and that the actuaries could not determine the true liability.
    c. If diagnostic testing were only reimbursed while the patient was in the hospital, there was a known maximum liability for all of the insurance companies in an area: the total number of inpatient beds multiplied by the average cost per bed-day, plus ancillary costs.
    d. This strategy enabled the insurance companies to set a finite and predictable limit to their financial liability.
6. **Health care markets**
    a. For health care to function like other markets, certain assumptions from a free market economic model must be invoked:
        (1) There must be many sellers and many buyers for the buyers to have free choice.
        (2) The buyers must have good information regarding the products from which they choose.
        (3) There must not be monopolies; that is, no seller and no buyer should be able to dominate the market.
    b. The health care market usually does not meet any of these characteristics.
        (1) In many areas, there are few choices of provider, either physician or hospital, and the federal government is often the dominant purchaser of care in the market.
        (2) Patients usually do not have enough information about the quality and cost of medical care for them to make a valid market-based decision concerning their medical needs.
        (3) Increasingly large market forces, such as hospital chains or huge HMOs, dominate the market in their primary regions.
    c. Market forces are inadequate to maximize efficiency and control the costs of medical care, because of the complexity of the care.

D. New cost-containment strategies
1. Introduction
    a. If resources for medical care are inadequate to meet demand, there are three basic methods of responding:
        (1) Resources may be increased.
        (2) Demand may be decreased (or at least utilization may be decreased).
        (3) Efficiency may be increased.

b. The recent emphasis is on decreasing demand and increasing efficiency.
2. **Prospective payment system based on diagnosis-related groups**
   a. The prospective payment system (PPS) based on diagnosis-related groups was developed in the 1970s but first applied nationally in the USA to Medicare reimbursement in 1983.
   b. Under this system, each hospital admission is classified into one of 23 major diagnostic categories based on organ systems, and then these diagnostic categories are further subdivided into diagnosis-related groups (DRGs).
   c. A DRG may consist of a single diagnosis or procedure, or it may consist of several diagnoses or procedures that, on average, have similar hospital costs per admission.
   d. DRGs were first developed to enable hospitals to look for cost "outliers," but the federal government decided to use the DRG system to pay hospitals on the basis of a prospectively determined average cost for each of the more than 470 DRGs.
   e. Although there is no federal requirement that hospital payers other than Medicare use the DRG system for reimbursement, several states requested and received federal permission to incorporate DRGs into their own prospectively determined rate-setting programs.
      (1) When this happened, all third-party payers in the state had to conform to the same prospectively determined rates.
      (2) This is referred to as an **all-payer system.**
   f. Note that the hospital is actually reimbursed *after* a specific type of care is given; however, the amount of payment for the specific type of care is decided prospectively (in advance).
   g. Medicare reimburses hospitals with the predetermined fixed amount for the entire hospital stay of a Medicare patient, based on his or her DRG, regardless of whether it actually cost the hospital more or less than the prospectively determined DRG-specific payment to provide that care.
   h. If a hospital can find a way to reduce the costs and provide the care for less than the amount reimbursed by the PPS, it can retain the excess amount; if a hospital is inefficient and has higher than average costs for a hospital admission, it will lose money on that admission.
   i. Because hospitals with the strongest administrative teams and data systems are best able to keep costs below PPS reimbursements, there is a tendency for the strong hospitals to get stronger and the weak hospitals to get weaker.
   j. Good results from the PPS include the following:
      (1) There are more and better hospital data than before.
      (2) There is a greater ability than before to find unnecessary costs.
   k. After a period of reduced costs resulting from PPS, the upward pressure on the cost of medical care has resumed.
   l. The full impact of the PPS on the quality of medical care has not been determined.
      (1) There is evidence that some patients are being discharged sooner than desirable, but no major change in medical care quality has been clearly discernible.
      (2) Often, early discharge merely passes the medical care problems (and therefore costs) down the line to the care institutions receiving the patients from the hospital: the home, home care agencies, and nursing homes.
   m. Because the PPS does not apply to ambulatory procedures, providers in ambulatory settings can set their own rates.
   n. Many hospitals and staff model HMOs began to develop infirmaries, where patients who did not need acute, intensive care could be given moderate supervision and some treatment at a much lower cost than if they were in hospitals.
3. **Managed care**
   a. Managed care is a system of administrative controls, the goal of which is to reduce the costs of medical care.
   b. With the advent of managed care, also known as **utilization management,** the trend appears to be away from physician autonomy in some aspects of medical practice, such as deciding which patients can be admitted to the hospital and how long they may remain there.
   c. **Preadmission review and certification** requires that some designated person in the managed care office (usually a nurse) must approve a nonemergent hospital admission before the admission occurs; otherwise, the hospital is not guaranteed payment from the patient's third-party payer.
   d. **Emergency department admission review** requires that an admission from the emergency department have a case review within a day to be sure it is a justified admission. If the reviewer does not consider the admission justified, the hospital will not be paid, and the patient will receive the bill.
   e. **Concurrent (continued stay) review** requires that the attending physician justify keeping a patient in the hospital longer than the number of days expected for that patient's DRG.
   f. To avoid discharging patients to inappropriate places for their needs and to avoid delays in outplacement, **discharge planning** should begin the day the patient is admitted to the hospital.
   g. A **second** (physician's) **opinion** concerning the patient's need for surgery must be obtained before a major elective invasive procedure is performed. Requiring a second opinion markedly reduces the rates for certain types of surgery, such as hysterectomy.
   h. Many managed care plans, particularly staff model HMOs, require that all referrals to specialists be approved by the patient's primary care practitioner. In this role, the primary care practitioners are functioning as **gatekeepers.**
   i. **High-cost case management**
      (1) If the hospital care for a particular patient is costing or potentially will cost the third-party payer a large amount of money, an administrator is appointed to look hard for a less costly alternative, such as ambulatory care or home care.
      (2) This strategy tends to work best in cases in which the patient has complex or multiple medical problems.
   j. **Benefit design**
      (1) Every benefit plan offered by a third-party payer, including HMOs of various types, seeks provisions to attract the patients they want to recruit to the plan while at the same time limiting the financial exposure of the insurer.

(2) The plan may try to reduce premiums and costs by enlisting the patients themselves in reducing costs by means of such traditional methods as deductibles and copayments.

(3) A common practice is to exclude or at least limit the amount of certain benefits from the policy.

(a) Plans frequently limit or exclude mental health and dental health benefits.

(b) Where it is legal to do so, some plans limit the coverage of costs related to human immunodeficiency virus (HIV) infection or substance abuse.

k. **Financial incentives for physicians**

(1) If physicians are paid on a salary or capitation basis rather than on a fee-for-service basis, they have no incentive to provide unnecessary services to their patients.

(2) In theory, there should be an incentive for preventive measures when a physician or a health plan is paid on a capitation basis, because effective prevention may reduce the time and effort the physician must spend on the average patient.

(3) A strategy that may be pursued to reduce unnecessary care is a system of bonuses for physicians if they use efficient practice techniques or if their group practice makes a sufficient profit in a given year.

E. National health insurance

1. The financial problems created by "cream skimming" (see above) and other techniques for enticing low-risk populations into third-party payer networks could be reduced by taking any large natural population (either the entire US population or each state's population) as a risk pool. If this were done for the entire US population, it would be a form of nationwide (national) health insurance.

2. National health insurance has been vigorously opposed by much of organized medicine, which still fears government control over medical practice as well as possible limitations of physician charges.

3. National health insurance is unlikely to be enacted unless a reliable way is found to control the costs of medical care.

4. In 1994, President Clinton proposed a new system designed to ensure universal coverage and yet provide people freedom of choice among health plans and physicians.

a. The centerpiece of this plan was the development of **health insurance purchasing cooperatives** (HIPCs), which the President termed **health alliances.**

b. Within defined regions, each of which could be as large as a state, the health alliances would (with minor exceptions) be the sole organizations to receive the health insurance premium payments from companies, individuals, and the government.

c. In turn, the health alliances would have the responsibility for approving a number of providers (mostly HMOs or IPAs) who met certain criteria.

d. Individuals or families could then choose one of the health plans approved by the health alliance in their area, and their premiums would be channeled through the health alliance to the health plan of their choice, which would provide their care.

e. Because the providers would be in competition with one another to attract people but would also be regulated by the government through the health alliances, the entire process was called **managed competition.**

f. This plan appeared so complex that the public and Congress were generally swayed by the arguments of opponents, and the plan failed politically.

5. The prospects for the foreseeable future are for continuing reform from within the medical care system. Strategies for reform include the following:

a. Efforts can be made by the states to create their own solutions.

b. Efforts can be made by the federal government to modify Medicare, Medicaid, and third-party insurance.

c. Efforts can be made to develop some form of national health insurance for pregnant women and young children.

d. Efforts can be made to develop some form of catastrophic health insurance plan.

F. Long-term care

1. A fairly high proportion (about 43%) of people who reach the age of 65 years eventually will spend some time in a nursing home.

2. Medicare (Title 18) will cover nursing home costs for a limited period (up to 100 days) only if two conditions are met:

a. The patient is released from a hospital directly to the nursing home.

b. This immediate posthospital nursing home care is considered likely to improve the patient's condition.

3. Medicare will not pay for nursing home care beyond the 100 days after discharge from a hospital.

4. Medicare will not pay for nursing home care if a nonhospitalized person requires nursing home care because of failing strength or inability to care for himself or herself in the activities of daily living.

5. After Medicare coverage of nursing home (or home care) services is used up, the patient must assume the costs for these services until his or her financial resources have been **"spent down"** to a prescribed level (e.g., ownership of a home may be maintained as long as a spouse or dependent survives, and up to about $4000 in personal resources may be maintained).

6. When the spend-down is complete, Medicaid (Title 19) begins to cover the costs, and in essence the patient is a welfare recipient.

a. Because part of the Medicaid costs are borne by the state in which the patient enters a nursing home, the state could put a lien on the patient's house and property, although it would not be taken until the patient's spouse had died.

b. Some elderly people, suspecting that they would likely need a long stay in a nursing home soon, gave their wealth to their children to avoid having their property and savings used to pay nursing home costs.

c. For this reason, state governments have stipulated that if the gift giver is placed in a nursing home and requires Medicaid within a certain number of years of giving the gift, the gift recipient must reimburse the state for the nursing home costs, up to an amount equal to the gift.

7. A number of experiments with long-term care insurance have been initiated.

a. **Long-term care insurance** will pay all or part of the costs of a nursing home but will not pay for an acute hospitalization or for physicians' charges.

b. This type of insurance tends to be very costly because good nursing home care now costs more than $1000 per week in many areas of the USA and the need for nursing home care can extend for many years; therefore, long-term care insurance has not gained general acceptance thus far.

G. **Assessment of the quality of medical care**
 1. The quality of medical practice has been a major concern since early in this century.
    a. In 1910, for example, the **Flexner Report** was especially concerned with the need to improve medical education.
    b. Quality of medical care became a bigger issue after World War II, when Donabedian and other investigators began to define more clearly the dimensions of quality. In 1969, Donabedian indicated that quality should be examined in terms of the following factors:
       (1) **Structure** (the physical resources and human resources that a hospital or HMO possessed for providing care) should be examined.
       (2) **Process** (the way in which the physical and human resources were joined in the activities of physicians and other health care providers) should be considered.
       (3) **Outcome** (the end results of care, such as whether the patients actually did as well as would be expected, given the severity of their problems) should be examined.
 2. State accreditation of facilities usually focuses on structural issues, with some evaluation of process.
 3. Quality review programs of the past, including the programs of professional review organizations, tended to focus on particular aspects of process called **procedural end points** and offered a detailed review of the methods of care provided and an analysis of how well certain disease-specific criteria were met.
 4. Unless outcomes are adjusted for the severity of the patients' illnesses, hospitals treating the sickest patients would be at an unfair disadvantage. The process of adjusting for the severity of illness is usually referred to as **case-mix adjustment.**
 5. The federal government now rates hospitals by giving a **case-mix adjusted mortality rate** for each hospital.

H. **Current trends in medical care**
 1. US companies are paying huge amounts for the medical care insurance they provide to their workers, and are exploring a variety of means to reduce their costs. These include self-insuring, repackaging their insurance policies, and requiring their employees to pay a portion of the medical care premiums.
 2. Cost-saving packages often come through special financial arrangements with providers in a preferred provider organization (see above).
 3. In the USA, current medical care policy is largely determined by what care is reimbursed, and this is highly variable by region, insurer, and patient, and is in a constant state of flux.
 4. Many of the current trends in medical care have evolved from efforts to reduce costs.
    a. One trend is the increasing use of ambulatory (outpatient) facilities for hernia repair, cataract removal, and a whole host of operative procedures that would have been unthinkable to perform on an outpatient basis a decade or two ago.
    b. Another trend is the increase in the amount of care provided by medical personnel other than physicians.
       (1) Experiences with medical corpsmen in World War II led to the development of the **physician assistant** (or **associate**), who could provide some primary care and could monitor ongoing care of more complex diseases once the diagnosis and treatment had been established.
       (2) **Nurse practitioners** are trained to provide primary care and to recognize when the care of a physician is needed.
       (3) **Certified nurse midwives** are licensed to provide prenatal care, labor and delivery care, and postpartum care for uncomplicated pregnancies and deliveries.
       (4) **Alternative practitioners,** particularly chiropractors, have became more common in the USA and are consulted by a larger proportion of the population. Recently, Congress established a national program to study alternative medicine approaches.
 5. For-profit, investor-owned hospital chains (e.g., the Hospital Corporation of America), nursing homes, diagnostic laboratories, radiologic facilities, home care programs, urgent care facilities, and renal hemodialysis units now represent a significant proportion of provider institutions. The percentages of medical care institutions run by for-profit organizations are 14% of community hospitals in 1985; 34% of psychiatric hospitals in 1984; 81% of nursing homes in 1980; 66% of HMO plans in 1987; 57% of PPOs in 1985; 32% of Medicare-certified home health agencies in 1985; 90% of freestanding surgery centers in 1986; 93% of primary care centers in 1986; and 63% of blood banks in 1986.
 6. There has been some vocal criticism of the increasing intrusion of for-profit ventures into the world of medical care (Relman 1991).

VI. **The Problem of the "Medical Commons"**
 A. In 1968, Garrett Hardin wrote "The Tragedy of the Commons," perhaps the most famous contribution to the population control debates of the 1960s.
 B. Pointing out that the shared resources of the earth (the commons) are limited, Hardin argued that the attempt by one individual or group to maximize its own welfare by using more than its fair share of the commons would necessarily diminish the good that others can derive from it.
 C. This logic can be applied to the use of medical resources in the USA.
    1. Unless Americans are able and willing to organize, finance, and regulate medical care in light of the needs of the entire population, then various individual groups (e.g., industries, hospitals, hospital chains, HMOs, insurance companies, nursing homes, and home care programs) will continue to seek to maximize their benefits (their share of the commons) at the expense of others.
    2. Apportioning resources from the medical commons equitably and efficiently is highly complex, for all of the reasons outlined above.

QUESTIONS

DIRECTIONS. (Items 1–4): For each numbered item below, select the ONE lettered option that is most closely associated with it. It is generally advisable to begin by reading the list of options. Then, for each item in the set, try to generate the correct answer and locate it in the option list, rather than evaluating each option individually. Each lettered option may be selected once, more than once, or not at all. Correct answers and explanations are given at the end of the chapter.

(A) limitation of functional ability as defined by a physician
(B) subjective experience of pathologic process
(C) medically defined pathophysiology
(D) social definition of physical limitations
(E) quality of life measure

1. disease
2. illness
3. impairment
4. disability

DIRECTIONS. (Items 5–10): Each of the numbered items or incomplete statements in this section is followed by answers or by completions of the statement. Select the ONE lettered answer or completion that is BEST in each case.

5. Which one of the following is a true statement regarding the "need" for medical care and the "demand" for medical care?

   (A) They are different because patients need care but physicians demand it.
   (B) They are different because "need" is professionally defined and "demand" is patient generated.
   (C) They are always the same.
   (D) They are different because "need" takes cost into consideration.
   (E) They are the same when barriers to care are minimal.

6. Which one of the following is a true statement regarding the utilization of medical resources?

   (A) It is equal to demand.
   (B) It is equal to need.
   (C) It is dependent on price elasticity.
   (D) It is greater than need.
   (E) It is influenced by unmet need.

7. The median age of a population is principally determined by

   (A) the crude mortality rate
   (B) the age-adjusted mortality rate
   (C) the birth rate
   (D) the prevalence of chronic disease
   (E) genetic factors

8. Which one of the following is a true statement regarding the US population over age 65 years?

   (A) The size of this population has been declining since World War II.
   (B) Rising cancer rates have shortened the life expectancy of this group.
   (C) It represents 30% of the total population.
   (D) It accounts for 50% of hospitalizations at any given time.
   (E) It enjoys a lower age-specific mortality rate than younger cohorts.

9. The medically indigent population is composed of

   (A) individuals receiving Medicare
   (B) individuals receiving Medicaid
   (C) individuals with limited or no medical insurance and too much income to qualify for Medicaid
   (D) individuals receiving welfare benefits
   (E) unemployed women with young children, exclusively

10. As of 1993, what percentage of the US population lacked medical insurance?

    (A) 5%
    (B) 15%
    (C) 20%
    (D) 25%
    (E) 37%

ANSWERS AND EXPLANATIONS

1. **The answer is C: medically defined pathophysiology.** A disease is a pathologic process occurring with or without symptoms. If a disease is asymptomatic, the physician might regard it more seriously than the patient. As an example of this, aortic root dilatation due to Marfan's syndrome, a disorder of connective tissue, might become life-threatening before symptoms are present, and a patient who feels well might need to be convinced of the need for cardiac surgery. The need to manage a disease must be balanced with the need to care for the patient, whose experience of the disease may differ markedly from that of the physician. *Epidemiology,* p. 261.

2. **The answer is B: subjective experience of pathologic process.** One may feel ill without having disease. One may harbor serious disease, as discussed in the answer to question 1, above, yet feel well. Illness is the patient's subjective experience of any disruption in normal health. *Epidemiology,* p. 261.

3. **The answer is A: limitation of functional ability as defined by a physician.** Impairment refers to a reduction in physi-

cal function resulting from illness or injury. The degree to which an individual can compensate for impairment determines the extent of the resultant disability. *Epidemiology,* p. 261.

4. The answer is D: social definition of physical limitations. Disability is loss of ability. It is the consequence of impairment. Disability is defined on the basis of what functions an individual can and cannot perform following illness or injury compared with the status of these functions before the impairment was acquired. Disability is a socially defined concept, largely based on a person's capacity to perform employment duties before and after an injury or illness. Disability can be partial or total, temporary or permanent. *Epidemiology,* p. 261.

5. The answer is B: They are are different because "need" is professionally defined and "demand" is patient generated. A patient with a "cold," or viral upper respiratory infection, may demand medical care the physician considers unnecessary (i.e., unneeded). Conversely, a patient using intravenous drugs may refuse help that the physician or other provider considers essential (i.e., needed). The need for medical care is professionally defined as the health care issues in a population for which medical intervention is indicated. Demand is the amount of care sought by the public under varying assumptions about the cost of care. *Epidemiology,* p. 262.

6. The answer is E: It is influenced by unmet need. Utilization of health care resources is effective demand, the overlap between the desire for care and the delivery of care. Both demand and need tend to be greater than utilization. The difference between need and utilization is unmet need. The greater the utilization of available resources, the smaller the disparity between utilization and need, and the smaller the unmet need is likely to become. *Epidemiology,* p. 262.

7. The answer is C: the birth rate. The birth rate varies from country to country far more than does either the crude death rate or the life expectancy. A high birth rate can rapidly bring down the median age of a population, even if the death rate is low among the elderly. A low birth rate will cause the median age of the population to climb steadily, even if life expectancy is limited by world standards and age-adjusted death rates are high. Often, a low birth rate and a low age-adjusted death rate occur together, as in Scandinavia, where the median age of the population is consequently high. *Epidemiology,* p. 262.

8. The answer is D: It accounts for 50% of hospitalizations at any given time. The age group over 65 years accounts for 11–12% of the US population at present but accounts for 50% of hospitalized patients at any given time. The utilization of health care resources by the elderly, therefore, is disproportionately great. This is to be expected, given the overwhelming impact of age on overall morbidity. The elderly population continues to increase in size and to represent a greater percentage of the total because of increasing life expectancy and, in particular, a low birth rate. The financial implications of this trend are discussed in the chapter. *Epidemiology,* p. 262.

9. The answer is C: individuals with limited or no medical insurance and too much income to qualify for Medicaid. Individuals with no financial resources often receive emergency medical care as wards of the state under Medicaid, Title 19 of the Social Security Act. These patients have no ability to pay medical bills and are not expected to do so. Individuals with inadequate resources to meet the financial demands of emergency medical care but with too many financial resources to qualify for Medicaid can be financially destroyed by illness or injury. This growing population in the USA is characterized as medically indigent and was the principal justification for the ill-fated attempt by the Clinton administration to guarantee universal health care insurance in this country. *Epidemiology,* p. 262.

10. The answer is B: 15%. This 15% of the population, as mentioned in the answer to question 9, above, was among the principal concerns of the Clinton administration in its attempt to institute health care reform. The US population exceeds 250 million, so that 15% represents more than 37 million people. The number of uninsured and underinsured adults is rising annually. Most members of this group are one health crisis away from poverty and are therefore considered medically indigent. *Epidemiology,* p. 262.

REFERENCES CITED

Committee on the Costs of Medical Care. Medical Care for the American People. Chicago, University of Chicago Press, 1932.

Donabedian, A. A Guide to Medical Care Administration. Vol. 2, Medical Care Appraisal. New York, American Public Health Association, 1969.

Flexner, A. Medical Education in the United States and Canada: A Report to the Carnegie Foundation for the Advancement of Teaching. Buffalo, N. Y., Heritage Press, 1910.

Hardin, G. The tragedy of the commons. Science 162:1243–1248, 1968.

Jekel, J. F. Epidemiology, Biostatistics, and Preventive Medicine. Philadelphia, W. B. Saunders, 1996.

Relman, A. S. The health care industry: where is it taking us? New England Journal of Medicine 325:854–859, 1991.

Sweet, V. Letter to the editor. New England Journal of Medicine 329:1655, 1993.

SECTION IV

Comprehensive Examination, Epidemiologic and Medical Glossary, and Appendix

Comprehensive Examination

QUESTIONS

DIRECTIONS. (Items 1–4): For each numbered item below, select the ONE lettered option that is most closely associated with it. It is generally advisable to begin each set by reading the list of options. Then, for each item in the set, try to generate the correct answer and locate it in the option list, rather than evaluating each option individually. Each lettered option may be selected once, more than once, or not at all. Correct answers and explanations are given at the end of this section.

(A) (1 − sensitivity)
(B) $d/(c + d)$
(C) (1 − specificity)
(D) $a/(a + b)$

Match the given value or rate with the correct formula.

1. false-positive error rate
2. false-negative error rate
3. positive predictive value
4. negative predictive value

DIRECTIONS. (Items 5–8): Each of the numbered items or incomplete statements in this section is followed by answers or by completions of the statement. Select the ONE lettered answer or completion that is BEST in each case.

5. The probability of disease in a patient with a negative test result, as expressed in a 2 × 2 table, is

(A) $a/(a + c)$
(B) $c/(a + c)$
(C) $c/(c + d)$
(D) $d/(c + d)$
(E) d

6. You wish to determine the proportion of myocardial infarctions that are fatal within the first 24 hours after they occur. You decide to examine the records of all local emergency rooms and doctors' offices. Then you will calculate the proportion of myocardial infarctions reported that resulted in death within 24 hours after the patient was seen initially. You briefly discuss your plan with a friendly neighborhood biostatistician, who immediately points out that your study is particularly subject to which one of the following forms of bias?

(A) selection bias
(B) observer bias
(C) measurement bias
(D) late look bias
(E) lead-time bias

7. An investigator is trying to determine whether medical screening programs using chest x-rays to detect lung cancer improve the survival time of the persons screened. She looks at the most recent year's data in the Connecticut Tumor Registry and discovers that the median survival time following the diagnosis of lung cancer is 6 months. She then performs a chest x-ray screening program in shopping centers and bowling alleys. She finds that in persons screened by this program, the median time from diagnosis to death is 9 months for the 95 cases of lung cancer discovered. The survival time difference is statistically significant. The investigator concludes that the screening program and subsequent treatment are adding an average of 3 months to the lives of lung cancer patients. You disagree on the basis of

(A) selection bias
(B) observer bias
(C) measurement bias
(D) late look bias
(E) lead-time bias

8. A company concerned about productivity lost because of health problems instituted an intensive medical treatment and support program for the 10% of its workers with the most time lost due to illness during a certain year. The treated group had much better attendance and productivity the next year. You know that the intensive program used was actually ineffective, and

189

you believe the most likely explanation for the outcome is

(A) random error
(B) the statistical regression effect
(C) effect modification
(D) type II error
(E) excessive power

DIRECTIONS. (Items 9–15): For each numbered item below, select the ONE lettered option that is most closely associated with it. It is generally advisable to begin each set by reading the list of options. Then, for each item in the set, try to generate the correct answer and locate it in the option list, rather than evaluating each option individually. Each lettered option may be selected once, more than once, or not at all.

(A) The interventions are properly selected and performed by trained professionals.
(B) There is public disclosure of financial records and quality standards.
(C) There are no geographic or financial barriers to care.
(D) The care provided is compatible with the patient's belief systems.
(E) The quality of medical care can be professionally reviewed.
(F) Care is provided during the times it is needed.
(G) The amount of care provided is commensurate with the need or demand, or both.

Match each of the criteria for high-quality health care with the corresponding definition or description.

9. availability of medical care
10. accessibility of medical care
11. adequacy of medical care
12. appropriateness of medical care
13. acceptability of medical care
14. accountability of medical care
15. assessability of medical care

DIRECTIONS. (Items 16–89): Each of the numbered items or incomplete statements in this section is followed by answers or by completions of the statement. Select the ONE lettered answer or completion that is BEST in each case.

16. A PPO, or preferred provider organization, is a variant of

(A) a DRG
(B) an HMO
(C) an ICF
(D) an IPA
(E) solo practice

Items 17–18

The following unpublished data concern the relationship between preventive treatment with chicken soup and the frequency with which colds develop. Assume that the data are from a randomized, double-blind trial of chicken soup versus placebo (e.g., gazpacho) over the course of 1 school year.

		OUTCOME		
		Colds	No Colds	Total
TREATMENT	Chicken Soup	36	21	57
	Placebo	35	11	46
	Total	71	32	103

17. When you begin to analyze the table, you decide to use percentages. Which of the following sets of percentages represents a correct and meaningful representation of the data above and provides a basis for making the comparison of interest?

(A) 63% and 76%
(B) 52% and 57%
(C) 65% and 51%
(D) 10% and 14%
(E) 5% and 95%

18. The appropriate significance test for these data is

(A) the Student's t-test
(B) the paired t-test
(C) the logrank test
(D) the Pearson correlation coefficient
(E) the chi-square analysis

Items 19–23

A test for the detection of baldness is devised. It is applied to a population of 200 adults in the USA who insist that they have hair. Based on a gold standard—namely, asking the subjects to remove their hats—20 of the 200 persons are actually bald. The new test detects 12 cases of baldness, of which 8 are positive by the gold standard and 4 are negative.

19. Based on these data, the sensitivity of the new test is

(A) 8%
(B) 12%
(C) 40%
(D) 92%
(E) 80%

20. The specificity of the new test is

(A) 40%
(B) 98%
(C) 66%
(D) 1%
(E) 12%

21. The positive predictive value of the new test in this population is

(A) 40%
(B) 98%
(C) 67%
(D) 1%
(E) 12%

22. The negative predictive value of the test is approximately

(A) 94%
(B) 98%

(C) 67%
(D) 1%
(E) 12%

23. The likelihood ratio positive for these data is

    (A) 2
    (B) 18
    (C) 8
    (D) 42
    (E) 12

Items 24–25

Assume that the sensitivity and specificity of the new test for baldness in questions 19–23 remain unchanged. Now, however, the gold standard is a consultant from Topknot Toupé, Inc., who decides that there are actually 150 cases of baldness instead of 20 cases.

24. The positive predictive value is now

    (A) 94%
    (B) 98%
    (C) 67%
    (D) 1%
    (E) 12%

25. The negative predictive value, given the revised prevalence, is

    (A) 35%
    (B) 98%
    (C) 67%
    (D) 1%
    (E) 12%

Item 26

Two radiologists review mammograms independently and recommend biopsy (+) or no biopsy (−). Of 200 mammograms reviewed, radiologist A recommends biopsies in 40 patients and radiologist B recommends biopsies in 32 patients. The distribution of their readings is as follows:

		RADIOLOGIST B		
		Positive	Negative	Total
RADIOLOGIST A	Positive	28	12	40
	Negative	4	156	160
	Total	32	168	200

26. The kappa test ratio for this table is

    (A) 0.50
    (B) 0.64
    (C) 0.18
    (D) 0.73
    (E) 0.88

Items 27–30

You are the medical director of a company that manufactures little plastic caps for the ends of shoelaces, a position to which you have aspired from early childhood. You wish to screen the workers for the dreaded *capus plasticus pedo-reversilosus*, a condition that causes the worker to place caps for the laces of the right shoe on the laces of the left, and vice versa. The condition can be treated effectively with ice-water immersion and concomitant hypnosis if it is detected early. From the literature on the subject, you determine that the sensitivity of the screening test is 94% and that the specificity is 90%. You estimate the prevalence of the condition in the workers to be 5%.

27. What percentage of positive screening test results will be false-positive results?

    (A) 23%
    (B) 48%
    (C) 67%
    (D) 18%
    (E) 81%

28. What is the positive predictive value of the screening test?

    (A) 33%
    (B) 48%
    (C) 67%
    (D) 28%
    (E) 81%

29. How many cases of the disease will be missed if 1000 workers are screened?

    (A) 95
    (B) 47
    (C) 8
    (D) 16
    (E) 3

30. Which one of the following facts would be the most important to establish before deciding whether or not to implement the screening test?

    (A) the prevalence of other diseases in the population
    (B) the cost of screening, follow-up, and treatment
    (C) the disease incidence
    (D) the size of the population at risk
    (E) the genetic risk factors for the disease

31. A screening test is applied to a population of 1000 in which the prevalence of disease Y is 10%. The sensitivity of this test is 96%, and the specificity is 92%. The diagnostic workup for each person found to have a true-positive result in the screening test will cost $50. A newer screening test has a specificity of 96% but costs $0.50 more per test than the older screening test. How much money would be saved or lost by choosing the newer test?

    (A) $1300 saved
    (B) $1300 lost
    (C) $ 500 saved
    (D) $1800 saved
    (E) $ 500 lost

Items 32–35

A friend of yours thinks that she may be pregnant. She purchased a product to test for pregnancy and found the following data provided in the product brochure:

		PREGNANCY TEST RESULT		
		Positive	Negative	Total
TRUE STATUS	Pregnant	253	24	277
	Not Pregnant	8	93	101
	Total	261	117	378

32. Turning to you for advice, your friend asks, "If the test says that I am pregnant, what is the probability that I really am pregnant?" Your answer is

   (A) 85%
   (B) 79%
   (C) 67%
   (D) 97%
   (E) 100%

33. Your friend then asks, "If the test says that I am not pregnant, what is the probability that I really am not?" Your answer is

   (A) 85%
   (B) 79%
   (C) 67%
   (D) 97%
   (E) 100%

34. Your friend is convinced that you actually know what you are talking about, so she asks, "If I really am pregnant, what is the probability that the test will discover that fact?" You confidently reply

   (A) 88%
   (B) 79%
   (C) 57%
   (D) 97%
   (E) 91%

35. You caution your friend that the numbers provided in the product brochure may not apply to her because

   (A) the test sensitivity is unknown
   (B) the test specificity is unknown
   (C) the prevalence in the population tested may not match her prior probability
   (D) the prevalence in the population tested is unknown
   (E) the false-positive error rate of the test is high

Items 36–38

Assume you are in charge of a state committee responsible for recommending whether or not to screen for congenital disease M, which is expected to be found in 1 per 10,000 births. If this disease is treated early, there will be a lifetime savings for the state of $250,000 (discounted for inflation) because of reductions in the cost of institutionalizing the affected infants. You determine from the literature that the sensitivity of the screening test is 95%, the specificity is 99%, the cost per screening test is $10, and the cost to evaluate each infant with a positive screening test result is $100.

36. What percentage of the infants with a positive screening test result will have disease M?

   (A) 1%
   (B) 5%
   (C) 9%
   (D) 20%
   (E) 68%

37. What will be the net yearly savings or loss to the state of screening if 100,000 infants are born each year?

   (A) $200,000 lost
   (B) $1.3 million saved
   (C) $1.5 million lost
   (D) $25,000 saved
   (E) $5 million saved

38. Approximately what percentage of the positive cases identified by screening would be diseased if disease M occurred once in each 1000 births rather than once in each 10,000 births?

   (A) 1%
   (B) 5.3%
   (C) 8.7%
   (D) 20%
   (E) 68.8%

39. You have just completed a history and physical examination on a patient for whom you consider the diagnosis of systemic lupus erythematosus possible but not very likely. You order a battery of six different tests for the disease, and four are positive and two are negative. To determine the probability of disease at this point, you would need to know

   (A) the prevalence, incidence, and natural history of systemic lupus erythematosus
   (B) the relative risk of systemic lupus erythematosus in this patient
   (C) the incidence of systemic lupus erythematosus and the predictive value of each test
   (D) the prior probability of systemic lupus erythematosus and the sensitivity and specificity of each test
   (E) the patient's risk factor profile, the family history, and the predictive value of each test

Items 40–41

The resting heart rates of 10 subjects are 70, 68, 84, 76, 88, 66, 56, 60, 80, and 70 beats per minute.

40. What is the variance of this data set?

   (A) 12.2
   (B) 104.4
   (C) 10.2
   (D) 86.6
   (E) 8.6

41. What is the standard deviation of this data set?

   (A) 12.2
   (B) 104.4
   (C) 10.2
   (D) 86.6
   (E) 8.6

Items 42–44

The following story is taken from the New King James Version of the Bible, Book of Daniel, chapter 1, verses 1–16:

> Daniel said to the steward whom the chief of the eunuchs had set over Daniel, Hananiah, Mishael, and Azariah, "Please test your servants for ten days, and let them give us vegetables to eat and water to drink. Then let our countenances be examined before you, and the countenances of the young men who eat the portion of the king's delicacies; and as you see fit, so deal with your servants."
>
> So he consented with them in this matter, and tested them [i.e., the Israelites] for ten days. And at the end of ten days their countenance appeared better and fatter in flesh than all the young men who ate the portion of the king's delicacies. Thus, the steward took away their portion of delicacies and the wine that they were to drink, and gave them vegetables.

42. What type of study design is described?

    (A) field trial
    (B) case-control study
    (C) retrospective cohort study
    (D) prospective cohort study
    (E) cross-sectional study

43. Although this study is in some ways quite primitive, a strength it demonstrates is

    (A) random allocation
    (B) random sampling
    (C) matching
    (D) use of controls
    (E) blinding

44. This study is susceptible to all of the following sources of error EXCEPT

    (A) selection bias
    (B) recall bias
    (C) measurement bias
    (D) confounding
    (E) observer bias

45. At a given level of alpha, the z value associated with a one-tailed test is smaller than the z value associated with a two-tailed test because

    (A) the rejection region is smaller for a one-tailed test
    (B) the rejection region is smaller for a two-tailed test
    (C) there is 1 less degree of freedom with a one-tailed test
    (D) the entire rejection region is on one side of the mean with a one-tailed test
    (E) the entire rejection region is on one side of the mean with a two-tailed test

46. A trial of an antihypertensive agent is performed by administering the drug or a placebo, with a "washout period" in between, to each study subject. The treatments are administered in random order, and each subject serves as his or her own control. The trial is double-blind. The appropriate significance test for the change in blood pressure with drug versus placebo is the

    (A) chi-square analysis
    (B) Pearson correlation coefficient
    (C) paired t-test
    (D) ANOVA
    (E) logrank test

Items 47–48

Two different groups of investigators perform clinical trials of the same therapy. The trials have approximately the same sample size. In the first trial, the investigators find more successes in the treatment group and report a p value of 0.04. Based on these results, the investigators recommend the therapy. In the second trial, the investigators find more successes in the treatment group but report a p value of 0.08. Based on these results, the investigators do not recommend the therapy.

47. After reviewing the studies, which one of the following statements could a clinician make that would best describe the statistical significance of the data?

    (A) If the data achieve statistical significance, the studies will be clinically important.
    (B) If the data do not achieve statistical significance, the studies cannot be clinically important.
    (C) The reason that statistical significance has not been achieved is probably because alpha has been set too high.
    (D) The reason that statistical significance has not been achieved is probably because beta has been set too low.
    (E) Data that are not statistically significant may nonetheless be clinically important.

48. An appropriate means for reconciling the conflicting recommendations would be to

    (A) conduct a case-control study
    (B) obtain expert opinion
    (C) analyze the data using multivariable methods
    (D) pool the data
    (E) perform intention-to-treat analysis

49. You are preparing a before-after trial of a hypoglycemic agent in the management of type II diabetes mellitus. In the trial, each subject will serve as his or her own control. You plan to set a two-tailed alpha at 0.05, to look for a difference $(\bar{d})$ of 10 mg/dL in the serum glucose level, and to ignore beta. You select a standard deviation (s) of 22 mg/dL from prior studies. The total sample size needed for your study is best represented by which of the following formulas?

    (A) $N = \dfrac{(z_\alpha)^2 \cdot (s)^2}{(\bar{d})^2}$

    (B) $N = \dfrac{(z_\alpha)^2 \cdot 2 \cdot (s)^2}{(\bar{d})^2}$

    (C) $N = \dfrac{2 \cdot (z_\alpha)^2 \cdot 2 \cdot (s)^2}{(\bar{d})^2}$

    (D) $N = \dfrac{\sqrt{(z_\alpha - \bar{d})^2}}{s^2}$

    (E) $N = \dfrac{(z_\alpha)^2 \cdot (\bar{d})^2}{s^2}$

50. Diagnosis-related groups (DRGs) are used to

    (A) assign patients to appropriate hospital wards
    (B) assess the quality of subspecialty care in hospitals
    (C) provide support for patients following hospital discharge
    (D) stipulate prospective payment to hospitals
    (E) review interobserver agreement in radiology

51. Epidemiology is best defined as the study of

    (A) disease outbreaks
    (B) the incidence of human disease
    (C) causal factors in disease
    (D) the factors that influence the occurrence and distribution of disease in human populations
    (E) the transmission of infectious diseases and exposure to environmental toxins

52. The mosquito involved in the transmission of malaria assumes the role of

    (A) animal reservoir
    (B) surrogate host
    (C) vector
    (D) parasite
    (E) commensal

53. When the number of cancer cases diagnosed annually in the USA is compared with the expected number of cases if the lowest rates applied, there are

    (A) 5 times as many cases as would be expected
    (B) 2 times as many cases as would be expected
    (C) half as many cases as would be expected
    (D) 10 times as many cases as would be expected
    (E) this comparison is not valid

54. The bacterium responsible for legionnaire's disease usually replicates in and is spread from

    (A) swimming pools
    (B) drinking water
    (C) unpasteurized beer
    (D) warm-water systems such as air-conditioning cooling towers
    (E) food, especially meat, cooked at inadequate temperatures

55. Mormons and Seventh-Day Adventists have lower-than-average age-adjusted death rates. The avoidance of alcohol and tobacco and the adherence to a healthful diet by members of these groups are factors

    (A) unrelated to mortality rates because the rates are largely the result of genetic factors
    (B) unrelated to age-adjusted death rates, which are largely influenced by the age distribution of the population
    (C) causing the lower mortality rates
    (D) associated with the lower mortality rates but may not be causal factors
    (E) associated with reduced mortality rates from cardiovascular disease but not from cancer

56. Which of the following best describes the effect of vaccination against diphtheria when the disease was more common than it is currently?

    (A) Frequent exposure resulted in infection, against which vaccination was not protective.
    (B) Frequent exposure resulted in boosted immunity following immunization.
    (C) Vaccination resulted in unusually severe clinical disease when immunity was not achieved.
    (D) Susceptibility to severe infection redeveloped several years following vaccination.
    (E) Vaccination resulted in hypersensitivity to the infectious agent.

57. Which one of the following is a true statement about the Sabin polio vaccine?

    (A) The vaccine does not confer immunity.
    (B) The vaccine is administered intramuscularly.
    (C) Use of the vaccine contributes to herd immunity.
    (D) The vaccine is a killed virus vaccine.
    (E) The Sabin polio vaccine is always superior to the Salk polio vaccine.

58. The demographic gap refers to

    (A) socioeconomic differences among countries
    (B) disruption of the nuclear family
    (C) persons living below the poverty line in the USA
    (D) the difference between Medicare reimbursement and hospital charges
    (E) the difference between the birth and death rates of a society

59. Construction of the Aswan High Dam in Egypt resulted in increased transmission of which one of the following diseases?

    (A) schistosomiasis
    (B) tuberculosis
    (C) leishmaniasis
    (D) yellow fever
    (E) necrotizing fasciitis

60. Which one of the following is a true statement regarding the quality of data reported for cases of relatively mild diseases such as chickenpox?

    (A) The data are considered valueless because mild diseases are underreported.
    (B) The data are highly accurate because of active surveillance.
    (C) The data are useful for establishing prevalence but not incidence.
    (D) The data are valueless because most cases do not present for medical care.
    (E) The data are valuable for detecting changes in the pattern of a disease.

61. Which of the following is a true statement regarding immunization against poliomyelitis with the Salk injectable vaccine?

    (A) It contributes to herd immunity.
    (B) It is never indicated.
    (C) It always protects the individual from poliomyelitis.
    (D) It results in both cell-mediated and humoral immunity.
    (E) It is useful for immunocompromised hosts.

62. An ecologic perspective is necessary to understand all of the following EXCEPT

    (A) cephalopelvic disproportion during labor in some Muslim and Indian women
    (B) the genetic basis of phenylketonuria
    (C) the transmission of Lyme disease
    (D) etiologic factors in atherogenesis
    (E) the epidemiology of obesity

63. Which one of the following types of data does epidemiology derive directly from the US census?

    (A) data for direct standardization
    (B) data for indirect standardization
    (C) denominator data for population-based studies
    (D) time trend data
    (E) the standardized mortality ratio

64. What percentage of death certificate diagnoses are thought to be potentially inaccurate?

    (A) 5%
    (B) 15–20%
    (C) 30%
    (D) 50%
    (E) 80–85%

65. The number of cases of viral influenza in a population is divided by the population at the beginning of the observation period. This is reported as the "risk" of influenza. Which one of the following is a true statement regarding this measure?

    (A) This is an acceptable risk estimate, even though the best denominator would include only subjects without protective immunity to influenza.
    (B) This is an acceptable risk estimate, although the best denominator would be the midyear population at risk.
    (C) This is an acceptable risk estimate because the risk is similar to the rate when the disease is common.
    (D) This is an unacceptable estimate of risk, because too many cases of influenza go unreported.
    (E) This is a rate, not a risk.

66. Which one of the following is a true statement regarding rates as compared with risks?

    (A) Rates are generally lower.
    (B) Rates are generally higher.
    (C) Rates are based on a more accurate denominator.
    (D) Rates better demonstrate the force of mortality.
    (E) Rates rely on more constant denominators.

67. Which one of the following is a valid rate?

    (A) uterine cancer in the population in the USA
    (B) lung cancer in men in the USA
    (C) poliomyelitis in grade school children
    (D) prostatitis in adults over age 65 years
    (E) rubella in pregnant women

68. A particularly useful measure of events that can occur repeatedly in individual subjects under observation is

    (A) the period prevalence
    (B) the incidence density
    (C) the risk
    (D) the incidence rate
    (E) the frequency

Items 69–78

For populations A and B, the following information about deaths in certain age groups is known:

Age Group (years)	Population A Population	Population A Deaths	Population B Population	Population B Deaths
< 30	13,000	13	4,000	3
30–65	15,000	45	15,000	30
> 65	4,000	120	13,000	145

69. The crude death rate for population A may best be expressed as

    (A) 178
    (B) 1.8 per 10,000
    (C) 0.005
    (D) 5.6 per 1000
    (E) 60 per 100,000

70. In selecting an answer for question 69, the following assumption was made:

    (A) The force of mortality was constant over the period of interest.
    (B) All deaths occurred during the period of interest.
    (C) The population figures are constant over time.
    (D) All deaths were due to chronic disease.
    (E) Diagnostic criteria for cause of death were consistent during the study period.

71. The crude death rate for population B is

    (A) 5.6 per 1000
    (B) 178
    (C) 60 per 100,000
    (D) 0.005
    (E) 1.8 per 10,000

72. Crude death rates in populations are determined by

    (A) the causes of death
    (B) the size of the population and the total number of deaths
    (C) seasonal variation in mortality rates and diagnostic consistency
    (D) disease surveillance and reporting
    (E) the age distribution

73. The age-specific death rate for those over 65 years in population A is

    (A) 3%
    (B) 120
    (C) 3 per 1000
    (D) always less than the crude rate
    (E) dependent upon the size of the standard population

74. Direct standardization is needed to compare the mortality rates of populations A and B because

    (A) the numerators are not comparable
    (B) the causes of death are different
    (C) the period of observation is unknown
    (D) the force of mortality is variable
    (E) the age distributions are different

75. Any one of the following could serve as a reference population for use in calculating standardized mortality rates by the direct method EXCEPT

    (A) the population in the USA
    (B) the sum of populations A and B
    (C) a population of nursing home residents
    (D) the sum of populations A and B in each age group
    (E) the population of a country other than the USA

76. The standardized mortality rate for population A is

    (A) 17
    (B) 0.003
    (C) 9.64 per 1000
    (D) 0.034
    (E) 5.1 per 1000

77. The standardized mortality rate for population B is

    (A) 9.64 per 1000
    (B) 0.007
    (C) 0.023
    (D) 4.06 per 1000
    (E) exactly 1%

78. The reason standardized mortality rates differ for populations A and B is that

    (A) population B is dying sooner
    (B) population B is older
    (C) population A is smaller
    (D) the crude rates differ
    (E) the crude rates are the same

79. All of the following are correct statements regarding control measures during an outbreak investigation EXCEPT

    (A) they include sanitation, prophylaxis, diagnosis and treatment, and vector control
    (B) they usually require an understanding of the modes of disease transmission
    (C) they must always await hypothesis testing
    (D) they may create public relief or alarm
    (E) they may be effective even if based on erroneous hypotheses

80. Obesity has been labeled an epidemic in the USA. This is

    (A) incorrect, because obesity is not a communicable disease
    (B) incorrect, because obesity takes years to develop
    (C) correct, because one in three adults is considered obese
    (D) correct, because obesity has a genetic component
    (E) correct, because current prevalence exceeds the historical baseline

81. The following deaths are reported for a population of 2807 men, age 35–68 years: 4 deaths in nonsmokers; 11 in those smoking up to one-half pack of cigarettes per day; 18 in those smoking one-half to one pack per day; and 43 in those smoking over one pack per day. From this, one may conclude that

    (A) smoking increases mortality
    (B) the relationship between mortality rates and smoking is linear
    (C) there is a dose-response relationship between smoking and death
    (D) age at death is required for each case in order to interpret the data
    (E) the data must be expressed as rates to be interpreted

82. Among elderly subjects who are fit, vigorous exercise reduces the risk of heart disease. Among elderly subjects who are unfit, the initiation of vigorous exercise might precipitate a myocardial infarction. Fitness may be considered

    (A) a risk factor
    (B) an effect modifier
    (C) a confounder
    (D) a necessary cause
    (E) a sufficient cause

83. In the relationship between obesity and cardiovascular disease, hyperlipidemia and hyperglycemia are

    (A) effect modifiers
    (B) confounders
    (C) necessary causes
    (D) intervening variables
    (E) unrelated

84. When questioned years after an illness, cases reported more severe subjective symptoms than did controls. This is probably an example of

    (A) misclassification
    (B) random error
    (C) spectrum bias
    (D) confounding
    (E) recall bias

85. What is the relationship of asbestos to asbestosis?

    (A) necessary cause
    (B) sufficient cause
    (C) confounder
    (D) effect modifier
    (E) synergism

86. A study is conducted in which the rate of cigarette smoking and the prevalence of tuberculosis are assessed in the same population at the same time. Both are found to be high, and the conclusion is drawn that cigarette smoking is causally related to tuberculosis. Based on this study only, all of the following are true statements regarding this conclusion EXCEPT

    (A) it may be incorrect due to confounding
    (B) it is merely a hypothesis
    (C) it could serve as the basis for a case-control or cohort study

(D) it is correct but not generalizable without further study
(E) it may be incorrect due to the ecologic fallacy

87. The National Health and Nutrition Examination Surveys (NHANES) are conducted at intervals by the government to establish trends in health-related behaviors and dietary practices. For example, the recently conducted NHANES III (Phase 1, 1988–1991) shows a decline in mean fat intake for the population from 37% of calories to 34% of total calories. This type of research is an example of

(A) a retrospective cohort study
(B) a longitudinal ecologic study
(C) a randomized field trial
(D) a longitudinal cohort study
(E) hypothesis testing

88. Which one of the following examples might represent a retrospective cohort study?

(A) cases of lung cancer assessed for prior exposures
(B) subjects with current angina followed for the development of myocardial infarction
(C) subjects with prior radiation exposure followed for the development of lymphoproliferative cancers
(D) subjects randomly assigned to active varicella vaccine versus placebo, followed for chickenpox
(E) subjects with skin cancer assessed for lifelong, cumulative sun exposure

89. Randomization and double-blinding in a controlled trial are intended to accomplish all of the following EXCEPT

(A) minimize bias
(B) optimize external validity
(C) reduce or eliminate known confounders
(D) reduce or eliminate unknown confounders
(E) produce similar groups that differ by the intervention only

DIRECTIONS. (Items 90–94): This section consists of a list of lettered options followed by numbered items. For each numbered item, select the ONE lettered option that is most closely associated with it. It is usually helpful to begin by reading the list of options. Then, for each numbered item, try to generate the correct answer and locate it in the option list, rather than evaluating each option individually. Each lettered option may be selected once, more than once, or not at all.

Items 90–91

(A) case-control study
(B) longitudinal ecologic study
(C) prospective cohort study

Match the study design with the appropriate description.

90. Multiple disease outcomes can be studied, but only the risk factors defined and measured at the beginning of the study can be assessed.

91. Only one outcome may be studied, but many risk factors or exposures may be assessed.

Items 92–94

(A) statistical analysis
(B) randomization
(C) hypothesis testing
(D) double-blinding
(E) increasing the sample size

Match each of the terms related to study design and interpretation with the appropriate description.

92. reduces confounding by both known and unknown factors

93. minimizes measurement bias

94. In a prospective randomized trial, the two subject groups differ significantly with regard to several factors other than the intervention under investigation. In order to compensate for this, what is required?

DIRECTIONS. (Items 95–101): Each of the numbered items or incomplete statements in this section is followed by answers or by completions of the statement. Select the ONE lettered answer or completion that is BEST in each case.

95. The Physicians' Health Study, which is designed to test the effects of aspirin and beta carotene on cardiovascular disease and cancer, is a randomized, double-blind, placebo-controlled trial (Physicians' Health Study Steering Committee 1989). The participants in the trial are approximately 22,000 male physicians in the USA who are 40–75 years of age. The design of the study as described helps to achieve which of the following?

(A) prevention of confounding by known and unknown factors
(B) elimination of bias
(C) internal validity
(D) external validity
(E) statistical significance

96. In a cohort study, the groups are compared on the basis of

(A) disease status
(B) exposure status
(C) inclusion criteria
(D) exclusion criteria
(E) age

Items 97–99

In the US population, approximately 50 million people have hypertension to varying degrees (Joint National Committee on Detection, Evaluation, and Treatment of High Blood Pressure 1993). Hypertension is the most important risk factor for cerebrovascular disease, which accounted for 144,000 deaths in 1990 (McGinnis and Foege 1993). The relative risk for cerebrovascular accident rises with the magnitude of hypertension, and has ranged in published studies from slightly over 1 to nearly 50 (Fletcher and Bulpitt 1994). To answer the following items, assume that the annual death rate from cerebrovascular accident is 150,000, that the total US population is 265 million, and that the rate of death due to cerebrovascular accident among hypertensives is 3.5 times higher than that among the general population.

97. Among the general population, how much of the total risk for death caused by cerebrovascular accident is due to hypertension?

    (A) 12%
    (B) 1/8
    (C) 5.7
    (D) 11.6 per 10,000 deaths
    (E) 3.4 per 10,000 deaths

98. Among the general population, what percentage of the total risk for cerebrovascular accident mortality is due to hypertension?

    (A) 3.4%
    (B) 59.6%
    (C) 72%
    (D) 8.4%
    (E) 0%

99. The risk for cerebrovascular accident mortality attributable to hypertension is

    (A) 28%
    (B) 32%
    (C) 17.7 per 10,000
    (D) 1.8 per 100,000
    (E) 3.1 per 10,000

100. Which one of the following measures is most useful in the generation of public health policy?

    (A) PAR%
    (B) AR
    (C) RR
    (D) OR
    (E) AR%$_{(exposed)}$

101. An investigator selects a sample size formula based on the study design. To determine the sample size required, the investigator must know all of the following EXCEPT

    (A) variance
    (B) the level of alpha
    (C) the number of steps to be compared
    (D) the p value
    (E) the smallest important difference $(\bar{d})$

DIRECTIONS. (Items 102–105): For each numbered item, select the ONE lettered option that is most closely associated with it. It is generally advisable to begin each set by reading the list of options. Then, for each item in the set, try to generate the correct answer and locate it in the option list, rather than evaluating each option individually. Each lettered option may be selected once, more than once, or not at all.

(A) $N = \dfrac{(z_\alpha + z_\beta)^2 \cdot 2 \cdot \bar{p}(1-\bar{p})}{(\bar{d})^2}$

(B) $N = \dfrac{(z_\alpha)^2 \cdot (s)^2}{(\bar{d})^2}$

(C) $N = \dfrac{(z_\alpha)^2 \cdot 2 \cdot (s)^2}{(\bar{d})^2}$

(D) $N = \dfrac{(z_\alpha + z_\beta)^2 \cdot 2 \cdot (s)^2}{(\bar{d})^2}$

Match the description of the trial with the appropriate sample size formula.

102. randomized controlled trial of the effects of chocolate versus placebo on the development of complete inner peace, with the outcome expressed as the proportion of subjects in each group

103. randomized controlled trial of the effects of sauteed gravel (unpretentious, with a lingering crunch) versus placebo on serum cholesterol levels, with experimental and control groups of comparable size (sample size considers type 1 error but not type 2 error)

104. before and after study in one group of subjects of the effects of watching Brady Bunch reruns on IQ

105. randomized controlled trial of soluble fiber versus placebo, and effects on postprandial glycemia, with comparable control and experimental groups (both type 1 error and type 2 error are considered)

DIRECTIONS. (Items 106–150): Each of the numbered items or incomplete statements in this section is followed by answers or by completions of the statement. Select the ONE lettered answer or completion that is BEST in each case.

106. The square of the correlation coefficient, the r^2 value, represents

    (A) the covariance of y
    (B) the degrees of freedom associated with x
    (C) the risk ratio
    (D) (1 − specificity)
    (E) the strength of the association between x and y

107. Linear regression would be an appropriate method for which of the following scenarios?

    (A) predicting serum cholesterol levels when dietary intake data are known
    (B) predicting blood pressure levels when the family history of the presence or absence of hypertension is known
    (C) calculating the probability of achieving remission of leukemia when the prior response to chemotherapy is known
    (D) predicting the blood type of a patient when the blood types of the parents are known
    (E) predicting the peak air flow rates when the pack-years of cigarettes smoked are known

108. In using the chi-square test for statistical analysis of a 2 × 2 contingency table, the null hypothesis is

    (A) that the two means do not differ
    (B) that the row and column totals are the same

(C) that the two variables are independent of each other
(D) that the two variances do not differ
(E) that p is < 0.05

Items 109–118

A study is conducted to determine the rates of iron deficiency anemia among 400 adult male manganese miners, some of whom use dental floss and some of whom do not. The following data are gathered:

		IRON DEFICIENCY ANEMIA	
		Positive	Negative
EXPOSURE TO DENTAL FLOSS	Positive	150	50
	Negative	20	180

109. What type of study is this?

    (A) ecologic study
    (B) cohort trial
    (C) case-control study
    (D) field trial
    (E) randomized study

110. Which one of the following can be determined from the data provided?

    (A) the mean value for each population
    (B) the odds ratio
    (C) the incidence of anemia
    (D) the relative risk
    (E) the prevalence

111. The number of subjects using dental floss and expected to have iron deficiency anemia is

    (A) 50
    (B) 20
    (C) 200
    (D) 75
    (E) 85

112. The number of subjects not using dental floss and expected to have iron deficiency anemia is

    (A) 50
    (B) 20
    (C) 200
    (D) 75
    (E) 85

113. The number of subjects not using dental floss and expected not to have iron deficiency anemia is

    (A) 40
    (B) 26
    (C) 115
    (D) 130
    (E) 230

114. The null hypothesis for this study is best expressed by which of the following?

    (A) Mining manganese causes iron deficiency anemia.
    (B) Using dental floss causes iron deficiency anemia.
    (C) Iron deficiency anemia is associated with both manganese mining and using dental floss.
    (D) Use of dental floss and iron deficiency anemia are not associated.
    (E) Manganese mining and use of dental floss are not associated.

115. Cell c of the table represents

    (A) those with the exposure but not the outcome
    (B) those with the outcome but not the exposure
    (C) those without the exposure but with the outcome
    (D) those with the exposure and the outcome
    (E) those with neither outcome nor exposure

116. The chi-square value for this table is

    (A) 12.5
    (B) 172.8
    (C) 400
    (D) 0.125
    (E) 16.4

117. How many degrees of freedom are associated with the chi-square value calculated in question 116?

    (A) 0
    (B) 1
    (C) 2
    (D) 3
    (E) 4

118. To determine the statistical significance of a given chi-square value, one must

    (A) determine the 95% confidence limits
    (B) find the corresponding p value in a chi-square table
    (C) determine the variance
    (D) perform power calculations
    (E) perform sample size calculations

119. Which of the following is most likely to contribute to a statistically significant but clinically meaningless outcome?

    (A) setting alpha at < 0.05
    (B) obtaining a p value larger than alpha
    (C) performing a two-tailed test instead of a one-tailed test
    (D) performing subgroup analysis
    (E) using a very large sample

Items 120–121

A study is conducted to determine the survival benefit of chemotherapy in subjects with pancreatic carcinoma. Subjects are enrolled at diagnosis and treated immediately after staging. The study is conducted over a period of 4 years. You observe 10 subjects for 6 months, and 6 of these subjects die and 4 are lost to follow-up. You observe 30 subjects for 4 months; 16 have died and 4 are alive when the study is terminated, and 10 are lost to follow-up. You observe 2 subjects for 3 years; 1 dies and 1 survives.

120. The total observation period

    (A) is best expressed as 4 years
    (B) is best expressed as 48 person-months
    (C) is best expressed as 4 person-years
    (D) is best expressed as 21 person-years
    (E) cannot be determined from the available information

121. The mortality rate in this study

    (A) is approximately 1.1 deaths per person-year of observation
    (B) is approximately 23 deaths per person-year of observation
    (C) is approximately 5.8 deaths per person-month of observation
    (D) is 5.75 deaths per year of observation
    (E) cannot be determined because of losses to follow-up

122. The principal way in which the Kaplan-Meier life table method differs from the actuarial method is that

    (A) the Kaplan-Meier method can be used only if death is the outcome of interest
    (B) the Kaplan-Meier method requires a larger sample
    (C) the Kaplan-Meier method is not limited to dichotomous outcomes
    (D) the Kaplan-Meier method counts censored subjects as deaths
    (E) the Kaplan-Meier method uses uneven death-free intervals

123. All of the following should be true of a disease for which screening is planned EXCEPT

    (A) the disease must be treatable
    (B) the natural history of the disease must be understood
    (C) the disease should be rare
    (D) the disease should be serious
    (E) a population at risk should be identifiable

124. An enormously successful hypertension screening campaign is repeated after 1 year in a retirement community. This time, the results are disappointing because very few new cases of hypertension are detected. The likely explanation for this outcome is

    (A) a new cutoff level for diagnosing hypertension
    (B) the phenomenon of regression to the mean
    (C) a decline in the prevalence of hypertension
    (D) a high rate of mortality among elderly individuals with hypertension
    (E) the fact that only incident cases are detected

125. Patients may be more amenable to tertiary prevention than to either primary or secondary prevention because

    (A) the onset of symptoms may cause concern and raise motivation
    (B) tertiary prevention is more cost-effective
    (C) only tertiary prevention limits disability in older age groups
    (D) only tertiary prevention occurs after the onset of disease
    (E) tertiary prevention produces the greatest net benefit

126. The effect of obesity on the risk of cardiovascular disease is known to be related to

    (A) gender
    (B) family history
    (C) age
    (D) body fat distribution
    (E) cigarette smoking

127. Data from the Framingham Study suggest that risk factors for coronary artery disease are

    (A) additive
    (B) common only in men
    (C) synergistic
    (D) too common to justify screening
    (E) unimportant in women

128. The circulating level of which one of the following lipid moieties is inversely related to the risk of myocardial infarction?

    (A) VLDL
    (B) LDL
    (C) triglycerides
    (D) HDL
    (E) total cholesterol

129. Which class of drugs disrupts cholesterol biosynthesis in the liver?

    (A) HMG co-A reductase inhibitors (i.e., statins)
    (B) bile acid sequestrants
    (C) fibrates
    (D) ACE inhibitors
    (E) selective serotonin reuptake inhibitors

130. According to the generally accepted definition of hypertension, approximately how many prevalent cases are there in the USA?

    (A) 5 million
    (B) 10 million
    (C) 25 million
    (D) 500,000
    (E) 50 million

131. Which of the following characteristics is true of most cases of hypertension?

    (A) They are caused by renal artery stenosis.
    (B) They are moderate to severe.
    (C) They are asymptomatic.
    (D) They are secondary.
    (E) They are unresponsive to life-style modification.

132. The only routine screening programs for diabetes that have proved cost-effective are those in

    (A) elderly persons
    (B) obese persons
    (C) pregnant women
    (D) neonates
    (E) school-age children

133. The difference between impairment and disability is that

    (A) impairment is socially defined
    (B) disability is socially defined
    (C) disability is medically defined
    (D) impairment is permanent
    (E) disability is permanent

134. The fundamental responsibility for the health of the US public lies with

    (A) the states
    (B) the federal government
    (C) the local governments
    (D) the Surgeon General
    (E) the Institute of Medicine

135. In contrast to the view of public health espoused in the USA, the Ottawa Charter (1986) emphasizes

    (A) individual responsibility for health
    (B) corporate responsibility for health
    (C) nuclear disarmament
    (D) consideration of the health impact of all public policy
    (E) population control and renewable energy sources

136. The targeted dietary fat intake in *Healthy People 2000* is

    (A) 80 g/d
    (B) 10% of total daily calories
    (C) 20% of total daily calories
    (D) 30% of total daily calories
    (E) 1 g/kg body weight per day

137. A 42-year-old woman presents to her physician for a routine examination. She has a family history of heart disease but not breast cancer. She smokes 1 pack of cigarettes daily and has done so for 24 years, and she is sedentary. Her diet is of the "typical American" variety, which is to say her intake of dietary fat and sodium exceeds recommendations. Her systolic blood pressure is 146 mm Hg and her diastolic blood pressure is 92 mm Hg. You estimate her weight to be about 35 lb above the ideal weight (she is 5 feet 5 inches tall and weighs 170 lb). Your highest priority for this encounter is to

    (A) order a mammogram
    (B) encourage physical activity
    (C) encourage smoking cessation
    (D) initiate antihypertensive therapy
    (E) obtain a lipid profile

138. The year 2000 goal for deaths resulting from coronary artery disease is to reduce the annual age-adjusted death rate to

    (A) 100 per 100,000 persons
    (B) 100 per 10,000 persons
    (C) 35 per 100,000 persons
    (D) 35 per 10,000 persons
    (E) 12 per 1000 persons

139. During a 3-month period, 9 cases of Hantavirus pulmonary syndrome are reported, and 3 of the cases die. Which of the following calculations would be 33%?

    (A) the attack rate
    (B) the crude death rate
    (C) the case fatality ratio
    (D) the standardized mortality ratio
    (E) the adjusted death rate

140. The epidemiologic year for hepatitis A runs from January to January. This is because

    (A) the calendar year runs from January to January
    (B) the incidence rate of hepatitis A peaks in January
    (C) hepatitis A is spread through respiratory droplets
    (D) fecal-oral transmission peaks in the summer
    (E) the incidence of hepatitis A does not vary seasonally

141. One disease for which surveillance in the USA is particularly sophisticated is

    (A) HIV infection and acquired immunodeficiency syndrome (AIDS)
    (B) syphilis
    (C) varicella
    (D) influenza
    (E) measles

142. The most appropriate sequence of action in an outbreak investigation is to

    (A) determine whether an epidemic exists; characterize the epidemic by time, place, and person; establish the diagnosis; and initiate control measures
    (B) establish the case definition; test hypotheses; characterize the epidemic by time, place, and person; and establish the diagnosis
    (C) determine whether an epidemic exists; initiate control measures; characterize the epidemic by time, place, and person; and establish the diagnosis
    (D) establish the case definition; determine whether an epidemic exists; characterize the epidemic by time, place, and person; develop hypotheses regarding spread; and initiate control measures
    (E) establish a diagnosis; conduct a case-control study; and initiate control measures

Items 143–150

Infection of the oropharynx by *Streptococcus pyogenes* (group A streptococcus) is a risk factor for rheumatic fever and may cause injury to the heart valves (carditis). Although the incidence of rheumatic fever declined in the USA from the early part of the twentieth century up until the 1980s, there has been a recent resurgence of this disease (Bisno 1991). You see a patient complaining of a sore throat for a routine office visit. The patient has no significant medical problems and denies having any allergies to antibiotics. You have recently read the article cited above and are fearful of leaving "strep throat" untreated, but you are also concerned about the unnecessary use of antibiotics and the consequent propagation of antibiotic-resistant strains of bacteria. Suddenly overcome with uncertainty, you resort to the following decision tree:

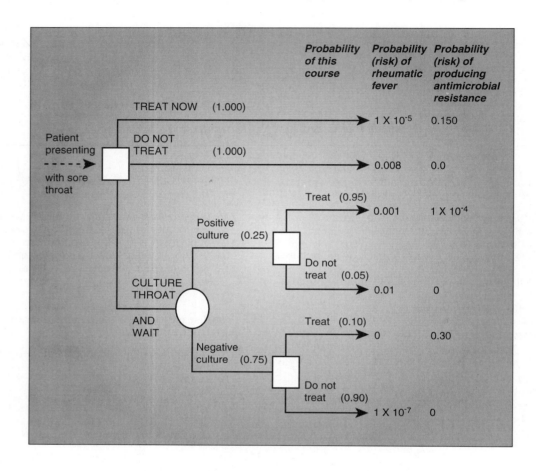

143. If you decide to treat your patient immediately, the risk of rheumatic fever is

    (A) 0.150
    (B) $1/10^7$
    (C) 0.44
    (D) $1/10^5$
    (E) 0.009

144. If you decide not to treat your patient, the risk of rheumatic fever is

    (A) 0.011
    (B) 0.008
    (C) 8%
    (D) 0.0063
    (E) 1

145. If you decide to obtain a throat culture and the result is positive, and you then decide to treat your patient, the risk of rheumatic fever is

    (A) 2.4/10,000
    (B) 9.5/10,000
    (C) 1%
    (D) 2.2%
    (E) 0.001

146. If you decide to obtain a throat culture and the result is positive, and you then decide to treat your patient, the risk of producing antimicrobial resistance is

    (A) $1/10^7$
    (B) 9.5/10,000
    (C) 0
    (D) $1/10^5$
    (E) $1/10^4$

147. The overall risk of rheumatic fever in all of the treated patients is

    (A) 1
    (B) $1.01/10^4$
    (C) $1/10^4$
    (D) 0
    (E) 0.30

148. The overall risk of rheumatic fever in all of the untreated patients is approximately

    (A) $1/10^4$
    (B) $1.01/10^4$
    (C) 1.8%
    (D) 0
    (E) 0.30

149. The overall risk of antimicrobial resistance developing in all of the treated patients is approximately

    (A) 1%
    (B) 10%
    (C) 33%
    (D) 45%
    (E) 1.8/1000

150. The overall risk of rheumatic fever in all of the patients for whom a throat culture is obtained is

    (A) $2.375/10^4$
    (B) $1.25/10^4$
    (C) $6.75/10^8$
    (D) $3.6/10^4$
    (E) 0%

ANSWERS AND EXPLANATIONS

1. The answer is C: (1 − specificity). In a 2 × 2 table, specificity is $d/(b + d)$, or the number of true-negative test results. Specificity is the ability of a test to exclude a disease when it is truly absent. Test results for those who do not have a disease can be either negative (i.e., true-negative results) or positive (i.e., false-positive results). The (1 − specificity) is the number of negative cases remaining after the true-negative cases have been subtracted (i.e., the false-positive results). This is shown algebraically as $1 - [d/(b + d)] = b/(b + d)$. This is the false-positive error rate. *Epidemiology,* p. 90.

2. The answer is A: (1 − sensitivity). In a 2 × 2 table, sensitivity is $a/(a + c)$. Sensitivity is the ability of a test to show a positive result when a condition is truly present. False-negative results occur when the test gives a negative result even though the condition is actually present. This is (1 − sensitivity), or $c/(a + c)$. *Epidemiology,* p. 90.

3. The answer is D: $a/(a + b)$. The positive predictive value is the probability of disease in a patient with a positive test result. This is the proportion of all patients with positive test results $(a + b)$ who truly have the condition (cell a), or $a/(a + b)$. *Epidemiology,* p. 90.

4. The answer is B: $d/(c + d)$. The negative predictive value is the probability of the absence of disease in a patient with a negative test result. This is the proportion of all patients with negative test results $(c + d)$ who truly do not have the disease (cell d), or $d/(c + d)$. *Epidemiology,* p. 90.

5. The answer is C: $c/(c + d)$. The probability that disease is truly absent given a negative test result is the negative predictive value, or $d/(c + d)$. This question, however, asks for the probability that disease is present even though the test result is negative. A disease may be either present or absent when a test result is negative, and, therefore, the probability that either one or the other of these options will be found is 1. The probability that a disease is present must be 1 minus the probability that the disease is absent. This is (1 − negative predictive value), or $1 - [d/(c + d)]$, which equals $c/(c + d)$. This question of probability is frequently asked in clinical medicine. A test is ordered on the basis of suspicion that a disease is present. If the test result is negative, one must decide whether to continue to suspect the diagnosis or to rule it out and suspect another diagnosis. This depends on the probability of the disease being present given the negative test result. Depending on the strength of one's suspicion (i.e., the prior probability) and the operating characteristics of the test (i.e., the sensitivity and specificity), the probability of disease following a negative result may be reduced to nearly 0 or remain very high. *Epidemiology,* p. 90.

6. The answer is D: late look bias. Late look bias is a tendency to detect only those cases of a serious disease that were mild enough to be detected prior to death. Many cases of massive myocardial infarction result in death before a patient can reach the doctor's office or emergency department. Consequently, your study might underestimate the probability of early death following infarction. Other errors, such as classification error, which is the result of inconsistency in the diagnosis of myocardial infarction, could create bias in your findings as well. Selection bias is a problem when subjects are recruited for study participation or assigned to treatment in a nonrandom manner, and neither of these conditions is operative here. Observer bias and measurement bias occur when there is distorted interpretation of the outcome in two study groups. In this case, the outcome is death, which is not particularly subject to interpretive error. Lead-time bias is the detection of a disease by screening earlier in its natural history than it would have otherwise been detected, resulting in a longer survival time after diagnosis but no change in actual survival time. *Epidemiology,* p. 67.

7. The answer is E: lead-time bias. Screening for lung cancer with chest x-rays has proved ineffective not because cases are never found but because when they are, the natural history of the disease is not changed from what it would have been if the disease were found and treated after symptoms developed. Lead-time bias is the prolongation of survival time after diagnosis not because death comes later but because diagnosis comes earlier. *Epidemiology,* p. 216.

8. The answer is B: the statistical regression effect. When subjects with the most extreme values for any measure are observed over time, the values tend to become less extreme. This is the statistical regression effect, or regression to the mean. This would be the most likely reason to see moderation of extreme absenteeism even though the treatment program has been ineffective. *Epidemiology,* p. 165.

9. The answer is F: Care is provided during the times it is needed. The provision of care does not necessarily make it available to the community. For people who cannot take time off from work to see a doctor, care is available only if it is provided in the mornings or evenings or on weekends. The provision of care during the times it can be utilized by those in need of it is availability. *Epidemiology,* p. 263.

10. The answer is C: There are no geographic or financial barriers to care. Medical care may be available but not accessible if patients lack transportation to the site where care is given or lack the means to pay for care. Care is accessible when financial and geographic barriers are negligible. *Epidemiology,* p. 263.

11. The answer is G: The amount of care provided is commensurate with the need or demand, or both. Care is adequate in supply when it meets demand. In health care, need and demand differ (see Chapter 21). Demand is patient-generated, while need is a professional assessment of the requirement for health care services. Care may be adequate when it is less than the demand for it, provided that the need is met. *Epidemiology,* p. 263.

12. The answer is A: The interventions are properly selected and performed by trained professionals. Care is appropriate when the right thing is done in the right way by the right person for the right reason. These are criteria for professional competency and comprehensiveness. Appropriateness is therefore defined by professional standards of care. *Epidemiology,* p. 263.

13. The answer is D: The care provided is compatible with the patient's belief systems. Care may be accessible and available but will not be utilized if it is unacceptable to patients. Acceptability requires that patients and providers can communicate in the same language, that care is delivered with compassion, and that cultural beliefs and practices of patients are respected. If the quantity of available care is commensurate with need but the care is unacceptable to patients in the community, the care provided cannot be considered adequate. *Epidemiology,* p. 263.

14. The answer is B: There is public disclosure of financial records and quality standards. Accountability of care is the disclosure of standards of care to the public and an acceptance of responsibility by providers for the judgment of the public. *Epidemiology,* p. 263.

15. The answer is E: The quality of medical care can be professionally reviewed. In contrast to accountability, assessability is contingent on professional rather than public scrutiny. The quality of medical care is assessable when it is open to review by impartial professionals. The care is accountable when the results of such a review are disclosed to the public. *Epidemiology,* p. 263.

16. The answer is D: an IPA. A PPO (preferred provider organization) is formed when an insurer establishes a network of contracts with independent practitioners. A patient may see a physician on the approved panel and receive full insurance benefits or the patient may consult a physician who is not on the panel and pay a surcharge. An IPA (independent practice association), like a PPO, is a group of providers associated by contract with an insurer. In the IPA, patients are not permitted to seek care outside of the network. An HMO (health maintenance organization) is a prepaid group practice that comes in several varieties (see Chapter 21). An ICF (intermediate care facility) is usually a type of nursing home. A DRG (diagnosis-related group) is a system of designating prepayment standards for hospital care. *Epidemiology,* p. 266.

17. The answer is A: 63% and 76%. When data from a table are expressed as percentages, the percentage of the dependent variable is usually shown as a function of the independent variable. For this table, the assignment to either chicken soup or gazpacho is the independent variable and the frequency of colds is the dependent variable. To express the frequency of colds as a function of treatment assignment, the percentage of subjects in each group developing (or not developing) colds should be shown. The percentage of subjects in the chicken soup group developing 1 or more colds is 63% and in the gazpacho group is 76%. The only other meaningful fact that could be learned from the data is the percentage in each group that did not develop colds (37% and 24%, respectively). This was not among the choices provided. *Epidemiology,* p. 145.

18. The answer is E: the chi-square analysis. The chi-square analysis is the appropriate significance test for the comparison of two dichotomous data sets. The dichotomous outcome here is having a cold or not having a cold. The two groups are chicken soup or placebo. The t-tests are used for comparing means and require two groups of continuous, not dichotomous, data. The log-rank test analyzes the significance of life table analysis. The Pearson correlation coefficient is used to assess the strength of association between two continuous variables. The chi-square value for this table is 1.985 at 1 degree of freedom and is not statistically significant at an alpha level of 0.05. *Epidemiology,* pp. 146–147.

19. The answer is C: 40%. Sensitivity is the proportion of positive cases that also have a positive test result. We have been told that there are 20 cases. Of these 20, 8 had positive test results with the new test. The sensitivity is 8/20, or 40%. The sensitivity can also be calculated by setting up the following 2×2 table:

		TRUE STATUS		
		Positive	Negative	Total
TEST STATUS	Positive	8	4	12
	Negative	12	176	188
	Total	20	180	200

The cells in the table contain the data we have been given. We know that there are 20 cases, so the total for the first column in the table, the disease (positive) column, must be 20. Cell a shows the positive cases detected by the test; we know that there are 8 of these. If cell a is 8 and column 1 adds to 20, cell c must be 12. We know that there are 12 positive test results, representing the total in row 1 of the table. If cell a is 8 and the row 1 total is 12, cell b must be 4. The total for the table must be 200. Cells a, b, and c add to 24, so cell d must be 176. The sensitivity is $a/(a + c)$, or $8/(8 + 12)$, which is 40%, the same answer we obtained above. *Epidemiology,* pp. 89–90.

20. The answer is B: 98%. The specificity of a test is the proportion of negative cases that it identifies as negative cases. There are 180 negative cases in this population, according to the available gold standard. The test identifies 4 of these as positive and the remainder, or 176, as negative. The proportion of negative cases correctly identified as such by the test is 176/180, or approximately 98%. From the table provided in the expla-

nation to item 19, above, the specificity is $d/(b + d)$, or $176/(4 + 176)$, which is 98%. *Epidemiology*, pp. 89–90.

21. The answer is C: 67%. The positive predictive value is the probability that the condition is present given a positive test result. This is the proportion of all positive results (cell a plus cell b) that are true-positive results (cell a). The formula for the positive predictive value is therefore $a/(a + b)$. In this case, the calculation is $8/(8 + 4)$, or about 67%. Recall that the predictive value, which is calculated as a percentage of subjects with a particular test result, includes subjects with and without the condition and is therefore dependent on the prevalence. Neither sensitivity nor specificity is dependent on the prevalence. *Epidemiology*, p. 90.

22. The answer is A: 94%. The negative predictive value is the probability that a disease is absent given a negative test result. This is the proportion of all negative test results (cell c plus cell d) that are truly negative (cell d). The negative predictive value is therefore $d/(c + d)$, which here is $176/(12 + 176)$. This is 93.6%, or approximately 94%. The negative predictive value is much higher than the positive predictive value in this instance because there are few cases of the condition in the population (i.e., the prevalence is low). The probability that anyone in the population has the condition is only 20 out of 200, or 10%. Therefore, the probability that a negative test result is correct is fairly high because of the low prevalence and the high specificity of the test. *Epidemiology*, p. 90.

23. The answer is B: 18. The likelihood ratio positive (LR+) indicates how much more likely it is for a positive test result to be true than to be false. The LR+ is the sensitivity, or $a/(a + c)$, divided by the false-positive error rate, or $b/(b + d)$. This becomes $8/20$ divided by $4/180$, which equals 18. The likelihood that disease is present in a subject with a positive test result is 18 times as great as the likelihood that disease is, in fact, absent. *Epidemiology*, p. 91.

24. The answer is B: 98%. The positive predictive value is profoundly influenced by the prevalence, even when the operating characteristics of the test remain unchanged. The sensitivity of the new test is 40%, and the specificity is 98%. We have now been told that there are 150 actual cases in the study population. Cell a in a new 2×2 table shows the number of true-positive results, or the sensitivity multiplied by the prevalence. This is 0.40×150, or 60. The total for column 1, the positive cases, must be 150, so cell c is 90. The number of noncases is equal to the total population, or 200, minus the 150 cases, or 50. The number of true-negative cases is equal to the specificity, or 0.98, multiplied by the number of noncases, or 50. This is 49, and this number is placed in cell d. Cells b and d must add to 50, so cell b is 1. The table is as follows:

TRUE STATUS

		Positive	Negative	Total
TEST STATUS	Positive	60	1	61
	Negative	90	49	139
	Total	150	50	200

The positive predictive value is $a/(a + b)$, or $60/(60 + 1)$, which is slightly greater than 98%. *Epidemiology*, p. 90.

25. The answer is A: 35%. The negative predictive value, as discussed in the explanation to question 22, above, is $d/(c + d)$. This is $49/(90 + 49)$, or approximately 35%. The negative predictive value has fallen as the prevalence has risen. The probability that disease is absent given a negative test result declines as the overall probability of disease being absent declines. The extreme example of this would be the use of a very sensitive and specific test in a population in which the condition being studied was universal. Despite the high sensitivity of the test, any negative test result would be false. Cell d in the table would be 0, and the negative predictive value would also be 0. *Epidemiology*, p. 90.

26. The answer is D: 0.73. The kappa test ratio is a means of assessing the degree to which agreement in paired data exceeds what would be expected by chance. The formula for kappa is $(A_o - A_c)/(N - A_c)$, where A_o is the observed agreement, A_c is the agreement expected by chance, and N is the total sample. The observed agreement is equal to the sum of cell a, in which both radiologists recommend biopsy, and cell d, in which both recommend no biopsy. The agreement expected by chance is the sum of the expected values in cells c and d, which are derived as they would be for a chi-square value. The expected value in any cell is the row total multiplied by the column total, divided by the sample total. The expected value for cell a is 40 multiplied by 32, divided by 200, or 6.4. The expected value for cell d is 160 multiplied by 168, divided by 200, or 134.4. When these figures are entered in the kappa formula, the following calculation can be made:

$$\text{kappa} = \frac{(28 + 156) - (6.4 + 134.4)}{200 - (6.4 + 134.4)}$$

$$= \frac{184 - 140.8}{200 - 140.8}$$

$$= \frac{43.2}{59.2}$$

$$= 0.73$$

A kappa ratio between 0.60 and 0.80 (60–80%) is usually considered good agreement, and a kappa ratio greater than 0.80 is considered excellent agreement. *Epidemiology*, p. 96.

27. The answer is C: 67%. The easiest approach to this question is to construct a 2×2 table based on an arbitrary sample size. The prevalence is 5%. If the sample size is 1000, then 50 workers will have the disease. The sensitivity of the test is 94%. Of the 50 workers, 94%, or 47, will be detected by the test. This is cell a. Cells a and c must add to the prevalence of 50, so cell c is 3. Of the 1000 workers, 950 are disease-free. Since the test specificity is 90%, the test will correctly identify 90%, or 855, of these workers; this is cell d. Cells b and d must add to 950, so cell b is 95. The table is drawn up as follows:

DISEASE STATUS

		Positive	Negative	Total
TEST STATUS	Positive	47	95	142
	Negative	3	855	858
	Total	50	950	1000

The percentage of positive test results (cells a and b) that are false-positive results (cell b) is $b/(a + b)$, or

95/(47 + 95). This is approximately 67%. The high percentage of false-positive results obtained despite the fairly good specificity of the test is the result of low prevalence. *Epidemiology*, pp. 89–90.

28. The answer is A: 33%. The positive predictive value is $a/(a + b)$. Based on the table shown in the explanation to question 27, above, this is 47/(47 + 95), or approximately 33%. Note that the probability that disease is absent when the test result is positive (calculated for question 27) and the positive predictive value (the probability that disease is present when the test result is positive) add to 100%, or 1. *Epidemiology*, p. 90.

29. The answer is E: 3. The prevalence of the disease is given as 5% of the population, and we are told that the population is 1000 workers. Therefore, we can expect 5% of 1000, or 50 persons, to have the disease. Of these 50 cases, 94%, or 47 cases, will be detected by our test. This is what sensitivity tells us. Therefore, the positive cases not detected, or the remaining 3 cases, are the false-negative cases. The number of false-negative cases is (1 − sensitivity) multiplied by the prevalence. In this example, that is [(1 − 94%) × 50], or (6% × 50), which is the 3 cases in cell *c* of the table. *Epidemiology*, p. 90.

30. The answer is B: the cost of screening, follow-up, and treatment. We have already been told that the disease in question can be treated. We also know the operating characteristics of our screening test. If the screening test is prohibitively expensive, we cannot implement it successfully. We need to consider the cost of follow-up testing because 67% of our positive test results will be false-positive results, and all of these subjects will need additional testing to demonstrate whether they do not, in fact, have the disease. Lastly, the cost of treatment for the true-positive cases must not be prohibitive or it will not be possible to provide therapy. The prevalence of other diseases in the population is not critical information, although if other diseases are considered more important and more prevalent, they might be a higher company priority. The disease incidence would be an important consideration in deciding how often to repeat the screening test after the initial round of testing. Once the prevalent cases have been detected, only incident cases remain to be found at subsequent screenings. The size of the population might influence the overall cost of screening, but a greater concern is the cost-effectiveness of the program for the company (i.e., whether the company profits by treating the disease and improving the work force). If the company profits, the size of the population will not be a limiting factor, and if the company loses, the program will not be cost-effective in even a small population. Lastly, identifying a genetic risk factor is usually important only in an effort to prevent disease or to determine which persons are at increased risk for a disease. This issue is not important in a program of screening for established disease. *Epidemiology*, p. 216.

31. The answer is A: $1300 saved. Regardless of which test is used, the entire population, or 1000 people, will be screened. We do not know the cost of the original test, but we know that the new test costs $0.50 more per application. Therefore, screening the population of 1000 will cost ($0.50 × 1000), or $500 more with the new test than the old; however, the specificity of the new test (i.e., 96%) is higher than the specificity of the old test (i.e., 92%). We know that the prevalence of disease Y is 10%, so 90% of the population, or 900 people, are disease-free. The original test will correctly identify 92%, or 828, of these 900 subjects, and there will be 72 false-positive results. Each person whose test result is falsely positive will need the diagnostic workup that costs $50 per person. The newer test will correctly identify 96%, or 864, of the 900 subjects who are disease-free, and there will be 36 false-positive results. There are 36 fewer false-positive results with the new test than with the old test. The number of true-positive results remains unchanged because the sensitivity of the new test is the same as that of the old test (i.e., 96%). All of the subjects that test positive will require the $50 workup, and there will be 36 fewer of these with the new test. The savings resulting from not having to do the workups for 36 people is (36 × $50), or $1800. When the $500 in additional costs associated with the new test is subtracted from $1800, there is a net savings of $1300 if the new test is used. *Epidemiology*, pp. 89–90.

32. The answer is D: 97%. This question is asking for the positive predictive value of the test, or the probability that the condition is present if the test result is positive. The positive predictive value is equal to the true-positive results (cell *a*) divided by all the positive test results (usually cells *a* and *b*, but this table is shown with the true disease status in rows rather than columns, and the data usually shown in cell *b* of a 2 × 2 table are in the position of cell *c* in this table). The positive predictive value is therefore 253/(253 + 8), or 253/261, which is approximately 97%. *Epidemiology*, p. 90.

33. The answer is B: 79%. This question calls for the negative predictive value, or the proportion of all negative test results (24 + 93) that are truly negative (93). In this case, 93/(24 + 93) yields a result of approximately 79%. *Epidemiology*, p. 90.

34. The answer is E: 91%. The probability that a test will detect a condition when it is actually present is the sensitivity. This is equal to the true-positive test results (253) divided by the true-positive cases (253 + 24), or 253/277, which is approximately 91%. *Epidemiology*, p. 89.

35. The answer is C: the prevalence in the population tested may not match her prior probability. The predictive value, both positive and negative, is dependent on the prevalence. In the case of an individual, the prevalence is not applicable, but the probability of the condition in the individual, or the prior probability, is analogous to the prevalence. Both indicate the probability of the condition before testing and influence the probability after testing. The data in the product brochure are derived from a population in which the prevalence of pregnancy is 277 out of 378 women, or about 73% of the population. If your friend is unlikely to be pregnant, perhaps because of consistent use of effective contraceptives, her prior probability of pregnancy may be much lower than 73%. If this were true, the estimates of positive and negative predictive value calculated above would not be applicable. All data should be assessed on the basis of both internal validity (i.e., correctness) and external validity (i.e., generalizability to people other than the study participants). The data reported in the pregnancy test brochure are most likely correct, but whether or not they pertain to your friend is uncertain. *Epidemiology*, pp. 100–102.

36. The answer is A: 1%. A 2 × 2 table could be constructed to answer this question, but the information provided is more amenable to the use of Bayes' theorem. The formula for Bayes' theorem is shown in Box 8–1. The sensitivity provided is 95%. The prevalence is 1 per

10,000, or 0.0001. The specificity provided is 99%; therefore, the (1 − specificity) is 1%. The (1 − prevalence) is 0.9999. The calculation is as follows:

$$= \frac{(0.95)(0.0001)}{[(0.95)(0.0001)] + [(0.01)(0.9999)]}$$

$$= \frac{0.000095}{0.000095 + 0.009999}$$

$$= \frac{0.000095}{0.01}$$

$$= 0.0095, \text{ or approximately } 1\%.$$

Epidemiology, p. 99.

37. The answer is B: $1.3 million saved. The cost per screening test is $10, and screening 100,000 newborn infants will therefore cost $1 million. Given a prevalence of 1 case per 10,000 in a population of 100,000, we can expect 10 cases of disease M. Of these 10 cases, 95%, or 9.5, will be detected. There will be 90,000 disease-free infants born, of whom our test will correctly identify 99%, or 89,100. The remainder of those without disease, or 900 infants, will be false-positive cases. The cost of evaluating each case with a positive test result, of which there are 9.5 true-positive and 900 false-positive results, for a total of 909.5, is $100. The total evaluation costs, therefore, are $90,950. For each of the 9.5 true-positive test results, $250,000 will be saved, for a total savings of (9.5 × $250,000), or $2,375,000. Subtracting the cost of screening ($1 million) and the cost of evaluating ($90,950) from the gross savings yields the following: 2,375,000 − (1,000,000 + 90,950) = 1,284,050, or nearly $1.3 million in net savings. *Epidemiology,* pp. 89, 216.

38. The answer is C: 8.7%. This question is essentially the same as question 36, except that the prevalence of disease is different. When the formula for Bayes' theorem is used (see Box 8–1) and the new prevalence of 0.001 is entered, the calculations are as follows:

$$= \frac{(0.95)(0.001)}{[(0.95)(0.001)] + [(0.01)(0.999)]}$$

$$= \frac{0.00095}{0.00095 + 0.00999}$$

$$= \frac{0.00095}{0.01094}$$

$$= 0.0868, \text{ or approximately } 8.7\%.$$

Epidemiology, p. 99.

39. The answer is D: the prior probability of systemic lupus erythematosus and the sensitivity and specificity of each test. Bayes' theorem is in constant application in clinical medicine, whether or not clinicians are aware of it. After completing a history and physical examination, a physician has an impression of the patient's health status that may differ from the impression he or she had before the history and physical examination or from the impression after the history had been taken but before the physical examination had been done. What has happened? The initial impression is based on the small amount of information known about the patient before the medical history is obtained. This is a relatively nonspecific prior probability. Each answer to each question in the medical history may cause the physician to generate a posterior probability of disease and overall health that is different than the probability he or she was considering before the answer. Upon completion of the examination, the impression may be revised further. Any tests ordered to pursue a diagnosis entertained after the history and physical examination are intended to modify the physician's estimate of the probability of the particular diagnosis in question. To use Bayes' theorem for this purpose, the prior probability of disease before each test as well as the sensitivity and specificity of each test must be known. The posterior probability after one test becomes the prior probability for the next test. Interestingly, we know little about the operating characteristics (sensitivity and specificity for different diagnoses) of the most sophisticated diagnostic instrument to which patients are exposed: the physician's brain. This same instrument makes the original estimation of disease probability upon which all subsequent testing is dependent. *Epidemiology,* p. 99.

40. The answer is B: 104.4. Variance, a measure of dispersion in a data set, is the sum of the squared differences of each data point from the mean divided by the sample size minus 1:

$$\text{Variance} = s^2 = \frac{\sum(x_i - \bar{x})^2}{N - 1}$$

The mean for this data set is 71.8, which is the 10 measures summed and divided by 10. The mean is then subtracted from each entry, and the difference is squared. The first entry is 70. Subtracting the mean of 71.8 from 70 yields −1.8. This number squared is 3.24. This calculation is repeated for each entry, and the 10 values obtained are summed, yielding 939.6. This number is divided by the sample size minus 1, which represents the degrees of freedom associated with the data set. This yields 939.6/9, or 104.4. This is the variance. *Epidemiology,* p. 117.

41. The answer is C: 10.2. The standard deviation is the square root of the variance. The variance for this data set was calculated for question 40, above, as 104.4. The square root of 104.4 is 10.2. *Epidemiology,* p. 118.

42. The answer is D: prospective cohort study. The groups of subjects are assembled on the basis of exposure. One group has been exposed to the king's delicacies and the other to vegetables. The groups, or cohorts, are followed over 10 days for the outcome, a change in countenance. This is a prospective cohort study. *Epidemiology,* p. 68.

43. The answer is D: use of controls. The inclusion of a control group improves this study considerably. Two groups are followed over time. One group receives the active intervention (vegetables), while the other group, the control group, receives the "placebo" (the king's delicacies). The "investigator" might simply have followed a group eating vegetables for 10 days and reported some improvement in countenance, but this study is better than that (which is not to say it would be likely to get NIH funding). The study is not blinded or randomized, and we have not been given any evidence that the cohorts are matched. *Epidemiology,* p. 69.

44. The answer is B: recall bias. Recall bias is introduced in retrospective studies, usually case-control studies, when one group is more likely than another group to remember a previous exposure. Recall bias is virtually never a factor in a prospective study. Selection bias is a likely source of error in this study. We know little, for example, about how subjects were chosen to participate and nothing about the appearance of their counte-

nances at baseline. Measuring the quality of a countenance is a rather unscientific endeavor and likely to be performed differently each time, introducing measurement error. Factors other than diet may differ between the two groups, and the difference in outcome may be the result of one of these unstudied influences. If these unstudied influences were associated with both exposure and outcome, they would be confounders. Observer bias would be introduced if the investigator assessed countenances differently for the two groups, a likely possibility. This is related to measurement bias. *Epidemiology,* p. 60.

45. The answer is D: the entire rejection region is on one side of the mean with a one-tailed test. The distinction between a one-tailed test and a two-tailed test of significance is illustrated in Fig. 10–1 and discussed in Box 10–6. The value of z represents the number of standard deviations between a data point and the mean for the sample. When alpha is 0.05 and the significance test is two-tailed, the rejection region must be divided evenly between the high end and the low end of the distribution, placing 0.025 of the total curve area in each rejection region. To deviate from the mean as far as a two-tailed alpha of 0.05, a data point must move past 0.475 of the curve area, since half the curve lies on either side of the mean. A dispersion of 0.475 of the total curve area from the mean is associated with a z value of 1.96. When a one-tailed test is performed, the entire rejection region, or 0.05 of the total curve area, is placed on one side of the mean. To deviate from the mean that far requires that a data point be removed only 0.45 of the curve area from the mean. This smaller deviation from the mean is associated with a smaller value of z, or 1.645. Achieving statistical significance, therefore, is easier with a one-tailed test than with a two-tailed test, provided that the outcome occurs in the expected direction. *Epidemiology,* pp. 129 and 134.

46. The answer is C: paired t-test. Selecting the appropriate significance test is essential for ensuring valid data analysis. The choice of test is dependent on the study type and the nature of independent and dependent variables. For an extensive list of significance tests and the data for which they are appropriate, see Table 11–1. This is a before-after trial, so the data are paired. There are two groups (active treatment group and control group), so the independent variable is dichotomous. The outcome variable is blood pressure measurements, which are continuous data. The appropriate test for this situation is the paired t-test, as indicated in Table 11–1 and discussed in Chapter 10. *Epidemiology,* p. 131.

47. The answer is E: Data that are not statistically significant may nonetheless be clinically important. When statistical significance is not achieved, the implication is that there is more than a 5% probability (if alpha is set at 0.05) that the outcome difference is due to chance. However, a p value of 0.08 still indicates a 92% probability that the outcome difference is not due to chance. In other words, failure to reject the null hypothesis does not mean that the null hypothesis is true. If a therapy is desperately needed and looks promising on the basis of trials that fail to show statistical significance, there are circumstances under which judicious use of the therapy would be appropriate. Certainly, with large samples, statistical significance may be achieved for outcomes that are not important clinically. Setting alpha high or setting beta low increases rather than decreases the probability of achieving statistical significance. *Epidemiology,* p. 136.

48. The answer is D: pool the data. The two studies show similar results, one with and one without statistical significance. If the studies are sufficiently similar, a quantitative meta-analysis would be appropriate. Pooling data effectively increases the sample size, providing greater power to detect a significant outcome difference. An alternative would be to recruit more subjects, but this approach is obviously more expensive. The studies were perhaps both somewhat underfunded in the first place, resulting in restriction of sample size to the bare minimum. When sample sizes are small, power is reduced, and type II (false-negative) error increases. *Epidemiology,* p. 105.

49. The answer is A:

$$N = \frac{(z_\alpha)^2 \cdot (s)^2}{(\bar{d})^2}$$

The smaller the alpha (and, therefore, the larger the z_α), the larger the sample size must be. The larger the variance, the larger the sample size must be. That is why z_α and s appear in the numerator of the sample size formula. The larger the difference one is seeking, the easier it is to detect the difference and the smaller the necessary sample size will be. The difference sought is in the denominator. When only one group of subjects is required, the sample size calculation provides the total for the study. When one group is to be compared with another, the size of each group is doubled because of the introduction of more variance, and the total is doubled again to allow for two groups. These formulas are shown as choices B and C, respectively. Choices D and E are statistical gibberish. *Epidemiology,* p. 160.

50. The answer is D: stipulate prospective payment to hospitals. DRGs represent categories of diagnosis for which a standard hospital stay and resultant cost of care is anticipated (see Chapter 21). Hospitals are paid by insurers such as Medicare on the basis of the diagnostic group rather than the actual care delivered. Efficient care results in a profit for the hospital. Complications result in hospital costs exceeding the DRG reimbursement and a potential financial loss for the hospital. *Epidemiology,* p. 270.

51. The answer is D: the factors that influence the occurrence and distribution of disease in human populations. Epidemiology (from the Greek *epi,* meaning upon, and *demos,* meaning population) is the study of all factors influencing patterns of human health and disease. Epidemiology is not restricted to outbreaks. It also is the study of endemic, or prevailing, patterns of disease in populations. The science of epidemiology addresses many types of human health problems and is not restricted to infectious agents or environmental toxins. While causality is one area studied in epidemiology, it is not the sole focus of this science. *Epidemiology,* p. 4.

52. The answer is C: vector. A vector, which is usually an arthropod insect, facilitates contact between the pathogen of a human disease and the host. The vector is usually not affected by the pathogen. The vector does not serve as an animal reservoir, because the pathogen must have another definitive host in order to complete its life cycle. *Epidemiology,* p. 5.

53. The answer is A: 5 times as many cases as would be expected. The number of preventable cancers in a population is determined by defining the population with the lowest observed incidence rate of a particular type of cancer (see Chapter 1). This rate is applied to another population to determine how many cases would occur

in the second population if it had the same low rate as the first population. The actual rate in the second population is compared with this standard, and the number of excess incident cancers is noted. This process is repeated for all the relevant types of cancer. The population with the lowest known incidence rate of a particular type of cancer is always identified, and its rate is used as the basis for comparison. This method suggests that there are 5 times as many cancers in the USA as there would be if the lowest observed rates could be achieved. *Epidemiology*, p. 6.

54. The answer is D: warm-water systems such as air-conditioning cooling towers. The original outbreak of legionnaire's disease in 1976 was ultimately traced to the air-conditioning and ventilation systems of the hotel that was hosting a legionnaires' conference. The causative organism, *Legionella pneumophila,* thrives in warm water. Respiratory therapy equipment is a potential means of spreading this infection and must be thoroughly sterilized after it is used. *Epidemiology*, p. 7.

55. The answer is D: associated with the lower mortality rates but may not be causal factors. Mormons and Seventh-Day Adventists are relatively strict about avoiding a number of potentially harmful substances and also about eating a diet rich in vegetables, fruits, and grains and low in processed foods and fat. These factors are thought to contribute to the exceptionally good health and low mortality rates of these groups. However, these life-style practices are only part of a larger context in which the family, work, and social environment often differs from that found in other populations in this country. Other factors may therefore also contribute to the differences in age-adjusted mortality rates (Berkman and Breslow 1983). Health outcomes are usually multifactorial, and establishing direct causality is often difficult. *Epidemiology*, p. 9.

56. The answer is B: Frequent exposure resulted in boosted immunity following immunization. Levels of protective antibody that develop following immunization often drop as time passes. Eventually, levels fall below the threshold required to prevent symptomatic infection and susceptibility to the disease redevelops. This is true of diphtheria now that the disease is uncommon. In the past, frequent exposure to the bacterium resulted in periodic boosting of the antibody levels following the original immunization, so that protective immunity was preserved. The more effectively immunization programs prevent exposure to historically important pathogens, the greater is the potential susceptibility to such agents as time passes. The administration of periodic booster vaccines is recommended under such circumstances. The difficulty of immunizing a large population is compounded when immunization must be repeated over a period of time. *Epidemiology*, pp. 12–13.

57. The answer is C: Use of the vaccine contributes to herd immunity. The Salk polio vaccine is given by injection. The Sabin vaccine, which became available later, is given orally. The Sabin vaccine, which is a live attenuated vaccine, stimulates cell-mediated immunity in the intestine, where the poliovirus multiplies. Herd immunity is promoted by transmission of the attenuated virus to contacts of the immunized individual and by the intestinal immunity that not only prevents disease but also prevents transmission of the wild virus, interrupting the pattern of exposure. The Salk vaccine may be preferred when the host is immunocompromised. *Epidemiology*, p. 13.

58. The answer is E: the difference between the birth and death rates of a society. The demographic gap is the difference between the birth rate and the overall rate of death in a population and is used to specify the rate of population growth. *Epidemiology*, p. 14.

59. The answer is A: schistosomiasis. The schistosomal parasites require snails to serve as hosts during a portion of their life cycle, and snails must have standing water in which to reproduce. Epidemiologists had anticipated an increase in the incidence of schistosomiasis as a result of the Aswan Dam construction, and their predictions were fulfilled. Whenever ecologic systems are modified, changes in the pattern of human diseases are possible. *Epidemiology*, p. 14.

60. The answer is E: The data are valuable for detecting changes in the pattern of a disease. Underreporting of common and mild diseases is substantial. Therefore, the cases of chickenpox reported are only a portion of the total cases of this disease. If the proportion of cases that seek medical care is fairly constant and the proportion of cases that are reported is fairly constant, then a sudden change in the number of cases reported, even if this number is inaccurate, should reliably reflect an actual change in the pattern of disease. *Epidemiology*, p. 19.

61. The answer is E: It is useful for immunocompromised hosts. The Salk injectable vaccine protects the individual from poliomyelitis, but still permits spread to others through the gastrointestinal tract, thereby not providing herd immunity. No vaccine is completely effective, and some, although few, immunized individuals nonetheless acquire infection. The Salk vaccine results in humoral immunity but not cell-mediated immunity in the gastrointestinal tract, which is conferred by the oral Sabin vaccine. The Salk vaccine is nonetheless indicated in situations where sanitation is poor and gastrointestinal infection prevalent, because the oral vaccine may fail to provide immunity. In immunocompromised hosts, such as persons with AIDS, the live oral vaccine may produce acute illness and is contraindicated, and the Salk vaccine is therefore used instead. *Epidemiology*, p. 13.

62. The answer is B: the genetic basis of phenylketonuria. An ecologic perspective is fundamental in epidemiology. Health and illness are evaluated in a context that includes the contributions of host, agent, environment, and vector. The etiologic factors in the "BEINGS" model are considered. Such an expansive view is often required to understand complex, multifactorial health problems. Cephalopelvic disproportion during labor in Muslim women is due to the practice of purdah, in which women are completely veiled. Lack of exposure to sunlight may result in vitamin D deficiency, which in turn results in poor calcium absorption and osteomalacia. The interaction of cultural and physiologic factors results in complications of labor due to cephalopelvic disproportion. The transmission of Lyme disease is complex, involving host exposure to a vector (the tick) carrying the agent *(Borrelia burgdorferi)*. The etiology of atherogenesis has a genetic component and is influenced by life-style practices such as diet, activity level, and smoking history, as well as other illnesses such as diabetes. The epidemiology of obesity is similarly complex, and requires consideration of genetic, social, environmental, and even psychologic factors for an adequate understanding. The genetic basis of phenylketonuria is addressed through laboratory methods that do not benefit from an ecologic perspective. However, rec-

ommending appropriate screening and interventions for genetic illness would be an epidemiologic responsibility for which an ecologic perspective would be useful. *Epidemiology*, p. 11.

63. The answer is C: denominator data for population-based studies. The US census, conducted every 10 years in years ending in 0, provides the best available estimate of the total US population and the population of various subgroups. These data are used as the denominators in population-based epidemiologic studies. While rates based on census data may provide the basis for direct or indirect standardization, they do not derive directly from the census data. Similarly, census data may be put to use in studies of various time trends, but the data do not reveal these trends directly. *Epidemiology*, p. 18.

64. The answer is B: 15–20%. Inaccuracies may arise due to uncertainty about immediate versus underlying cause of death, lack of familiarity with the deceased, or some combination of these, along with sleep deprivation of the reporting physician, who is often a house officer. (See Moriyama 1966.) *Epidemiology*, p. 18.

65. The answer is A: This is an acceptable risk estimate, even though the best denominator would include only subjects without protective immunity to influenza. Risk is defined as the proportion of subjects unaffected at the beginning of the study period who undergo the event during the study period. The denominator should include only subjects at risk for the study event at the beginning of the study. In the case of infectious disease, the at-risk subjects are those who lack protective immunity. However, as this is generally not known, the population is often used as the denominator. *Epidemiology*, p. 22.

66. The answer is D: Rates better demonstrate the force of mortality. Risk will be reported as the same whether the event in question, such as death, occurs evenly throughout the study period, all at the beginning, or all at the end. The rate will vary with these changes and will be highest for the clustering of events at the beginning of the study period. When the event is death, rate reveals the force of mortality on a population by rising when most of the deaths occur early. The denominator for a rate is the midperiod population. Rates may be higher or lower than risks, or they may be virtually the same, depending on the distribution of events over the study period and on the proportion of subjects in the population considered to be at risk for the event. *Epidemiology*, p. 23.

67. The answer is B: lung cancer in men in the USA. To be valid, a rate must have a numerator representing events in subjects from the denominator and must have a denominator representing those at risk for the numerator. In all of the choices except B, the denominators contain subjects not at risk for the event in the numerator. Uterine cancer in the US population is not valid, because the US population includes men, who are not at risk. Poliomyelitis in grade school children is not valid, because most school children in this country have been immunized, and are not susceptible to polio virus. Prostatitis in adults over age 65 years is not valid, because only men in this group are at risk. Rubella in pregnant women is not valid, because only the unimmunized are at risk. Lung cancer in men is valid, as all men are potentially at risk for this disease. *Epidemiology*, p. 23.

68. The answer is B: the incidence density. The incidence density is the frequency of new events per person-time. The denominator is the cumulative time contributed by subjects under observation, rather than the number of subjects. Consequently, this measure is useful for characterizing the pattern of events that may recur in individuals. *Epidemiology*, p. 24.

69. The answer is D: 5.6 per 1000. The crude death rate is calculated by adding all of the deaths that occur in the studied population during the period of observation and dividing by the midyear population. The total deaths are 120 plus 45 plus 13, divided by the total population of 32,000. This yields a rate of 5.6/1000. The population figures provided are assumed to be the midperiod populations. *Epidemiology*, p. 26.

70. The answer is B: All deaths occurred during the period of interest. The crude death rate is the total deaths in the study population during the period of observation, divided by the midperiod population. To derive the crude death rate from the data provided, one assumes that all deaths occurred during the observation period and that the population figures represent the midperiod population. The force of mortality, or the temporal distribution of deaths during the observation period, does not influence the crude death rate. Crude mortality rates are independent of the causes of death. *Epidemiology*, p. 26.

71. The answer is A: 5.6 per 1000. The crude death rates are the same for populations A and B. They are calculated the same way, as described in the explanation for question 69, above. By convention, crude death rates are multiplied by a constant, in this case 1000, so that the rates are easier to manipulate and discuss. *Epidemiology*, p. 26.

72. The answer is B: the size of the population and the total number of deaths. The total number of deaths, divided by the midperiod population, is the crude death rate. As these two terms define the rate, nothing else influences it directly. Causes of death may affect the force of mortality, with some causes producing more rapid, and others slower, deaths. Seasonal variation in mortality would also influence the force of mortality. Diagnostic consistency is necessary only to ensure that comparable numerators are compared when cause of death is being considered. Disease surveillance and reporting does not affect the death rate, as few deaths are unregistered. The age distribution of a population will affect the crude death rate by influencing the total number of deaths that occur during a specified period. *Epidemiology*, p. 26.

73. The answer is A: 3%. The age-specific death rate is the total deaths in a particular age group, divided by the population in that age group. In this case, that is 120 divided by 4000, which is 0.03, or 3%. *Epidemiology*, p. 27.

74. The answer is E: the age distributions are different. Crude death rates will be high in an elderly population, even if health care is excellent and longevity great. Crude death rates will usually be low in young populations, even if health care is poor and life expectancy limited. To make valid comparisons, the rates must be manipulated to answer this question: What would mortality rates be in these two populations if they had the same age distribution? *Epidemiology*, p. 26.

75. The answer is C: a population of nursing home residents. Whenever possible, direct standardization is used to compare mortality rates between populations; as discussed in the text, indirect standardization can be used when age-specific mortality rates are not known for one or more of the populations to be compared. In direct standardization, age-specific death rates (ASDRs)

must be calculated, or provided, for the populations being considered. Then, a hypothetical age distribution is established, usually by finding the sum of the numbers in each age group of the study populations. The ASDRs are then applied to the new age distribution, generating an expected number of total deaths in that population. This is then divided by the total in the standard population (e.g., the total sum of the study populations). The resultant standardized mortality rates allow one to compare the populations as if they had the same age distribution. Any reference population with an adequate age distribution might be used. A population of nursing home residents would be unacceptable, because the entire population would be clustered in a narrow age range, and this would not permit calculation of standardized death rates for the younger groups in the study populations. *Epidemiology*, p. 28.

76. The answer is C: 9.64 per 1000. The table below will be used in making the calculations for questions 76–78. The age-specific death rate is calculated for each age group by dividing incident deaths in that age group by the population of that age group. A standard age distribution is created by obtaining the sum of populations A and B in each of the three age groups. The ASDRs are then applied to the standard population to generate the expected numbers of deaths in each age group if populations A and B had the same age distribution. The expected numbers of deaths in each age group are added to produce the total number of expected deaths; this figure is then divided by the total standard population to produce the standardized death rate for the population. By convention, the result is converted to a number with one digit to the left of the decimal point by use of a constant multiplier. *Epidemiology*, p. 29.

77. The answer is D: 4.06 per 1000. Please refer to the answer to question 76; the same discussion pertains in this case. Calculations are shown in the table below.

78. The answer is B: population B is older. The standardized death rates for two populations might be the same, even when crude death rates differ. The determinants of standardized rates are the crude rate of mortality and the age distribution. In the example provided, populations A and B have the same crude death rate, 5.06 per 1000; this does not explain why the standardized rates differ. We know nothing about the timing of death in either population during the observation period. The size of the population does not influence the mortality rate, except that in a very small population, rates might be unstable estimates due to random variation. The reason population B has less than half the standardized mortality rate of population A is that population B is much older, with a great many more members in the age group that is over 65 years of age. If population B had the same age distribution as population A, the death rate would be much lower in population B; this is what the standardized rate tells us. *Epidemiology*, p. 28.

79. The answer is C: they must always await hypothesis testing. The most famous example of the institution of control measures without a thorough understanding of disease etiology or transmission was the proverbial removal of the handle from the Broad Street pump in London by John Snow in the 1850s to interrupt the transmission of cholera. As in this now legendary example, the institution of control measures is often appropriate before understanding is complete or hypotheses formally tested. Clearly, the better the disease in question is understood, the more likely it is that effective control measures will receive priority. However, even a rudimentary understanding of an outbreak may be sufficient to justify attempts at control if the disease is serious and control methods are readily available. The announcement of a need for control measures may generate alarm if the public is thereby made aware of its susceptibility to the disease. If, however, as is often the case, the outbreak receives media attention, the public may welcome the implementation of control measures. The announcement of an outbreak requires professional discretion, so that the public is kept well informed but undue anxiety is avoided. *Epidemiology*, p. 50.

80. The answer is E: correct, because current prevalence exceeds the historical baseline. By definition, an epidemic is the occurrence of any adverse health event within a population at a rate that exceeds either precedent or expectation. The current prevalence of obesity is greater than the historical precedent, and defies the expectations of many in the weight control and nutrition fields, although it is not truly surprising given the lifestyle and dietary practices in this country (Kuczmarski et al. 1994; Stamler 1993). Obesity has a complex etiology and is partly the result of genetic factors (Lindpainter 1995; Bennett 1995). Approximately 1 in 3 adults in the USA is considered obese, but simple frequency does not define an epidemic. Although most individuals suffer an occasional headache, headache is endemic, occurring at predictably high frequency, rather than epidemic. If the high prevalence of obesity persists over time, it too may come to be recognized as an endemic state, and the current rise in prevalence may be recognized as a secular trend rather than an epidemic. *Epidemiology*, p. 37.

81. The answer is E: the data must be expressed as rates to be interpreted. The number of deaths associated with cigarette smoking at different levels cannot readily be interpreted, unless some information is provided about who in the population is smoking. Specifically, if

Age Group (years)	Population A			Population B		
	Population Size	Age-Specific Death Rate	Expected Number of Deaths	Population Size	Age-Specific Death Rate	Expected Number of Deaths
< 30	17,000	× 0.001	17	17,000	× 0.00075	12.75
30–65	30,000	× 0.003	90	30,000	× 0.002	60
> 65	17,000	× 0.03	510	17,000	× 0.011	187
Total	64,000		617	64,000		259.75
Standardized Death Rate	617/64,000 = 9.64 per 1000			259.75/64,000 = 4.06 per 1000		

a very large percentage of the population were smoking more than a pack of cigarettes daily, one would expect a comparably large percentage of all deaths in this group. If the older cohorts in the population were smoking more heavily than the younger cohorts, one would expect a higher percentage of deaths among the smokers due to the influence of age on mortality. The raw data in the question imply, but cannot be interpreted to show, that smoking increases mortality. Conversion of the data to rates would facilitate interpretation. Of course, smoking does contribute to overall mortality rates, but this cannot be known from the information provided in the question. *Epidemiology*, p. 23.

82. The answer is B: an effect modifier. An effect modifier is a third variable that alters the relationship between an independent and a dependent variable. For effect modification to occur, a true relationship must exist between the independent variable of interest (in this case, physical activity) and the dependent, or outcome, variable (in this case, the occurrence of myocardial infarction). The effect modifier neither explains nor obscures the relationship between the causal factor of interest and the outcome but rather alters the relationship so that under differing conditions of the effect modifier (in this case, fitness), the relationship between independent and dependent variables changes in magnitude, or, as in this case, even direction. *Epidemiology*, p. 62.

83. The answer is D: intervening variables. Obesity, particularly abdominal obesity, represents a risk for cardiovascular disease (Pi-Sunyer, 1991). When obesity precedes cardiovascular disease, however, it is often, although not always, associated with hyperlipidemia and hyperglycemia, as well as hypertension. There is less evidence that obesity contributes to heart disease incidence when these other conditions do not develop. Therefore, obesity may only increase cardiovascular disease risk when it results in the development of other risk factors that intervene. Heart disease can develop in the absence of either hyperlipidemia or hyperglycemia, so neither is a necessary cause of heart disease. *Epidemiology*, p. 55.

84. The answer is E: recall bias. Once an adverse outcome has occurred, events in the past that may have contributed to it take on greater significance. These same events, when the adverse outcome does not occur, may be easily forgotten. The tendency to recall events preferentially when they are related to an adverse outcome is recall bias. *Epidemiology*, p. 60.

85. The answer is A: necessary cause. Asbestosis, a disease in which asbestos fibers in the lung disrupt the pulmonary architecture and impair respiratory capacity, can only occur with exposure to such fibers. However, many individuals exposed to asbestos fibers do not develop asbestosis. Therefore, asbestos is a necessary, but not sufficient, cause of asbestosis. *Epidemiology*, p. 55.

86. The answer is D: it is correct but not generalizable without further study. In an ecologic study, populations rather than individuals are the subject of investigation. The ecologic fallacy is the tendency to draw conclusions about causal associations in individuals as a result of surveying a population. In an indigent population, rates of both cigarette smoking and tuberculosis are likely to be relatively high but may not correlate in individuals. Within this population, tuberculosis may be as likely in nonsmokers as in smokers. However, an association observed at the population level may be true in individuals. Recall that an ecologic study should serve only to develop, but not to test, hypotheses. In this case, the hypothetical association between smoking and tuberculosis should be tested in either a case-control study or a cohort study. In fact, cigarette smoking is not a specific risk factor for tuberculosis, although pulmonary infections in general are more common in smokers than nonsmokers (Bartecchi, MacKenzie, and Schrier 1994); both tuberculosis and cigarette smoking tend to occur at high rates in the same populations in the USA.

Based on the information provided, the conclusion is not known to be correct. Even if supported by the results of additional studies, the association between cigarette smoking and tuberculosis might or might not be generalizable to populations other than the ones studied. *Epidemiology*, p. 67.

87. The answer is B: a longitudinal ecologic study. The National Health and Nutrition Examination Surveys (NHANES) are an example of cross-sectional surveys conducted at regular intervals to establish secular trends in the behaviors studied. This is an example of the longitudinal ecologic study. *Epidemiology*, p. 67.

88. The answer is C: subjects with prior radiation exposure followed for the development of lymphoproliferative cancers. A retrospective cohort study is performed when an exposure or risk factor is established at some time in the past, before the outcome in question has occurred. Subjects are then interviewed or assessed through chart review to identify cases of the outcome that occurred after the exposure or risk factor was recorded. If a radiation exposure occurred in the distant past, and the subjects were followed for lymphoproliferative disease into the more recent past, this would be a retrospective cohort study. If the exposure were current, and subjects were followed forward in time, as in option B (subjects with current angina followed for the development of myocardial infarction), the study would be a prospective cohort design. Retrospective cohort studies are usually cheaper, and certainly quicker, than comparable prospective studies. Option A (cases of lung cancer assessed for prior exposures) and option E (subjects with skin cancer assessed for lifelong, cumulative sun exposure) are amenable to case-control studies. Option D (subjects randomly assigned to active varicella vaccine versus placebo, followed for chickenpox) is a randomized field trial. *Epidemiology*, p. 68.

89. The answer is B: optimize external validity. Randomization prevents selection bias on the part of either the subject or the investigator. Each subject is equally likely to enter each of the study groups, and groups that differ only on the basis of the intervention will result when large enough samples are assembled. With small trials, random differences may exist between groups, requiring statistical compensation in analysis. Randomization reduces both known and unknown confounders by producing groups that are similar in both measured and unmeasured ways. Blinding reduces bias in measurement.

External validity is the degree to which results pertain to the populations beyond the subjects in the study. This depends on many factors, particularly the degree to which study subjects resemble the general population. Neither randomization nor double-blinding ensures external validity. Random sampling, a technique used to obtain subjects representative of the general public (see Chapter 12) is helpful in optimizing external validity, also known as generalizability. *Epidemiology*, p. 72.

90. The answer is C: prospective cohort study. In a prospective cohort study, the cohort is assembled on the basis of risk factors or exposures. As these characteristics define the cohort, only those defined and measured at the beginning of the study may be analyzed. However, observed outcomes in the cohort may be assessed even if they were not the outcomes anticipated. The occurrence of an unanticipated outcome should serve as the basis for a new hypothesis, to be tested in another study. *Epidemiology,* p. 68.
91. The answer is A: case-control study. In a case-control study, cases are those with the outcome, and controls are those without the outcome. As the study groups are defined at entry on the basis of outcome, only this one outcome may be assessed. However, both groups may be evaluated for any number of past exposures or risk factors. *Epidemiology,* p. 69.
92. The answer is B: randomization. The process of randomization provides each study subject with an equal chance of entering each of the study groups. With a large enough sample, the characteristics of the subjects will be balanced among the study groups as a result of random distribution. This effect will influence the distribution of known confounders as well as factors not yet known to confound the association under investigation. The smaller the sample, the less likely that subject traits will be balanced among the study groups. Therefore, randomization cannot guarantee that the groups will be comparable, with the exception of the intervention under study, or that confounding will be completely eliminated. However, randomization will at least reduce confounding by known and unknown factors. *Epidemiology,* p. 70.
93. The answer is D: double-blinding. In a double-blind study, neither the investigator nor the subject knows which intervention is being administered. While measurement error may occur, it will not be influenced by knowledge of the intervention and will, therefore, be random. Random error reduces the power of a study to detect a true difference between control and intervention groups (see Chapter 12), but it does not introduce bias. *Epidemiology,* p. 71.
94. The answer is A: statistical analysis. Randomization ensures that all participants in a trial will be assigned to the different groups with equal probability. However, particularly when groups are small, randomization cannot ensure that all relevant characteristics will be evenly distributed among the groups. Bias is removed from the assignment by a randomization technique, but residual confounding by unevenly distributed factors may remain. This residual confounding is controlled in the statistical analysis of the data. *Epidemiology,* p. 71.
95. The answer is A: prevention of confounding by known and unknown factors. The goal in an intervention trial is to compare two groups that differ only with regard to the intervention of interest. In animal research, this is achieved by using genetically homogeneous populations that are, for all relevant intents and purposes, identical. Clearly, human populations cannot be manipulated in this way. The only opportunity to study genetically identical humans is afforded by studies of monozygotic twins. Randomization is performed so that all of the factors contributing to human variability are equally likely to appear in the control and intervention groups. When successful, this method results in groups that differ significantly only by the intervention, *although this is not ensured*. Randomization, therefore, minimizes both known and unknown confounders. A study that successfully tests its principal hypothesis in an unbiased manner is internally valid; randomized, prospective, double-blind trials often meet this condition. However, selection bias is not prevented by this design, and measurement error is always possible. Therefore, the study design described does not ensure internal validity. Randomization ensures unbiased distribution of participants to control and intervention groups but does not ensure that the study population represents the general population, although there are methods to achieve this as well. This study, restricted to male physicians, may not be generalizable to other groups and therefore may not be externally valid due to sampling and selection bias. Although a large sample size increases the likelihood of detecting a statistically significant difference in the outcome measure, this is never ensured, particularly if no such difference exists. *Epidemiology,* p. 71.
96. The answer is A: disease status. In a cohort study, the group to be followed is assembled on the basis of inclusion and exclusion criteria; an age range is usually included among such criteria. These factors define the cohort. The exposure in question should be included among the eligibility criteria. Similar groups that differ with regard to the exposure in question should be assembled. The comparison to be made, however, is the rate at which disease (or other outcome) occurs in those with and those without the exposure of interest. Therefore, disease status is the important comparison to be made at the conclusion of any cohort study. In contrast, in a case-control study, disease status is known at entry, and exposure status is compared between cases and controls. *Epidemiology,* p. 75.
97. The answer is E: 3.4 per 10,000 deaths. The proportion of the total population risk for deaths associated with cerebrovascular accident due to hypertension is the population attributable risk (PAR). This measure is calculated by subtracting the risk in the unexposed group from the risk in the total population. The question does not provide the risk in the unexposed group. To determine this, we must first calculate the risk in the total population and then in the exposed group. The total population risk for death due to cerebrovascular accident is 150,000 per 265 million = 5.7 per 10,000. This represents the total population cause-specific death rate for cerebrovascular disease. Given a risk for death due to cerebrovascular accident 3.5 times higher among those with hypertension compared with the general population, the death rate in the exposed group is (3.5×5.7) per 10,000 = 20 per 10,000. Given 50 million individuals in the exposed group, a cerebrovascular accident death rate of 20 per 10,000 yields 100,000 deaths among the exposed. Subtracting deaths among the exposed from total deaths yields 50,000 deaths among the unexposed. The unexposed population is the 265 million US residents minus the 50 million in the exposed group, or 215 million. The population attributable risk becomes:

$$PAR = Risk_{(total)} - Risk_{(unexposed)}$$
$$= (150{,}000/265 \text{ million}) - (50{,}000/215 \text{ million})$$
$$= (5.7/10{,}000) - (2.3/10{,}000)$$
$$= 3.4/10{,}000$$

Out of every 5.7 deaths due to cerebrovascular accident in the general population, 3.4 may be attributed to hypertension. *Epidemiology,* p. 79.

98. The answer is B: 59.6%. The percentage of total risk in the population due to a particular exposure is the population attributable risk percent (PAR%):

$$PAR\% = \frac{(Risk_{(total)} - Risk_{(unexposed)})}{Risk_{(total)}} \times 100$$

$$= \frac{(5.7/10,000) - (2.3/10,000)}{(5.7/10,000)} \times 100$$

Because the total risk minus the risk in the unexposed is the same as the risk in the exposed, a more direct approach to the same calculation is

$$PAR\% = \frac{Risk_{(exposed)}}{Risk_{(total)}} \times 100$$

Either way, the calculation becomes

$$PAR\% = \frac{3.4/10,000}{5.7/10,000} \times 100 = 0.596 \times 100 = 59.6\%$$

The answer, 59.6%, is the percentage of deaths from cerebrovascular disease in the general population attributable to hypertension.

Alternatively, the PAR% can be calculated using the following formula:

$$PAR\% = \frac{(Pe)(RR - 1)}{1 + (Pe)(RR - 1)} \times 100$$

where RR is the risk ratio (relative risk) and Pe is the proportion of the population with the exposure. Here, Pe is 50 million per 265 million = 0.19. RR is the risk of death due to cerebrovascular accident among those with hypertension (20 per 10,000) divided by the risk in those without hypertension (2.3 per 10,000) = 8.7. The formula therefore becomes

$$PAR\% = \frac{(0.19)(8.7 - 1)}{1 + (0.19)(8.7 - 1)} \times 100$$

$$= \frac{1.463}{2.463} \times 100 = 59.4\%$$

The slight discrepancy is due to the error introduced by rounding to the nearest decimal. *Epidemiology,* p. 79.

99. The answer is C: 17.7 per 10,000. This question calls for calculation of the attributable risk, the absolute risk of cerebrovascular accident mortality that may be attributed to hypertension. The formula and calculations for attributable risk (AR) in this case are as follows:

$$AR = Risk_{(exposed)} - Risk_{(unexposed)}$$

$$= (20/10,000) - (2.3/10,000)$$

$$= 17.7/10,000$$

Epidemiology, p. 76.

100. The answer is A: PAR%. The utility of attributable risk to health policy is its application to cost-benefit (or cost-utility) analysis, to cost-effectiveness analysis, or to both (see Chapter 14). The PAR% is the most useful of the measures in that it provides information about the impact of a particular exposure on the total population, and the degree to which that exposure influences total risk in the population. This measure is of greater interest to policymakers, as it considers the aggregate population risk. The AR, AR%$_{(exposed)}$, RR, or OR might be very high if an exposure were highly correlated with an adverse outcome, even in a very small segment of the population. For example, the risk of death due to Ebola virus may be impressively high among the exposed, but the rate of exposure in the USA may be too low to warrant the allocation of financial resources to a program aimed at prevention of the disease. Of the choices, only the PAR% would reveal that the impact of Ebola virus on the US population is negligible, despite the virulence of the pathogen. *Epidemiology,* p. 79.

101. The answer is D: the *p* value. The level of alpha must be specified before a study is conducted, in order to indicate the maximum risk the investigator is willing to take of spuriously rejecting the null hypothesis (i.e., the risk of false-positive error). The sample size is influenced by the level of alpha; the smaller the risk of false-positive error the investigator wishes to accept (alpha = 0.05 is the conventional level), the larger the sample size. Variance must be estimated, often based on an earlier study, before the sample size is determined. Beta, or the risk of false-negative error, can be left out of sample size calculations but should not be if one wishes to limit the risk of a false-negative outcome. The difference to be detected must be specified before the sample size can be determined; a large difference is detectable with a small sample, while smaller differences are only detectable with large samples. The *p* value is determined after data have been analyzed and, therefore, long after the sample size is established. The *p* value indicates the likelihood that a given outcome is the result of chance; if the *p* value is less than alpha, below the previously specified risk level for false-positive outcome, the null hypothesis is rejected. *Epidemiology,* p. 161.

102. The answer is A:

$$N = \frac{(z_\alpha + z_\beta)^2 \cdot 2 \cdot \bar{p}(1 - \bar{p})}{(\bar{d})^2}$$

The study described measures outcome as a proportion. The sample size formula discussed in Chapter 12 for use with proportional data is shown above. The use of this formula is displayed in Box 12–4. *Epidemiology,* p. 161.

103. The answer is C:

$$N = \frac{(z_\alpha)^2 \cdot 2 \cdot (s)^2}{(\bar{d})^2}$$

The study described uses separate control and intervention groups. Therefore, the sample size formula has a 2 in the numerator to double *N*. The use of this formula is detailed in Box 12–2. *Epidemiology,* p. 161.

104. The answer is B:

$$N = \frac{(z_\alpha)^2 \cdot (s)^2}{(\bar{d})^2}$$

This study uses subjects as their own controls, thereby reducing the number of subjects needed. The use of this formula is demonstrated in Box 12–1. *Epidemiology,* p. 161.

105. The answer is D:

$$N = \frac{(z_\alpha + z_\beta)^2 \cdot 2 \cdot (s)^2}{(\bar{d})^2}$$

In a study using separate control and intervention groups and considering both alpha and beta error, the appropriate sample size formula is shown above. N is increased both by adjusting for beta error and by the need for separate controls. The formula provides the number of subjects needed for each group. In order to recruit adequately for both control and intervention subjects, the number of participants needed would be $2N$. The use of this formula is shown in Box 12–3. *Epidemiology,* p. 161.

106. The answer is E: the strength of the association between x and y. The square of the correlation coefficient, r^2, is an expression of the degree of variation in the dependent variable (generally, y) explained by the variation in the independent variable (generally, x). For example, if the correlation coefficient for a set of data displaying height (the independent variable) and weight (the dependent variable) were 0.90, the square of this, or 0.81, would represent the degree of variation in height explained by variation in weight. The more the variation in the dependent variable is explained by variation in the independent variable, the greater the strength of the association between the two. *Epidemiology,* p. 141.

107. The answer is E: predicting the peak air flow rates when pack-years of cigarettes smoked are known. Linear regression is a form of statistical modeling. The model is used to predict y, given x and information about the relationship between x and y (i.e., the slope and y-intercept). The quality of the model is determined by how closely the predicted values of y approximate the actual measured values. Linear regression is related to correlation, and, like correlation, can be performed only when both the independent and dependent variables are continuous. In choice A (predicting serum cholesterol levels when dietary intake data are known), dietary intake data would provide multiple independent variables; linear regression analyzes the relationship between only one independent variable and one dependent variable. In choice B (predicting blood pressure levels when the family history of the presence or absence of hypertension is known) and choice C (calculating the probability of achieving remission of leukemia when the prior response to chemotherapy is known), the independent variables are not continuous. In choice D (predicting the blood type of a patient when the blood types of the parents are known), neither the independent nor the dependent variable is continuous. *Epidemiology,* p. 141.

108. The answer is C: that the two variables are independent of each other. A chi-square analysis of a 2×2 contingency table seeks evidence that the variation in one of the variables is due in part to variation in the other variable rather than to chance. The data are dichotomous and therefore cannot be characterized by a mean. The expected values for each cell, calculated by methods discussed in Chapter 11, are the values expected if the null hypothesis is true (i.e., if the variation in the two variables is independent). The significance test associated with the chi-square value determines whether or not the variation in one variable is explained by the variation in the other variable to a greater degree than would be expected by chance alone, at a conventional alpha level of 0.05. *Epidemiology,* p. 145.

109. The answer is B: cohort trial. A cohort trial is defined by the assembly of one or more groups (cohorts) characterized by an exposure and followed for an outcome. The cohorts here are manganese miners who have or have not been exposed to the benefits (and hazards) of dental floss. The cohorts are then followed for the occurrence of the specified outcome, which is iron deficiency anemia. *Epidemiology,* p. 68.

110. The answer is D: the relative risk. Data from a cohort trial may be used to calculate the relative risk, whereas data from a case-control trial may only be used to estimate the relative risk by calculation of the odds ratio (see Chapters 5 and 6). The odds ratio is a good estimate of the risk ratio only when the disease under investigation is rare. The relative risk of iron deficiency anemia is the risk in those who floss, which is $150/(150 + 50)$, divided by the risk in those who do not floss, which is $20/(20 + 180)$, or 7.5. To calculate the incidence, we would need to know the specific period of time during which new cases of iron deficiency anemia developed in our population. To define the prevalence, we would need to know the number of cases of iron deficiency anemia present in our population at a given point in time. A dichotomous outcome is not characterized by a mean. *Epidemiology,* pp. 68 and 76.

111. The answer is E: 85. The question refers to the expected count for cell a. To calculate the expected count (E) for any cell in a contingency table, you need to know the total for the row in which the cell is located (row total), the total for the column in which the cell is located (column total), and the total for the table (study total). For cell a, the row total is $150 + 50 = 200$; the column total is $150 + 20 = 170$; and the study total is $150 + 50 + 20 + 180 = 400$. The formula can be expressed as follows:

$$E = \frac{\text{Row total}}{\text{Study total}} \times \text{Column total}$$

The expected count is therefore $(200/400)(170) = 85$. Alternatively, the row total could be multiplied by the column total and then divided by the study total, and the expected count is the same: $(200 \times 170)/400 = 85$. For further review of the concept, see Box 11–3. *Epidemiology,* pp. 145–146.

112. The answer is E: 85. This question asks for the expected count in cell c of the contingency table. To calculate the expected count, the row total $(20 + 180)$ is multiplied by the column total $(150 + 20)$, and the product is divided by the total for the table, or 400. This becomes $(200 \times 170)/400$, or 85. The conceptual justification for this method is discussed in Box 11–3. *Epidemiology,* pp. 145–146.

113. The answer is C: 115. This question asks for the expected count in cell d of the contingency table. The row total is $(20 + 180)$. The column total is $(50 + 180)$. The table total is 400. The expected count is $(200 \times 230)/400$, or 115. See Box 11–3 for further discussion. *Epidemiology,* pp. 145–146.

114. The answer is D: Use of dental floss and iron deficiency anemia are not associated. The null hypothesis for a contingency table is that the variables of interest are independent. In a cohort study such as the one described, the independent variable is the exposure—in this case, use of dental floss. The dependent variable is the outcome—in this case, iron deficiency anemia. The null hypothesis stipulates that the exposure and outcome are independent, or that, in this case, the use of dental floss and the development of iron deficiency anemia are independent. *Epidemiology,* p. 145.

115. The answer is B: those with the outcome but not the exposure. Conceptual understanding of a contingency table requires identification of the type of data repre-

sented by each cell in the table. By convention, disease status is expressed in the columns of the table and exposure status is in the rows, as is true of the example provided for this set of questions. Cell a is subjects with both exposure and outcome; cell b is subjects with exposure but not the outcome; cell d is subjects with neither exposure nor outcome; and cell c, the correct answer, is subjects with the outcome (column 1) but not the exposure (row 2). *Epidemiology,* pp. 145–146.

116. The answer is B: 172.8. The method for calculating the chi-square is discussed at some length in Chapter 11. In brief, the chi-square value, $[(O - E)^2/E]$, is calculated for each of the four cells in the table and then these values are summed. For cell a, the observed value is 150 and the expected value is 85. The expected value for each cell is the row total multiplied by the column total divided by the total for the study. For cell a, this is $[(150 + 50) \times (150 + 20)]/400$; this equals 85. The chi-square value for the cell is $(150 - 85)^2/85$, which is 49.7. The chi-square value for each of the other cells is determined in the same way. The expected value for cell b is 115; for cell c, 85; and for cell d, 115. You may be surprised that the expected values are the same for cells a and c and for cells b and d. This will occur any time the sample is evenly divided into two groups with and without the exposure. Because half of the sample is in row 1 and half in row 2, half of each column total is expected to occur in the top row and half in the bottom row.

The chi-square value for cell b is 36.7; for cell c, 49.7; and for cell d, 36.7. The chi-square value for the table is therefore $49.7 + 36.7 + 49.7 + 36.7$, which is approximately 172.8. *Epidemiology,* p. 147.

117. The answer is B: 1. In any contingency table, there are $(R - 1) \times (C - 1)$ degrees of freedom, where R is the number of rows in the table and C is the number of columns. In a 2×2 table, therefore, there is always $(2 - 1) \times (2 - 1)$, or 1 degree of freedom. The conceptual basis for this is uncomplicated. Given the total number of data points for the table and the totals for each of the rows and columns, how many of the cells are "free" to vary? In a 2×2 table, if we know the total for row 1 and fill in a value for cell a, cell b is no longer free to vary; cell a plus cell b must add up to the known total for row 1. Now that cells a and b are filled in, cells c and d become fixed because cell a plus cell c must add up to the known total for column 1, and cell b plus cell d must add up to the known total for column 2. Therefore, once one value is entered into a 2×2 table, the other counts become fixed and are not "free." The degrees of freedom specifies the number of cells in the table that are free to vary. *Epidemiology,* p. 147.

118. The answer is B: find the corresponding p value in a chi-square table. Once the chi-square values for the table and the degrees of freedom have been established, no further calculation is necessary to determine the statistical significance of the model. A chi-square table is provided in the Appendix, as well as in most books of biostatistics. Across the top are p values associated with a given value of chi-square, with a specified number of degrees of freedom listed in a column. In this case, the corresponding p value is less than 0.0005, and this model is therefore highly statistically significant. Our conclusion from this study is that, at least in male manganese miners, use of dental floss predisposes to iron deficiency anemia. We have learned nothing of the effects of flossing in other populations. *Epidemiology,* p. 148.

119. The answer is E: using a very large sample. When sample size is sufficiently large, even trivial differences may achieve statistical significance. If, for example, a vaccine against Lyme disease were tested and found to reduce the rate of infection from 11.1 per 1000 among the unvaccinated to 10.9 per 1000 among the vaccinated (these figures are fictitious), the vaccine would be of virtually no clinical value. To prevent one case of Lyme disease with this vaccine, more than 5000 people would need to be vaccinated. If data were collected for many thousands of subjects over many years, even this trivial difference might be statistically significant. Statistical significance may be misleading when outcome differences are not clinically meaningful. *Epidemiology,* p. 136.

120. The answer is D: is best expressed as 21 person-years. Person-time methods are useful in survival analyses, when the observation period varies from subject to subject. By adding each participant's experience to the total, regardless of outcome, the interpretable observation period is maximized. The total observation period need not be measured in years but generally is whenever this provides whole numbers. In this study, 10 subjects under observation for 6 months contribute 60 months, or 5 years, of observation. The 30 subjects observed for 4 months each contribute 120 months, or 10 years. The remaining 2 subjects contribute 3 years apiece, for a total of 21 person-years. Those subjects lost to follow-up contribute the time they were observed prior to leaving the study. *Epidemiology,* p. 154.

121. The answer is A: is approximately 1.1 deaths per person-year of observation. As explained in the answer to question 120, the observation period in this study is best expressed as 21 person-years. A total of 23 deaths occurred during the observation period. The death rate for the study is simply total deaths divided by the units of observation, or 23 deaths per 21 person-years, which is approximately 1.1 deaths per person-year of observation. *Epidemiology,* p. 154.

122. The answer is E: the Kaplan-Meier method uses uneven death-free intervals. As discussed in Chapter 11, the actuarial method and the Kaplan-Meier method are both forms of life table analysis. While these methods are often used for survival analysis, the outcome of interest need not be death. The outcome must be dichotomous for both of these methods. The principal difference between the two techniques is that the Kaplan-Meier method begins a new observation interval with each death. Because deaths are generally not evenly distributed, the observation intervals, or so-called death-free intervals, are uneven. This approach compensates better for losses to follow-up and censored data when samples are small than does the actuarial method. The uneven intervals of the Kaplan-Meier method account for the uneven stairlike curves that result when the data are plotted. *Epidemiology,* pp. 154–155.

123. The answer is C: the disease should be rare. Screening is an effort to detect occult disease for which early diagnosis and treatment will improve the outcome. The disease must be treatable or there is no value in its detection. The natural history of the disease must be understood in order to know at what point detection and treatment will modify the outcome. The disease must usually be serious, in order for the costs of the screening program to be justifiable. Screening for dandruff, for example, would only make sense if you owned stock in a certain shampoo company. The population at risk for the disease must be identifiable or the screening

effort cannot be appropriately directed. In the case of extremely common diseases, screening may be less appropriate than devising prevention strategies, which would probably be more effective at containing costs and promoting the public health. Very rare diseases are not generally amenable to screening, because even a test of high sensitivity and specificity will produce more false-positive than true-positive results. It is worthwhile to screen for certain rare diseases of childhood, such as phenylketonuria, because the difference in outcome between treated and untreated individuals is so extreme. *Epidemiology,* p. 216.

124. The answer is E: the fact that only incident cases are detected. Whenever a new screening program is introduced, the number of cases detected includes new cases but also all the cases still living that have accumulated over time. If the same program is repeated in the same population at relatively frequent intervals, only the much smaller number of incident cases remains to be detected. *Epidemiology,* p. 218.

125. The answer is A: the onset of symptoms may cause concern and raise motivation. Adverse health events heighten awareness of and interest in disease prevention and health promotion. This is the proverbial case of not knowing what you've got til it's gone. The restoration of health following serious illness is often incomplete and therefore inferior to the preservation of health in the first place. The author has learned from reading the literature and from personal experience in treating patients that most patients are much more willing to modify potentially dangerous life-styles after they have suffered some of the adverse consequences of such life-styles. Tertiary prevention is generally less cost-effective than prevention at earlier stages, although this is not universally true. Illness may in fact limit compliance with medical advice, as might wellness, but this does not explain the greater influence of tertiary prevention. Both secondary and tertiary prevention occur after the onset of disease. Tertiary prevention usually produces the least net benefit of all preventive strategies. *Epidemiology,* p. 226.

126. The answer is D: body fat distribution. Gender, family history, age, and cigarette smoking influence the risk of cardiovascular disease whether or not obesity is present. Obesity is associated with increased risk of cardiovascular disease, especially, and perhaps exclusively, when fat is distributed predominantly in the abdomen rather than in the buttocks and lower extremities (Lemieux et al. 1993). The accumulation of adipose tissue in the upper body and abdomen has been labeled android (the "apple" pattern) because it predominates in men. This pattern may contribute to the higher risk of cardiovascular disease in men than in age-matched premenopausal women. The accumulation of adipose tissue in the lower extremities has been labeled the gynoid pattern. Neither pattern, of course, consistently respects the gender boundary. *Epidemiology,* p. 227.

127. The answer is C: synergistic. As risk factors for cardiovascular disease accumulate, the relative risk of the disease rises at a rate that is faster than the simple addition of the effects of each factor: this is synergy (Dawber, Meadors, and Moore 1951). Until recently, men have been preferentially selected as subjects in studies of heart disease. Cardiovascular disease is just as important in women, although the onset is later (Kuhn and Rackley 1993). *Epidemiology,* p. 227.

128. The answer is D: HDL. Low-density lipoprotein, particularly small, dense variants, and particularly when oxidized, is considered the most atherogenic of lipid particles in blood. Levels of very low density lipoprotein (VLDL) and triglycerides and the total cholesterol level, which incorporates all of the substances listed previously, correlate positively, to varying degrees, with the risk of coronary atherosclerosis. High-density lipoprotein (HDL) functions in reverse cholesterol transport, removing atherogenic lipid particles from the circulation. A detailed discussion of the mechanism is beyond the scope of this book, but the highly interested may refer to Gordon and Rifkind (1989). HDL levels, therefore, are inversely correlated with the risk of coronary artery disease. The ratio of total cholesterol to HDL, or of LDL to HDL, is preferable to any of the measures used alone in the estimation of heart disease risk (National Cholesterol Education Program 1994). *Epidemiology,* p. 227.

129. The answer is A: HMG-coA reductase inhibitors (i.e., statins). HMG-coA reductase is the rate-limiting enzyme in cholesterol biosynthesis in hepatocytes. The statin drugs competitively inhibit this enzyme, disrupting cholesterol biosynthesis. Bile acid sequestrants cause cholesterol to exit the enterohepatic circulation and be eliminated from the body in stool. Fibrates (fibric acid derivatives) such as gemfibrozil interfere with the metabolism and transfer of triglycerides (National Cholesterol Education Program 1994). There is recent noteworthy evidence that the statins reduce mortality when used as either secondary prevention (Scandinavian Simvastatin Survival Study Group 1994) or as primary prevention (Shepherd et al. 1995). Angiotensin-converting enzyme (ACE) inhibitors, which are antihypertensive drugs, are not used for reducing lipid levels. Selective serotonin reuptake inhibitors, such as fluoxetine, are antidepressants. *Epidemiology,* p. 229.

130. The answer is E: 50 million. Hypertension is defined as a mean systolic blood pressure of 140 mm Hg or higher or a mean diastolic blood pressure of 90 mm Hg or higher, or both. By this definition, there are approximately 50 million prevalent cases in the USA (National Institutes of Health 1995). *Epidemiology,* p. 229.

131. The answer is C: They are asymptomatic. Screening and case finding are important when disease can become harmful without producing overt symptoms. This is generally true of hypertension, which is rarely symptomatic before it reaches an advanced stage. Most patients have "essential hypertension," for which an underlying cause cannot be found. Most cases are relatively mild and considered stage 1 disease according to the new preferred nomenclature (National Institutes of Health 1995). Hypertension is usually at least somewhat responsive to life-style modification and may in fact be entirely reversible through dietary modification, increased physical activity, and weight loss when these can be achieved and sustained. *Epidemiology,* p. 229.

132. The answer is C: pregnant women. The usefulness of any screening program is its ability to detect a disease that would otherwise go undetected (at least temporarily) and to thereby improve the outcome. While diabetes is common, screening for it is not considered cost-effective because symptoms tend to unmask the disease early. The yield of screening in completely asymptomatic individuals is generally low. The one exception to this is pregnant women who have not been previously diagnosed with diabetes; in these women, counterregulatory hormones may unmask a susceptibility to the disease. Another factor that contributes to the cost-effectiveness of diabetes screening in pregnancy is the

severe congenital deformities associated with gestational diabetes, which can be completely prevented through the maintenance of normal serum glucose levels (euglycemia). Recall that the cost-effectiveness of a screening program is determined by the cost of screening and by the number of individuals who would need to be screened in order for one case to be identified that would have gone undetected without screening. The cost-effectiveness of screening may change over time, as lower-cost diagnostic methods are developed, treatment options change, or prevalence varies. *Epidemiology*, p. 230.

133. The answer is B: disability is socially defined. The severity and chronicity of disability and impairment are variable. Impairment is a medically defined change in physical status following injury or illness. The resultant decline in functional status (disability) is defined socially in terms of what the patient can and cannot do. *Epidemiology*, p. 231.

134. The answer is A: the states. The limited responsibility of the federal government for the administration of public health policy is predicated upon the interstate commerce clause and the general welfare clause from section 8 of the US Constitution. The fundamental responsibility for the health of the public lies with the states, as clarified in the Tenth Amendment to the US Constitution: "The powers not delegated to the United States by the Constitution, nor prohibited by it to the States, are reserved to the States respectively, or to the people." *Epidemiology*, p. 252.

135. The answer is D: consideration of the health impact of all public policy. As noted in the text, the public health perspective in the USA may be considered narrow by world standards. The mission statement by the Institute of Medicine in 1988 placed emphasis on the responsibilities of all public health agencies in the promotion of the public well-being. In contrast, the Ottawa Charter, released in Canada in 1986, considers the pursuit of well-being and health to be a universal responsibility. All public policy, according to the charter, is to be assessed for any potential impact on health and modified so as to promote health whenever possible. The notion of "healthy public policy" as public health policy is increasingly popular worldwide, and is advocated by many in the USA (Levin 1995). *Epidemiology*, p. 254.

136. The answer is D: 30% of total daily calories. The restriction of dietary fat to no more than 30% of total daily calories is the official guideline in the USA and is included among the nutrition goals in *Healthy People 2000*. At the time of the second National Health and Nutrition Examination Survey (NHANES II), the mean fat intake for adults in the USA was estimated to be 37% of calories. In the more recent NHANES III, the estimate fell to 34%. Given the rising prevalence of obesity and the caloric density of fat (9 kcal/g), some nutrition experts question whether this trend reflects dietary changes or merely changes in the way people respond to dietary surveys. The 30% goal is not considered optimal but is the lower limit of what authorities consider a feasible goal for the year 2000. Most nutritionists agree that if fat intake were much lower than 30%, there would be additional health benefits for adults. *Epidemiology*, p. 256.

137. The answer is C: encourage smoking cessation. Smoking is considered the most important cause of preventable morbidity and mortality in the USA, and smoking cessation would be the top priority for this encounter. In general, the treatment of any imminently dangerous condition is the highest priority for any clinical encounter. The treatment of symptomatic but not necessarily dangerous conditions would be next in order of priority. The management of conditions posing neither an immediate threat nor causing symptoms is an attempt to prevent such outcomes and is third in order of priority. Screening for additional risk factors is important but less so than managing those which are already apparent. Of the several risk factors for chronic disease revealed in this encounter, smoking is the single most important and might exacerbate other risk factors, such as the slightly elevated blood pressure. Of importance always is the patient's preferences. If this patient were interested in weight loss but not smoking cessation, it would be appropriate to encourage her to consider smoking cessation at a later date and then to give advice regarding diet and exercise to promote weight loss. Although it is difficult to make multiple life-style changes at one time, health-promoting behaviors are often interrelated. Interest in smoking cessation might arise after the initiation of an exercise program, for example, when the patient first becomes aware of activity-limiting dyspnea (breathlessness). *Epidemiology*, pp. 255–259.

138. The answer is A: 100 per 100,000 persons. As noted in Chapter 20, the age-adjusted death rate from coronary artery disease in the USA was 135 per 100,000 persons in 1987. The goal for the year 2000 is to reduce this to 100 per 100,000 persons. *Epidemiology*, p. 258.

139. The answer is C: the case fatality ratio. The case fatality rate, or ratio, is the percentage or proportion of cases of a particular disease that prove fatal. The attack rate is the proportion of susceptibles who become ill. The crude death rate is not cause-specific and would not accurately be derived from so small a population. The standardized mortality ratio derives from indirect standardization of death rates and is not relevant. The adjusted death rate is the same as the standardized death rate and is not relevant. *Epidemiology*, p. 39.

140. The answer is D: fecal-oral transmission peaks in the summer. Hepatitis A is spread through the fecal-oral route. Diseases associated with this mode of transmission tend to show peak incidence rates in the summer, due at least in part to proliferation of the organisms in contaminated water and food and in part to contamination by flies and other insects. The epidemiologic year is a plot of incident cases over a year and begins during the month of lowest incidence. The incidence of hepatitis A would peak in the summer and nadir in the winter, perhaps in January for a particular year. The epidemiologic year may mimic the calendar year but often does not. *Epidemiology*, p. 39.

141. The answer is D: influenza. Viral influenza A varies seasonally, with peak incidence in the USA during the fall and winter. Because of this variation, because of the known propensity of the virus to mutate (i.e., antigenic drift and shift), because of the extent of illness and death associated with the disease, and because of the availability of a protective vaccine, a sophisticated surveillance system has been established for influenza to facilitate recognition and control of outbreaks. *Epidemiology*, p. 39.

142. The answer is D: establish the case definition; determine whether an epidemic exists; characterize the epidemic by time, place, and person; develop hypotheses regarding spread; and initiate control measures. The appropriate sequence of actions in an outbreak investigation is delineated in Box 3–1. Of the answers provided, none is complete, but only D lists actions in the correct order.

The initial action is to make a diagnosis, albeit a tentative one, which specifies the disease to be investigated. The next step is to establish a case definition, so that cases may be distinguished from noncases. Only after these steps have been taken can the number of cases be established and a judgment made as to whether or not a true epidemic exists. Once the determination has been made that an epidemic does exist, it must be characterized by time, place, and person. This leads to hypotheses regarding the source, type, and route of spread. The hypotheses must be tested, such as in case-control studies and laboratory assays. Control measures are then implemented, and follow-up surveillance is devised. In reality, more than one of these steps may take place concomitantly. This sequence, however, provides the theoretical model upon which any true outbreak investigation should be based. *Epidemiology,* p. 44.

143. The answer is D: $1/10^5$. Each branch of the decision tree represents the outcomes associated with a particular decision. If you decide to treat your patient immediately, the probability of treatment is 1 (100%). The risk of rheumatic fever in patients treated immediately is given as $1/10^5$; $1/10^5$ multiplied by 1 is $1/10^5$, the correct answer. *Epidemiology,* p. 102.

144. The answer is B: 0.008. The probability of not treating is 100%, or 1, once the decision not to treat is made. The risk of rheumatic fever in the patients for whom neither a culture is obtained nor treatment is given is 0.008. The answer is 0.008×1, or 0.008. *Epidemiology,* p. 102.

145. The answer is E: 0.001. To answer this question, you need only follow the branch of the decision tree that corresponds to the management of this patient and find the associated risk. The risk of rheumatic fever in patients treated on the basis of a positive culture is given as 0.001, or 1 per thousand; this is the correct answer. While the probability of a positive culture in those patients cultured is only 1 in 4 (0.25) and the probability of treatment is only 0.95 in those patients with a positive culture, these numbers are not needed to answer this question, because we have been told that the patient's culture was positive and that the patient was then treated. Therefore, in this scenario, the probability of a positive culture is 1, and the probability of treatment is 1. The answer, then, may be thought of as the probability of a positive culture multiplied by the probability of treatment multiplied by the risk of rheumatic fever in a patient given treatment after a positive culture, or $1 \times 1 \times 0.001$, which is 0.001. *Epidemiology,* p. 102.

146. The answer is E: $1/10^4$. As in question 145, the scenario described represents the end of a decision tree branch, and, therefore, the risk is provided in the tree and does not require calculation. Start with the branch for obtaining a throat culture and waiting for the results. Follow it to the branch for positive cultures. Follow this to the branch for treatment. The risk of rheumatic fever in this group, 0.001, was the answer to question 145. The risk of antimicrobial resistance is provided as $1/10^4$, and this is the answer here. *Epidemiology,* p. 102.

147. The answer is B: $1.01/10^4$. The question asks us for the risk of rheumatic fever in all treated patients. The first step is to consider who is in this group. Treated patients include those for whom a culture has not been obtained, those treated after a positive culture, and those treated after a negative culture. The risk in the group treated without culture is $1/10^5$. The risk in those who have a positive culture and are then treated is 0.001. The risk in those who are treated after a negative culture is 0. The answer is the sum of these: $1/10^5 + 0.001 + 0 = 1.01/10^4$. *Epidemiology,* p. 102.

148. The answer is C: 1.8%. The untreated patients include those for whom neither culture nor treatment has been ordered, in whom the risk of rheumatic fever is 0.008; those with a positive culture but no treatment, in whom the risk is 0.01; and those with a negative culture and no treatment, in whom the risk is $1/10^7$. The aggregate risk in the untreated group is the sum of these, or 1.80001/100, or approximately 1.8%. *Epidemiology,* p. 102.

149. The answer is D: 45%. The risk of antimicrobial resistance in those who are treated but for whom no culture is ordered is 0.150; in those treated after a positive culture is obtained, $1/10^4$; and in those treated after a negative culture is obtained, 0.30. The sum of these is 0.4501, or approximately 45%. *Epidemiology,* p. 102.

150. The answer is D: $3.6/10^4$. In this question, we are not told whether the patients are treated or not. The risk in all patients for whom a culture was ordered includes those with positive and negative cultures, and those treated and untreated. In the group of patients for whom a culture was ordered, the probability of the culture being positive is 0.25. In the group with positive cultures, the probability of being treated is 0.95. The risk of rheumatic fever associated with positive culture and treatment is 0.001. Therefore, of the patients for whom a culture was ordered, those who are treated on the basis of a positive culture contribute a risk of ($0.25 \times 0.95 \times 0.001$), or $2.375/10^4$, to the overall risk in this group. Similarly, those who are not treated after a positive culture is obtained contribute a risk of ($0.25 \times 0.05 \times 0.01$), or $1.25/10^4$, to the overall risk.

The probability that patients for whom a culture is ordered will have a negative culture is 0.75. In this group, the treated patients contribute ($0.75 \times 0.10 \times 0$), or 0, to the overall risk; and the treated patients contribute [$0.75 \times 0.90 \times (1/10^7)$], or $6.75/10^8$, to the overall risk.

The risk of rheumatic fever in all cultured patients is the sum of these: $2.375/10^4 + 1.25/10^4 + 0 + 6.75/10^8 = 0.0003626$, or approximately $3.6/10^4$.

In actual practice, you would consider other characteristics in each patient when estimating the risk of rheumatic fever, and then follow the branches of the decision tree that would lead to the lowest risk of an adverse outcome for that individual. For example, the risk of treatment might be high in a patient with multiple antibiotic allergies, and this would indicate either that no treatment should be given or that it would be wise to obtain a culture and wait. In a patient known to have recurrent positive cultures for "strep throat" and a history of rheumatic fever, immediate treatment would be prudent. A decision tree, like a hammer, is a tool. A decision tree is neither necessary nor sufficient for building the foundation of a good decision, but, like any appropriate tool used well, it helps improve the construction process and the product. *Epidemiology,* p. 102.

REFERENCES CITED

Bartecchi, C. E., T. D. MacKenzie, and R. W. Schrier. The human costs of tobacco use. New England Journal of Medicine 330:907–912, 1994.

Bennett, W. I., ed. Beyond overeating. New England Journal of Medicine 332:673–674, 1995.

Berkman, L. F., and L. Breslow. Health and Ways of Living: The Alameda County Study. New York, Oxford University Press, 1983.

Bisno, A. L. Group A streptococcal infections and acute rheumatic fever. New England Journal of Medicine 325:783–793, 1991.

Dawber, T. R., G. F. Meadors, and F. E. Moore, Jr. Epidemiologic approaches to heart disease: the Framingham Study. American Journal of Public Health 41:279–286, 1951.

Fletcher, A. E. and C. J. Bulpitt. How far should blood pressure be lowered? New England Journal of Medicine 326:251–254, 1992.

Gordon, D. J., and B. M. Rifkind. High-density lipoprotein: the clinical implications of recent studies. New England Journal of Medicine 321:1311–1316, 1989.

Institute of Medicine. The Future of Public Health. Washington, D. C., National Academy Press, 1988.

Jekel, J. F. Epidemiology, Biostatistics, and Preventive Medicine. Philadelphia, W. B. Saunders Company, 1996.

Joint National Committee on Detection, Evaluation, and Treatment of High Blood Pressure. The Fifth Report of the Joint National Committee on Detection, Evaluation, and Treatment of High Blood Pressure (JNC V). Archives of Internal Medicine 153:154–183, 1993.

Kuczmarski, R. J., et al. Increasing prevalence of overweight among US adults. The National Health and Nutrition Examination Surveys, 1960 to 1991. Journal of the American Medical Association 272:205–211, 1994.

Kuhn, F., and C. E. Rackley. Coronary artery disease in women: risk factors, evaluation, treatment, and prevention. Archives of Internal Medicine 153:2626–2636, 1993.

Lemieux, S., et al. Sex differences in the relation of visceral adipose tissue accumulation to total body fatness. American Journal of Clinical Nutrition 58:463–467, 1993.

Levin, L. [Yale University School of Medicine, Department of Epidemiology and Public Health, International Health Division.] Personal communication, 1995.

Lindpainter, K. Finding an obesity gene—a tale of mice and men. New England Journal of Medicine 332:679–680, 1995.

McGinnis, J. M. and W. H. Foege. Actual causes of death in the United States. Journal of the American Medical Association 270:2207–2212, 1993.

Moriyama, I. M. Inquiring into the diagnostic evidence supporting medical certifications of death. In Lilienfeld, A. M. and A. J. Gifford, eds. Chronic Diseases and Public Health. Baltimore, Johns Hopkins University Press, 1966.

National Cholesterol Education Program. Second report of the expert panel on detection, evaluation, and treatment of high blood cholesterol in adults. Circulation 89:1329–1445, 1994.

National Institutes of Health. The Fifth Report of the Joint National Committee on Detection, Evaluation, and Treatment of High Blood Pressure. Publication No. (NIH)95-1088. Washington, D. C., Government Printing Office, 1995.

Ottawa Charter for Health Promotion. Report of an International Conference on Health Promotion, Sponsored by the World Health Organization, Health and Welfare Canada, and the Canadian Public Health Association, Ottawa, Ontario, Canada, November 17–21, 1986.

Physicians' Health Study Steering Committee. Final report on the aspirin component of the ongoing Physicians' Health Study. New England Journal of Medicine 321:129–135, 1989.

Pi-Sunyer, F. X. Health implications of obesity. American Journal of Clinical Nutrition 53:1595s–1603s, 1991.

Scandinavian Simvastatin Survival Study Group. Randomised trial of cholesterol lowering in 4444 patients with coronary heart disease. Lancet 344:1383–1389, 1994.

Shepherd, J., et al. Prevention of coronary heart disease with pravastatin in men with hypercholesterolemia. New England Journal of Medicine 333:1301–1307, 1995.

Stamler, J., ed. Epidemic obesity in the United States. Archives of Internal Medicine 153:1040–1044, 1993.

Epidemiologic and Medical Glossary

Acceptability of medical care: A measure of patient satisfaction with available medical care. This is influenced by such factors as whether the providers can communicate well with their patients, whether the care is seen as warm and humane and concerned with the whole person, and whether the patients believe in the confidentiality and privacy of information shared with their providers.

Accessibility of medical care: Care that patients can receive without undue geographic or financial obstacles.

Accountability of medical care: The degree to which the health care system takes public responsibility for its actions. This involves public representation on the board of directors of the health care facility, regular review of the financial records by certified public accountants, and appropriate public disclosure of financial records and of quality of care studies.

Accuracy: The ability of a test to obtain the correct measure, on average.

Acquired immunodeficiency syndrome (AIDS): A state of severe immunocompromise resulting from infection with the human immunodeficiency virus (HIV).

Active immunity: Immunity that is conferred by exposure to an antigen that stimulates antibody production by the host. Active immunity is far superior to passive immunity, because active immunity lasts longer (a lifetime in some cases) and is rapidly stimulated to high levels by a reexposure to the same or closely related antigens.

Active surveillance: When public health officials initiate contact with physicians, laboratories, or hospitals to obtain information about diseases of interest.

Actuarial method of life table analysis: One method of life table analysis, in which proportionate survival is assessed at fixed intervals, such as months, which have been established prior to data accrual. *See also* Life table analysis.

Acute sera: The first serum samples collected soon after symptoms of an infectious disease occur.

Acute tubular necrosis: Sudden severe injury to renal tubule cells, often resulting from transient hypoperfusion.

Adequacy of medical care: A sufficient volume of care to meet the need of a community.

Adjusted rates: *Same as* Standardized rates.

Advanced life support (ALS): Intervention protocols applied to resuscitate or stabilize (or both) the condition of critically ill or critically injured patients.

Air inversion: When cooler air settles close to the surface of the earth and warmer air rises above, so that the natural mixing of air does not occur and pollution is concentrated.

Alcohol abuse: *See* Chemical substance abuse.

Allostatic load: The ongoing level of demand for adaptation in an individual, which may be an important contributor to many chronic diseases.

Alpha error: *Same as* Type I error.

Alpha level: The maximum probability of making a false-positive error that the investigator is willing to accept.

Alternative hypothesis: The hypothesis that there is in fact a real (true) difference between means or proportions of groups being compared or that there is a real association between two variables.

Ames test: A quick and frequently used test to estimate the mutagenic potential of a chemical substance.

Analysis of covariance (ANCOVA): A method of significance testing based on the ratio of between-groups variance to within-groups variance. It is used in multivariable analysis if the dependent variable is continuous, some of the independent variables are categorical (i.e., nominal, dichotomous, or ordinal), and some of the independent variables are continuous.

Analysis of variance (ANOVA): A method of significance testing based on the ratio of between-groups variance to within-groups variance. It is used in statistical analysis if the dependent variable is continuous and the independent variable or variables are categorical (i.e., nominal, dichotomous, or ordinal). If there is only one independent variable, the technique is called one-way ANOVA. If there is more than one independent variable, the technique is called *N*-way ANOVA.

Anergy panel: A panel containing several prevalent antigens that are injected to assess immunocompetence. At least one of the antigens will elicit a reaction if the immune system is not impaired.

Angina pectoris: Chest pain resulting from periods of myocardial ischemia.

Antigenic drift: Relatively minor change in the surface antigens of a viral influenza strain.

Antigenic shift: Major change in the surface antigens of a viral influenza strain, with the potential to create worldwide epidemics.

Appropriateness of medical care: The procedures being performed are properly selected and carried out by trained personnel in the proper setting.

Asbestosis: Pulmonary compromise resulting from an accumulation of asbestos fibers in the lungs and from the associated inflammatory response, ultimately leading to pulmonary fibrosis.

Assessability of medical care: Care can be readily evaluated.

Atherogenesis: The development and accumulation of atherosclerotic plaque in arteries.

Attack rate: The proportion of exposed persons that become ill. This is the customary measure used to establish the severity of a disease outbreak.

Attributable risk (AR): The proportion of the total risk for a particular outcome attributable to a particular exposure. *Same as* Risk difference.

Attributable risk percent in the exposed [$AR\%_{(exposed)}$]: Answers the question: among those with the risk factor, what percentage of the total risk for the disease is due to the risk factor?

Availability of medical care: The provision of care during the hours and days when people need and can use it.

Bacillus Calmette-Guérin: A bacterial antigen vaccine used to provide partial immunity to *Mycobacterium tuberculosis*.

Bayes' theorem: Answers the two important questions that remain unanswered by sensitivity and specificity: (1) If the test results are positive, what is the probability that the patient has the disease? (2) If the test results are negative, what is the probability that the patient does not have the disease? The theorem stipulates that the probability of a given condition in an individual is related to the prevalence of that condition in the population of which the individual is a member.

Berylliosis: Poisoning by fumes or dust of the metal beryllium, usually resulting in pneumonitis.

Best estimate: Estimate achieved with the statistical model that produces the smallest sum of the squared error terms.

Beta error: *Same as* Type II error.

Between-groups mean square: *Same as* Between-groups variance.

Between-groups variance: Measurement of the variation between (or among) the means of more than one group based on the independent variables under study.

Bias: The introduction of error that produces deviations or distortions of data that are consistently in one direction, as opposed to random error. *Same as* Differential error.

Binary variables: *Same as* Dichotomous variables.

Biologic oxygen demand: The demand of aerobic bacteria in sewage for oxygen in water.

Bivariate analysis: Analysis of the relationship between one independent variable and one dependent variable.

Black lung disease: *Same as* Coal worker's pneumoconiosis.

Bonferroni procedure: A method for adjusting alpha when multiple hypotheses are being tested. To keep the risk of a false-positive finding in the entire study to no more than 0.05, the alpha level chosen for rejecting the null hypothesis is made more stringent by dividing alpha by the number of hypotheses to be tested.

Botulism: Poisoning by a neurotoxin produced by the bacterium *Clostridium botulinum*, usually through exposure to improperly canned or prepared food.

Bronchitis: Inflammation of the bronchi producing a clinical syndrome of cough, dyspnea, chest discomfort, and fever.

Byssinosis: An asthmalike pulmonary syndrome resulting from inhalation of textile dust (e.g., cotton dust).

Capitation basis for payment: Payment of primary care physicians on a "per head" basis. Regardless of the variation in patient need for and consumption of medical services, the physician receives the same amount of money per patient per year.

Case definition: Characterizes the clinical manifestations of the condition under investigation. It provides the inclusion and exclusion criteria that are used to determine which subjects are, and which are not, cases in an outbreak investigation.

Case fatality ratio: The proportion of clinically ill persons who die of the condition under study. This is a marker of virulence.

Case finding: The process of searching for asymptomatic diseases or risk factors among people while they are in a clinical setting (i.e., among people who are under medical care). The distinction between screening and case finding is frequently ignored in the literature and in practice; the distinction is important because many of the criteria for community screening do not need to be met during the process of case finding.

Case-control study: Study groups are defined on the basis of disease (or outcome) status. The frequency of the risk factor (exposure) in the cases (diseased) is compared with the frequency of the risk factor (exposure) in the controls (nondiseased).

Causality: A factor produces, or contributes to the production of, a specified outcome.

Cause-specific rates: Rates that provide numerators for comparison that are comparable with regard to diagnosis.

Cell-mediated immunity: A tissue-based cellular response to foreign antigens that involves mobilization of killer T cells.

Central limit theorem: For reasonably large samples, the distribution of the means of many samples is normal (gaussian), even though the data in individual samples may have skewness, kurtosis, or unevenness.

Chemical substance abuse: Both physical dependence (including tolerance) and psychologic dependence on the use of chemicals such as alcohol or illegal drugs to modify mood and performance and escape from anxiety. Cigarette smoking is also a form of chemical dependency or substance abuse.

Chickenpox: The common childhood illness characterized by fever and a papulovesicular rash resulting from infection by the varicella-zoster virus.

Chi-square test of independence: A statistical significance test used for nominal or dichotomous data in contingency tables. The chi-square value is the sum of the squares of the observed count (O) minus the expected count (E) divided by the expected count for each cell in the table. The standard chi-square formula is $\Sigma[(O - E)^2/E]$.

Cholera: An acute and, at times, fulminant diarrheal illness caused by an enterotoxin produced by the bacterium *Vibrio cholerae*.

Chronic renal disease: A nonspecific term referring to a gradual, as opposed to acute, decline in the functional capacity of the kidney as measured by creatinine clearance and glomerular filtration rate. The term is conventionally applied to renal insufficiency that is irreversible.

Coal worker's pneumoconiosis: Pulmonary inflammation or fibrosis, or both, resulting from chronic inhalation of coal dust. It is also known as black lung disease.

Coefficient: A weighting factor used in an equation, with the weight based on the relative importance of the factor in predicting prognosis.

Cohort: A clearly defined group of persons studied over time.

Cohort study: Study of a clearly identified group characterized by exposure and followed for the outcome.

Completeness of medical care: Adequate attention to all aspects of a medical problem, including prevention, early detection, diagnosis, treatment, follow-up measures, and rehabilitation.

Comprehensiveness of medical care: The extent to which care is provided for all types of health problems, including dental and mental health problems.

Confounding: Confusion of two supposedly causal variables, so that part or all of the purported effect of one variable is actually due to the other.

Contingency table: Used to determine whether the distribution of one variable is conditionally dependent (contingent) upon the other variable.

Continuity of medical care: Management of a patient's care over time is coordinated among providers.

Continuous variables: Data measured over the range of an uninterrupted numerical scale (e.g., height, weight, age).

Convalescent sera: Follow-up serum samples collected to allow sufficient time for antibody titers to rise after exposure to infection. Convalescent sera are compared with acute sera to identify evidence of recent infection.

Coronary artery disease: The accumulation of atherosclerotic plaque in the coronary arteries, resulting in reduced perfusion of the myocardium and, potentially, ischemia or infarction, or both.

Cost-benefit analysis: Measures and compares the costs and benefits of a proposed course of action in terms of the same units, usually monetary units such as dollars.

Cost-effectiveness analysis: Provides a way of comparing the cost of different proposed means to a particular end in terms of the most appropriate measurement units. Assesses the least costly means of achieving a fixed goal.

Covariance: The product of the deviation of an observation from the mean of the x variable, multiplied by the same observation's deviation from the mean of the y variable.

Cox method: *Same as* Proportional-hazards method.

Cox models: *Same as* Proportional-hazards models.

Crack cocaine: Freebase cocaine, the most potent, addictive form of cocaine, which is usually smoked.

Critical ratios: A class of tests of statistical significance that depend on dividing some parameter (such as a difference between means) by the standard error of that parameter.

Cross-sectional ecologic study: Assessment of the frequency with which some characteristic (e.g., smoking) and some outcome of interest (e.g., lung cancer) occur in the same geographically defined population at one particular time.

Cross-sectional survey: Survey of a population at a single point in time.

Crude rates: Rates that apply to an entire population, without reference to any characteristics of the individuals in it.

Cumulative incidence: The total number of incident cases over a specified period of time. Incident cases resulting in death would be included in the cumulative incidence measure but not in the prevalence.

Data dredging: The analysis of large data sets with modern computer techniques, permitting the assessment of hundreds of possible associations among the study variables. Unless alpha is adjusted, the testing of multiple hypotheses raises the risk of false-positive error.

Decision analysis: A type of analysis intended to improve clinical decision making under conditions of uncertainty.

Deductive reasoning: Reasoning that proceeds from the general (i.e., from assumptions, from propositions, and from formulas considered true) to the specific (i.e., to specific members belonging to the general category).

Degrees of freedom: The number of observations in a data set that are free to vary once the parameters of the data set (e.g., the mean) have been established.

Denominator data: Data that define the population at risk.

Diabetes mellitus: Impairment in glucose metabolism resulting from lack of insulin production (type I) or from insulin resistance (type II). Hyperglycemia is the principal but not only expression of the metabolic derangements associated with diabetes mellitus.

Dichotomous variables: Variables with only two levels.

Differential error: Nonrandom, systematic, or consistent error in which the values tend to be inaccurate in a particular direction. *Same as* Bias.

Dimensional variables: *Same as* Continuous variables.

Diphtheria: Acute infectious disease caused by *Corynebacterium diphtheriae* and acquired from a person with the disease or a carrier of the disease. It usually involves the upper respiratory tract and is characterized by the formation of a pseudomembrane attached to the underlying tissue.

Direct causality: When the factor under consideration exerts its effect without intermediary factors.

Direct standardization: Two populations to be compared are given the same age distribution, which is then applied to the observed age-specific death rates, providing the number of deaths that would have occurred in each of the two populations if they had been identical in age distribution.

Disability: A social definition of limitation, based on the degree of impairment. The four formal categories of disability used in most states for reimbursement of workers who have job-related injuries or illnesses covered under a workers' compensation program are permanent total disability, permanent partial disability, temporary total disability, and temporary partial disability.

Disability limitation: Medical and surgical measures aimed at controlling or correcting the anatomic and physiologic components of disease in symptomatic patients and preventing resultant limitations in functional ability.

Discounting: A reduction in the present value of delayed benefits (or an increase in their present costs) to account for the time value of money (i.e., inflation).

Discrete variables: Dichotomous variables and nominal variables are sometimes called discrete variables because the different categories are completely separate from each other.

Disease: A medically definable process characterized by pathophysiology and pathology. *Compare* Illness.

Dose-response relationship: Increases in the intensity or duration of exposure increase the risk of an adverse outcome. The relationship is often demonstrated by chronic exposures, such as that between the quantity of cigarette smoking and the risk of lung cancer. In research involving a potential carcinogen, the dose-response relationship is measured beyond the usual latent period for the carcinogen, which is the period between the onset of exposure and the development of cancer.

Double-blind study: A study in which neither the subjects nor the investigators are aware of the treatment assignment (active agent or placebo) until the trial is terminated.

Drug abuse: *See* Chemical substance abuse.

Dystress: Harmful form of stress that must be present in only limited amounts in order for an individual to have good health.

Early fetal death: Delivery of a dead fetus during the first 20 weeks of gestation.

Ebola virus infection: A virulent hemorrhagic disease with a high case fatality ratio.

Ecologic fallacy: The use of ecologic data to draw inferences about causal relationships in individuals. Although the frequency of exposure and outcome is determined in the same population, no information regarding the occurrence of exposure and outcome in the same individual is provided by these studies.

Effect modification: When the strength, or even direction, of the influence of a causal factor on outcome is altered by a third variable, the effect modifier.

Endemic disease: Disease that is occurring regularly in a defined population.

Enzootic disease: Regularly occurring disease in animal populations.

Eosinophilia-myalgia syndrome: A syndrome of muscle pain and hypereosinophilia caused by a contaminant in one commercially prepared brand of the amino acid L-tryptophan.

Epidemic: The occurrence of any disease at a frequency that is unusual (compared with baseline data) or unexpected.

Epidemic threshold: The necessary degree of variation from usual patterns required for a disease to qualify as an outbreak.

Epidemiologic year: Runs from the month of lowest incidence of a particular condition in one year to the same month in the next year.

Epidemiology: The study of factors that influence the occurrence and distribution of disease in human populations.

Epizootic disease: Disease outbreaks in animals.

Error term: A term that is needed to make an equation true if the prediction is not perfect. The error term is the portion of variation in the independent variable not explained by the statistical model.

Eustress: A helpful form of stress that must be present in order for an individual to have good health.

External validity: When the results of a study are true and meaningful for the larger population beyond the study participants.

False-negative error: *Same as* Type II error.

False-positive error: *Same as* Type I error.

Fee-for-service method of payment: Physicians are paid for each major item of service provided.

Fetal death: Delivery of a dead fetus after 28 weeks' gestation.

Fisher exact probability test: A statistical significance test used for contingency tables, in which one or more of the expected counts are too small (i.e., <2) to satisfy conditions for use of the chi-square analysis.

Frequency distribution: A plot of data displaying the value of each data point on one axis and the frequency with which that value occurs on the other axis.

F-test: The test of statistical significance used with ANOVA. The F ratio is the test statistic or critical ratio. It is the ratio of between-groups variance to within-groups variance.

General linear model: The general model depicting the linear (first-order) relationship between multiple independent variables and one dependent (outcome) variable. ANOVA, ANCOVA, and other multivariable techniques are derivations of this basic model.

German measles: *Same as* Rubella.

Gohn complex: A characteristic abnormality seen on chest x-ray following resolution of initial infection with tuberculosis.

Gonorrhea: Various clinical manifestations of infection with the sexually transmitted pathogen *Neisseria gonorrhoeae*.

Goodness-of-fit test: A general term used to describe the comparison of actual data with the results predicted by a particular statistical model, such as the expected counts generated by use of the chi-square analysis.

Granuloma: A collection of inflammatory cells in a nodular formation isolating the inflammatory agent (e.g., pathogen) within the complex.

Ground water: Water that is found in underground spaces called aquifers, which when adequately protected from surface pollutants, represent an important source of potable water.

Hantavirus pulmonary syndrome: A newly emergent respiratory infection caused by particular serotypes of the Hantavirus. Infection can lead to respiratory distress syndrome and death.

Health: A difficult term to define. The World Health Organization defines it as "a state of complete physical, mental, and social well-being and not merely the absence of disease or infirmity."

Health belief model: Before seeking preventive measures, people generally must believe that the disease at issue is serious, if acquired; that they or their children are personally at risk for the disease; that the preventive measure is effective in warding off the disease; and that there are no serious risks or barriers involved in obtaining the preventive measure. In addition, there need to be cues to action, consisting of information regarding how and when to obtain the preventive measure, as well as the encouragement from or support of other people.

Health maintenance organizations (HMOs): Prepaid group practices consisting of a legal and fiscal entity that does the contracting and financial transactions; a group of physicians that provides the outpatient and inpatient medical care; and an associated hospital or hospitals.

Health risk assessments (HRAs): Use of questionnaires or computer programs to elicit and evaluate information concerning individuals in a clinical or industrial medical practice. Each assessed person receives information concerning estimates of his or her life expectancy and the types of interventions that are likely to have a positive impact on health or longevity.

Healthy worker effect: People with jobs must be in reasonably good health to remain employed, and, therefore, they have a lower risk of death and illness than the population as a whole. A carrier that insures such a group benefits from this effect.

Hepatitis: A nonspecific term referring to inflammation of the liver. Common causes include viruses and excess alcohol consumption.

Herd immunity: Immunity that results when a vaccine not only prevents the vaccinated person from contracting the disease but also prevents him or her from spreading the disease, protecting even the unimmunized persons in the population.

Heritability: The degree to which acquisition of a disease is influenced by genetic predisposition.

Herpes zoster: A painful dermatomal rash resulting from reactivation of the latent varicella-zoster virus, often many years after a case of chickenpox. It is also known as shingles.

Hospice: A nursing home that specializes in providing terminal care, especially for patients with cancer or AIDS.

Human immunodeficiency virus (HIV): A retrovirus with a particular trophism for CD4 helper cells. It is the infectious agent responsible for AIDS.

Hyperlipidemia: A level of circulating lipoprotein particles or total cholesterol that exceeds the established reference range for a given population.

Hypersensitivity pneumonitis: Pulmonary inflammation resulting from an allergic response to an inspired antigen.

Hypertension: An average systolic blood pressure of 140 mm Hg or higher or an average diastolic blood pressure of 90 mm Hg or higher (or both) in an otherwise healthy person.

Iatrogenic diseases and injuries: Diseases and injuries generated during the process of treatment.

Iceberg phenomenon: The fact that the earliest identified cases of a new disease are often fatal or severe (representing the tip of the iceberg), but as more becomes known about the disease, less severe cases and asymptomatic cases are usually discovered.

Illness: What the patient experiences when he or she is sick. *Compare* Disease.

Immunodeficiency: A deficiency of the immune system, which may be long-term (as in AIDS) or transient (lasting for a short period following some types of infections or the administration of chemotherapy for certain types of cancer).

Impairment: A limitation of capacity or functional ability, usually as determined by a licensed physician.

Incidence: The frequency (number) of new occurrences of disease, injury, or death in the study population during the time period being examined.

Incidence density: The frequency (density) of new events per person-time.

Incidence density measures: Measures of the frequency of adverse health events that are often used to determine the optimal timing for administration of a new vaccine and the duration of the immunity produced.

Incidence rate: The number of incident cases over a defined study period, divided by the population at risk at the midpoint of that study period.

Independent practice association (IPA): Patients enrolled in this program can choose a primary care physician from a list of physicians who have contracted to provide services for the IPA. The IPA pays each physician on a fee-for-service basis whenever a member uses that physician's services. The IPA physicians limit their fees to the rates specified in the contract, and they agree to certain kinds of quality review and practice controls that are often similar to those of managed care.

Index case: The case in which the condition under investigation was first identified.

Indirect causality: When one factor influences one or more other factors that are, in turn, directly causal.

Indirect standardization: Used if age-specific death rates are not available in the study population, or if the study population is small and would therefore yield age-specific death rates that would be statistically unstable. Standard rates from the population are applied to the known age distribution of the study group.

Individual practice association: *See* Independent practice association.

Inductive reasoning: Reasoning that proceeds from the specific (i.e., from data) to the general (i.e., to formulas or conclusions).

Infant death: The death of a live-born child before that child's first birthday.

Infectiousness of organism: The proportion of exposed persons who become infected (this is also influenced by the conditions of exposure).

Influenza: Infection of the upper respiratory tract by the influenza virus, resulting in an illness generally characterized by fever, sore throat, dry cough, and severe myalgia.

Interaction: *See* Effect modification.

Intermediate care facility (ICF): A facility that is suitable if the patient's primary need is for help with the activities of daily living. An ICF is not required to have a registered (skilled) nurse on duty at all times.

Intermediate fetal death: Delivery of a dead fetus between 20 and 28 weeks' gestation.

Intermediate hospital: A medium to large community hospital that has a considerable amount of the latest technology but less research and investigational activity than a tertiary medical center.

Internal validity: When the results of a study are true and meaningful for the participants.

Interobserver variability: A measure of disagreement between or among different observers.

Intervening variables: Intermediary factors involved in indirect causality.

Interview survey: A type of cross-sectional survey.

Intraobserver variability: A measure of inconsistency in repeated assessments by a single observer.

Joint distribution graph: A plot of two continuous variables that can be used to visualize the relationship between the two variables (if one exists) and to determine the direction (positive or negative) and linearity of such a relationship.

Kaplan-Meier method of life table analysis: The most commonly used approach to survival analysis in medicine. The Kaplan-Meier method is different from the actuarial method in that the occurrence of each death defines the end of one observation period and the beginning of the next. Therefore, with this method, the duration of observation periods for which survival is determined varies throughout. *See also* Life table analysis.

Kappa test: A measure of the extent to which agreement between two observers improves on chance agreement.

Kendall rank test: A nonparametric significance test of correlation used for ordinal data. It is analogous to the Pearson correlation used for continuous data.

Kruskal-Wallis one-way ANOVA: A nonparametric significance test used to compare three or more groups of ordinal data. It is analogous to the ANOVA used for three or more groups of continuous data.

Kurtosis: Vertical stretching of a frequency distribution.

Kwashiorkor: A disease that tends to occur in children at the time of weaning, when starchy foods replace breast milk and there is protein deficiency despite nearly adequate calorie intake. The development of ascites (fluid in the abdominal cavity) produces a distended abdomen, which suggests obesity but is actually due to severe undernutrition.

Late look bias: The preferential detection of mild, slowly progressive cases of disease because such cases live longer and can be interviewed, whereas more severe cases of illness result in death and go undetected by the survey.

Lead-time bias: Bias that occurs when screening detects disease earlier in its natural history than would otherwise have happened, so that the time from diagnosis to death is lengthened. Having additional lead time does not alter the natural history of the disease and, therefore, does not extend the length of life.

Leavell's levels: Three levels of preventive health care (primary, secondary, and tertiary levels) based on the premise that all of the activities of physicians and other health professionals have the goal of prevention. What is to be prevented depends on the stage of health or disease in the individual receiving preventive care.

Length bias: Bias that occurs when milder, more indolent cases of disease are detected disproportionately in population screening programs. More aggressive cases have already resulted in death or in symptoms requiring medical intervention.

Life expectancy: Traditionally defined as the average number of years of life remaining at a given age.

Life table analysis: A statistical analysis of survival (or another dichotomous outcome) in which proportionate survival is assessed repeatedly over the intervals of observation. The pattern of mortality, as well as the overall rates of death and survival, are captured. *See also* Actuarial life table analysis and Kaplan-Meier life table analysis, the two methods in common use.

Likelihood ratio negative: The ratio of the false-negative error rate of a test to the specificity of the test.

Likelihood ratio positive: The ratio of the sensitivity of a test to the false-positive error rate of the test.

Linear regression analysis: A statistical test of the strength of the linear relationship between one independent and one dependent variable, both of which must be continuous.

Live attenuated vaccines: Created by altering infectious organisms so that they are no longer pathogenic but are still viable and antigenic.

Live birth: Delivery of a product of conception that shows any sign of life after complete removal from the mother.

Local community hospital: Hospital that provides services such as routine diagnosis, treatment, and surgery but lacks the personnel and facilities for complex procedures.

Lockjaw: *Same as* Tetanus.

Logrank test: A significance test used to compare the rates of survival as determined by life table analysis.

Longitudinal ecologic studies: Use of ongoing surveillance or frequent cross-sectional studies to measure trends in disease rates over years in a defined population.

Lyme disease: A complex disease affecting multiple systems of the body that can result from exposure to the tickborne pathogen *Borrelia burgdorferi*.

Malaria: A febrile illness caused by infection with a mosquito-borne protozoan of the genus *Plasmodium*. The pathogens are obligate intracellular parasites.

Managed care: A system of administrative controls, the goal of which is to reduce the costs of medical care.

Mann-Whitney *U* test: A nonparametric significance test used to compare two groups of ordinal data. It is analogous to the Student's *t*-test for continuous data.

Marasmus: A severe wasting syndrome in infants resulting from malnutrition, occurring when all nutrients in the diet are deficient and causing almost total growth retardation.

McNemar chi-square test: A modification of the chi-square test for use with paired data. The formula is $(|b - c| - 1)^2/(b + c)$.

Mean: The average value, calculated as the sum of all of the observed values divided by the total number of observations.

Mean deviation: The average of the absolute value of the deviations of all observations from the mean.

Mean square: Another name for variance, which is defined as a sum of squares divided by the appropriate number of degrees of freedom.

Measles: A highly contagious infection caused by a paramyxovirus and recognized particularly by a characteristic maculopapular rash. It is effectively prevented by vaccination.

Measurement bias: Bias resulting in distorted quantification of exposures or outcome because of improper technique or subjectivity of the measurement scale.

Median: The middle observation when data have been arranged in order from the lowest to the highest value.

Medicaid: A program funded by general tax revenues of the federal and state governments. The benefits are considered to be social welfare, instead of social insurance.

Medical commons: Shared medical resources. An attempt by one individual or group to maximize its own welfare by using more than its fair share of the medical commons would necessarily diminish the good that others can derive from it.

Medically indigent persons: People whose incomes are too high to be eligible for Medicaid, who do not receive medical insurance in their jobs, and who are not able to pay for individual medical care insurance policies.

Medicare: A program administered by the federal government that pays health benefits for most individuals who are 65 years of age or older and those who receive Social Security benefits due to disability.

Meningitis: Inflammation of the meninges, which are the sheaths of connective tissue surrounding the brain and spinal cord.

Meta-analysis: Used increasingly in medicine to try to obtain a pooled quantitative or qualitative (methodologic) analysis of the research literature on a particular subject.

Metal fume: Gaseous metal oxide that comes primarily from activities in occupational settings, such as welding without adequate ventilation.

Metal fume fever: An acute syndrome usually having flu-like symptoms that may occur a few hours following exposure to metal fumes.

Miscarriage: *Same as* Early fetal death.

Mode: The most commonly observed value in a distribution (i.e., the value with the highest number of occurrences).

Multiple drug–resistant tuberculosis (MDRTB): An increasingly common type of tuberculosis that is resistant to more than one antimicrobial agent as a result of incomplete courses of antituberculous therapy.

Multiple linear regression: Method used in multivariable analysis if the dependent variable and all of the independent variables are continuous.

Multivariable analysis: Analysis of the relationship of more than one independent variable to a single dependent variable.

Multivariable models: Statistical models that have one dependent (outcome) variable but include more than one independent variable.

Multivariate analysis: A term that is frequently used incorrectly. It refers to methods for analyzing more than one dependent variable as well as more than one independent variable.

Mumps: An infectious disease, usually of childhood, caused by a paramyxovirus and characterized in particular by parotitis (painful swelling of the parotid glands). It is effectively prevented by vaccination.

Myocardial infarction: Death of heart muscle resulting from protracted ischemia.

Natural booster phenomenon: Augmentation of immunity that occurs with periodic exposure to an infectious agent. This effect may be lost when immunization prevents exposure.

Necessary cause: Precedes a disease and has the following relationship with it: If the cause is absent, the disease cannot occur. If the cause is present, the disease may or may not occur.

Negative predictive value: Indicates what proportion of subjects with negative test results are truly free of the disease.

Neyman bias: *Same as* Late look bias.

Nominal variables: "Naming" or categorical variables that have no measurement scale.

Noncausal association: The relationship between two variables is statistically significant, but no causal relationship exists, either because the temporal relationship is incorrect (the presumed cause comes after, rather than before, the presumed effect) or because another factor is responsible for both the presumed cause and the presumed effect.

Nondifferential error: Produces findings that are too high and too low in approximately equal amounts, owing to random factors.

Nonparametric data: Data for which descriptive parameters such as the mean and standard deviation are lacking, leaving the underlying frequency distribution undefined.

Nosocomial infections: Hospital-acquired infections, which are more common than often supposed.

Null hypothesis: The hypothesis that there is no real (true) difference between means or proportions of the groups being compared or that there is no real association between two continuous variables.

Numerator data: Data that define the events or conditions of concern.

Nursing home: *Same as* Skilled nursing facility.

N-way ANOVA: *See* Analysis of variance (ANOVA).

Occupational asthma: A syndrome of bronchospastic disease occurring only during periods of exposure to respiratory irritants in the workplace. Symptoms are usually less pronounced during periods away from work, such as weekends.

Odds ratio (OR): The odds of exposure in the diseased group divided by the odds of exposure in the nondiseased group.

One-way ANOVA: *See* Analysis of variance (ANOVA).

Ordinal variables: Medical data that can be characterized in terms of more than two values and have a clearly implied direction from better to worse but are not measured on a continuous measurement scale.

Outbreak: Often used to denote a local epidemic (may be considered interchangeable with the term *epidemic*).

Outliers: Extreme values that are widely deviant from the mean.

Overall percent agreement: The percentage of the total observations found in cells a and d of a 2×2 table.

Qualitative characteristic: A characteristic that must be described in detail but is not truly measurable.

Quantitative characteristic: A characteristic that can be described by a rigid, dimensional measurement scale.

p value: The probability that the observed difference could have been obtained by chance alone, given random variation and a single test of the null hypothesis.

Parametric data: Data for which descriptive parameters, typically the mean and standard deviation, are known, and define the underlying frequency distribution of the data. The underlying distribution is often assumed to be normal, as provided in the central limit theorem.

Particulate matter: Small solid particles dispersed in air, such as matter resulting from cigarette smoking or from fuel combustion, which often contains carcinogenic substances.

Passive immunity: Protection against an infectious disease provided by circulating antibodies made in another organism.

Passive surveillance: Reporting of disease by physicians, laboratories, and hospitals on a routine basis.

Pathogenicity of organism: The proportion of infected individuals who are clinically ill.

Pathognomonic test: A test that elicits a reaction that is synonymous with having the disease.

Pearson correlation coefficient: A measure of strength of the linear relationship between two continuous variables.

Peptic ulcer disease: Erosions of the stomach or duodenum related to hypersecretion of gastric acid and often associated with infection by the bacterium *Helicobacter pylori*.

Period prevalence: The number of persons who had the disease of interest at any time during the specified time interval. It is the sum of the point prevalence at the beginning of the interval plus the incidence during the interval.

Person-time: A unit of time during which one person is observed for that particular interval, or two persons for half the interval, and so on.

Person-time methods of survival analysis: Methods that control for the fact that the length of observation varies from subject to subject. They are used in calculating risks and rates of death.

Pertussis: An acute, highly contagious respiratory infection caused by the bacterium *Bordetella pertussis*. It is commonly known as whooping cough because of the paroxysmal coughing.

Pneumococcal infection: Infection with the bacterium *Streptococcus pneumoniae*, which is a common cause of community-acquired pneumonia.

Pneumoconiosis: Pulmonary injury caused by deposition of respirable particulate matter in the lungs and by the resulting inflammatory response.

Point prevalence: The number of cases in the study population at one point in time.

Poliomyelitis: A febrile viral disease that can cause paralysis in severe cases. It is preventable by vaccination and is targeted for eradication by the World Health Organization.

Population attributable risk: The risk in the total population minus the risk in the unexposed group. The population attributable risk answers this question: in the general population, how much of the total risk for the disease of interest is due to the risk factor (exposure) of interest?

Population attributable risk percent (PAR%): The PAR% answers the question: among the general population, what percentage of the total risk for the disease of interest is due to the risk factor (exposure) of interest?

Positive predictive value: The proportion of subjects with positive test results who actually have the disease of interest.

Posterior probability: Revised estimate of the probability of disease in a given patient following a diagnostic test or intervention.

Precision: The ability of an instrument to provide the same or a very similar result with repeated measurements.

Prediction model: A model, complete with coefficients, for use in predicting a particular outcome given the presence of a variety of independent variables.

Preferred provider organization (PPO): Formed when a third-party payer (e.g., an insurance plan or a company) establishes a network of contracts with independent practitioners. The patients in a PPO can see physicians who are not on the panel, although they will have to pay a surcharge for their services.

Prevalence: The number of persons in a defined population who have a specified disease or condition at a point in time, usually the time a survey is done.

Prevalence rate: The proportion of persons with a defined disease or condition at the time they are studied. This is not truly a rate, although it is conventionally referred to as one.

Preventive medicine: A medical specialty emphasizing practices that help individuals and populations promote and preserve health and avoid injury and illness.

Prior probability: The probability of disease in a given patient estimated prior to the performance of laboratory tests and based on the estimated prevalence of a particular disease among patients with similar signs and symptoms.

Product-limit method of life table analysis: *Same as* Kaplan-Meier method of life table analysis.

Prognostic stratification: *Same as* Stratified allocation.

Proportional hazard model: A modification of multiple logistic regression to permit multivariable modeling of the data in a life table analysis.

Proportional-hazards method: Used to test for differences between Kaplan-Meier survival curves while controlling for other variables. It is also used to determine which variables are associated with better survival.

Prospective cohort study: The investigator assembles the study groups in the present time on the basis of exposure, collects baseline data on them, and follows subjects over time for the outcomes of interest.

Prospective payment system (PPS) based on diagnosis-related groups (DRGs): Each hospital admission is classified into one of 23 major diagnostic categories based on organ systems, and these diagnostic categories are further subdivided into DRGs. A DRG may consist of a single diagnosis or procedure, or it may consist of several diagnoses or procedures that, on average, have similar hospital costs per admission.

Public health: (1) The health status of the public (i.e., of a defined population). (2) The organized social efforts made to preserve and improve the health of a defined population.

Quality-adjusted life years (QALY): A health status index that incorporates both life expectancy and the perceived impact of illness and disability on the quality of life.

Rabies: A highly virulent viral infection of the central nervous system that is spread to humans from infected animals. The disease is preventable by postexposure prophylaxis.

Random error: Produces findings that are too high and too low in approximately equal amounts, owing to random factors.

Random sampling: Entails selecting a small group for study from a much larger group of potential study subjects.

Randomization: Entails allocating the available subjects to one or another study group and is generally used in clinical studies.

Randomized controlled clinical trials (RCCTs): Subjects are randomly assigned to one of the following groups: (1) the intervention group, which will receive the experimental treatment, or (2) the control group, which will receive the nonexperimental treatment, consisting either of a placebo (inert substance) or a standard method of treatment.

Randomized controlled field trials: Similar to RCCTs but are designed to test a preventive intervention such as a vaccine.

Randomized field trial: The standard way to measure the effectiveness of a new vaccine. Susceptible persons are randomized into two groups and are then given the vaccine or a placebo, usually at the beginning of the high-risk season of the year. Testing the efficacy of vaccines by randomized field trials is very costly, but it may be required the first time a new vaccine is introduced.

Ranked variables: *Same as* Ordinal variables.

Rapid sand filtration: A technique for purifying water by adding a flocculent (usually aluminum sulfate, called alum) to the water before filtration. The flocculent coagulates and traps suspended materials, preventing them from passing through the sand with the filtered water. The flocculent is removed periodically by back flushing, and new flocculent is added to the next batch of water.

Rate: The frequency (number) of events that occur in a defined time period, divided by the average population at risk.

Rate difference: *Same as* Risk difference.

Ratio variables: The variables derived from a continuous scale that has a true 0 point.

Recall bias: Bias resulting from differential recall of exposure to causal factors among those who have the disease as compared with those who do not have it.

Regression toward the mean: Patients chosen to participate in a study precisely because they had an extreme measurement on some variable are likely to have a measurement that is closer to average at a later time for reasons unrelated to the type or efficacy of the treatment they are given. It is also known as the statistical regression effect.

Rehabilitation: An attempt to mitigate the effects of disease and thereby prevent or limit social and functional disability.

Relative risk (RR): The ratio of the risk in the exposed group to the risk in the unexposed group, which is expressed as $RR = Risk_{(exposed)}/Risk_{(unexposed)} = [a/(a + b)]/[c/(c + d)]$.

Retrospective cohort study: Cohorts are identified on the basis of *past* exposure and followed forward to the present for the occurrence of the outcome of interest.

Rheumatic fever: Immune-mediated inflammatory condition following streptococcal infection, which can result in permanent damage to the valves of the heart.

Risk: The proportion of persons who are unaffected at the beginning of a study period but who undergo the risk event during the study period.

Risk difference: The risk in the exposed group minus the risk in the unexposed group.

Risk event: Death, disease, or injury.

Risk factor: A characteristic which, if present and active, clearly increases the probability of a particular disease in a group of persons who have the factor compared with an otherwise similar group of persons who do not. A risk factor is neither a necessary cause nor a sufficient cause of the disease.

Risk ratio: *Same as* Relative risk.

Rubella: An acute respiratory infection caused by a togavirus. Although the infection is usually benign and self-limited, it can cause death or severe injury to an embryo if maternal infection occurs during the first trimester. It is preventable with vaccination. It is also known as German measles.

Sabin oral polio vaccine (OPV): A vaccine against poliomyelitis that provides herd immunity and has resulted in the apparent eradication of wild virus from the western hemisphere. It is made from live virus and has produced clinical illness under certain circumstances.

Salk polio vaccine: An injectable vaccine against poliomyelitis that provides individual immunity to poliomyelitis but does not interrupt transmission (i.e., does not confer herd immunity).

Schmutzdecke: *See* Slow sand filtration.

Screening: The process of identifying a subgroup of people who are at high risk for having asymptomatic disease or who have a risk factor that puts them at high risk for developing a disease or becoming injured. Screening takes place in a community setting and is applied to a community population, such as students in a school or workers in an industry.

Selection bias: Bias resulting when allocation of participants to a study, or a particular study group, is influenced by characteristics of the participants that also influence the probability of the outcome.

Sensitivity: The ability of a test to detect a disease when it is present.

Sentinel health event: An adverse health event (death, disease, or impairment) that serves to identify a potential threat to the public health.

Shingles: *Same as* Herpes zoster.

Sick building syndrome: A constellation of symptoms such as headache, watery eyes, and wheezing ascribed by people to a building in which they work. The syndrome often, but not always, occurs in buildings that are tightly sealed for energy conservation or are not well ventilated. The cause seems to vary from one building to the next but is, in essence, unknown.

Sign of life: Any spontaneous movement, a breath, a cry, or pulsation of the umbilical cord at the time of birth.

Sign test: A nonparametric significance test for use with continuous, ordinal, or dichotomous data. It is used to determine whether or not, on average, one group experienced a significantly better outcome than the other.

Silicosis: A pneumoconiosis resulting from inhalation of the dust of stones or sand containing silicon dioxide. *See* Pneumoconiosis.

Skewness: Horizontal stretching of a frequency distribution, so that one tail of the plot is longer and contains more observations than the other.

Skilled nursing facility (SNF): Usually provides specialized care, such as intravenous fluids and medicines. A registered (skilled) nurse must be on duty at all times.

Sleeping sickness: A form of trypanosomiasis endemic in parts of Africa. It is spread by the bite of the tsetse fly.

Slow sand filtration: A technique for purifying water by filtering it through a large bed of packed sand, on which an organic layer (the *Schmutzdecke,* German for "dirt layer") forms and assists in the filtration process.

Smallpox: An acute infectious disease characterized by distinctive skin eruptions. This disease has now been eradicated worldwide as a result of successful vaccination.

Spearman rank test: *Same as* Kendall rank test.

Specific rates: Rates that pertain to some homogeneous subgroup of the population such as an age group, gender group, or ethnic group.

Specificity: The ability of a test to indicate nondisease when no disease is present.

Staff model health maintenance organization (HMO): Most of the physicians are salaried, full-time employees who work exclusively in the health plan or belong to a physician group that contracts to provide all of the medical services to the plan. Some specialists may be retained on part-time contracts. The HMO may have its own hospitals or may hospitalize its patients in one or more local hospitals, in which case the local hospitals are not usually a formal part of the HMO.

Standard deviation: The square root of the variance.

Standard error: The standard deviation of a population of sample means, rather than of individual observations.

Standard error of the difference between the means (SED): The square root of the sum of the respective population variances, each divided by its own sample size.

Standardized mortality ratio (SMR): The observed total events in the study group, divided by the expected number of events based on the standard population rates applied to the study group. The constant multiplier for this measure is 100.

Standardized rates: Crude rates that have been modified to allow for valid comparisons. This is usually necessary to correct for differing age distributions in different populations.

Statistical regression effect: *Same as* Regression toward the mean.

Statistical significance: Whenever a significance test produces a p value less than the preset value of alpha, conventionally 0.05. The implication of statistical significance at alpha of 0.05 is that chance would produce such a difference between comparison groups no more often than 5 times out of 100. This is taken to mean that chance is not responsible for the outcome.

Stillbirth: *Same as* Fetal death.

Stratified allocation: Assignment of patients to different risk groups depending on such baseline variables as the severity of disease (e.g., stage of cancer) and age.

Strength of association: The degree to which variation in one variable explains variation in another. The greater the strength of association between variables, the more completely variation in one predicts variation in the other.

Strep throat: Infection of the oropharynx by group A beta-hemolytic streptococci *(Streptococcus pyogenes).*

Sufficient cause: Precedes a disease and has the following relationship with it: if the cause is present, the disease will always occur.

Surface water: Surface sources of potable water, including protected surface reservoirs, lakes, and rivers.

Synergy: An impact of two or more factors on an individual or population that is greater than the sum of the separate effects of each individual factor.

Syphilis: A sexually transmitted infectious disease caused by the spirochete *Treponema pallidum.* The disease is most communicable in its early stages. Early treatment with penicillin prevents disease progression to late stages but consequently permits reinfection. Reinfection may lead to greater rates of transmission if the highly infectious early stages of the disease occur repeatedly in the same individual.

Tertiary medical center: Hospital that has most or all of the latest technology and usually participates actively in medical education and clinical research.

Tetanus: Acute infection with the bacterium *Clostridium tetani.* The bacterium produces a neurotoxin that can cause severe muscle spasm and paralysis. It is also known as lockjaw.

Threshold level: The level below which the body can adapt to an adverse exposure successfully and with no harm resulting. There is such a threshold level for most chemical and physical agents and even for most microbes. Usually, nonthreshold exposures are limited to those which alter genetic material, producing genetic mutations and, potentially, cancer. Ionizing radiation is currently considered a nonthreshold exposure.

Total variation: Equal to the sum of the squared deviations, which is usually called the total sum of squares but is sometimes referred to as the sum of squares.

Toxoids: Inactivated or altered bacterial exotoxins, such as diphtheria vaccine and tetanus vaccine.

Trypanosomiasis: Infection with protozoa of the genus *Trypanosoma.*

***t*-tests:** Tests that compare differences between means.

Tuberculosis: Diverse clinical manifestations resulting from infection by *Mycobacterium tuberculosis.*

Type I error: Error that occurs when data lead one to conclude that something is true when in reality it is not true.

Type II error: Error that occurs when something is said to be false when in reality it is true.

Typhoid: An acute febrile illness caused by ingestion of food contaminated with *Salmonella typhi.* It is also known as typhoid fever.

Unit of observation: The source of data in a medical study.

Utilization management: *Same as* Managed care.
Variable: A measure of a single characteristic.
Variance: The sum of the squared deviations from the mean, divided by the number of observations minus 1.
Varicella: *Same as* Chickenpox.
Vector: A factor in disease transmission, often an insect, which carries the agent to the host.
Visceral protein malnutrition: *Same as* Kwashiorkor.
Whooping cough: *Same as* Pertussis.
Wilcoxon matched-pairs signed-ranks test: A nonparametric significance test used to compare paired, ordinal data. It is analogous to the paired t-test for continuous data.
Within-groups mean square: *Same as* Within-groups variance.

Within-groups variance: Measurement based on the variation within each group (i.e., variation around a single group mean).
Workers' compensation laws: Laws stipulating that people with a job-related injury or illness have their medical and rehabilitation expenses paid and also receive a certain amount of cash payments in lieu of wages while they are recuperating.
Yates correction for continuity: An adjustment to the chi-square value recommended when counts in the contingency table are small, because the binomial distribution is discontinuous.
Yellow fever: An acute, mosquito-borne infectious disease caused by a flavivirus. The clinical manifestations are variable.
z-tests: Significance tests used to compare differences between proportions.

APPENDIX

TABLE A. RANDOM NUMBERS*

53872	34774	19087	81775	71440	12082	75092	34608	75448	13148
04226	62404	71577	00984	56056	32404	87641	53392	92561	33388
28666	44190	75524	62038	21423	46281	92238	96306	72606	80601
63817	30279	14088	86434	16183	06401	90586	80292	54555	47371
22359	16442	83879	47486	19838	32252	39560	95851	36758	36141
50968	28728	83525	16031	77583	65578	84794	51367	32535	83834
39652	24248	96617	91200	10769	52386	39559	75921	49375	22847
35493	00529	69632	29684	80284	87828	72418	80950	86311	34016
75687	53919	80439	20534	96185	72345	96391	52625	50866	45132
31509	93521	10681	44124	88345	84969	88768	48819	22311	41235
40389	76282	37506	60661	23295	67357	95419	10864	87833	09152
59244	54664	63424	97899	44153	69251	08781	18604	02312	21658
99876	17075	40934	08912	96196	58503	63613	24486	98092	45672
06457	50072	18060	71023	84349	40984	59487	77782	32107	53770
14297	07687	05517	10362	35783	62236	63764	45542	68889	03862
51661	57130	97442	29590	21634	79772	73801	70122	46467	47152
53455	41788	16117	09698	24409	05079	76603	57563	33461	46791
48086	31512	62819	27689	63744	11023	11184	87679	22218	70139
19108	01602	96950	41536	39974	88287	83546	69187	45539	78263
39001	77727	33095	58785	29179	45421	71416	20418	38558	78700
72346	55617	14714	21930	14851	38209	52202	03979	05970	74483
19094	64359	89829	10942	53101	37758	29583	26792	42840	45872
82247	77127	01652	50774	04970	83300	33760	22172	67516	62135
75968	18386	31874	52249	21015	20365	57475	32756	58268	75739
01963	38095	99960	91307	99654	74279	80145	53303	11870	50485
64828	15817	80923	55226	51893	93362	15757	47430	84855	95822
64347	61578	44160	06266	35118	52558	56436	96155	10293	67506
54746	52337	84826	39012	59118	19851	10156	78167	41473	99025
22241	41501	02993	99340	91044	67268	51088	12751	74008	33773
11906	20043	10415	44425	31712	54831	85591	62237	88797	14382
76637	07609	95378	95580	86909	50609	99008	99042	50364	36664
93896	47120	98926	30636	28136	49458	84145	79205	79517	93446
75292	88232	14360	12455	13656	65736	70428	66917	64412	38502
98792	29828	10577	48184	29433	98278	22543	76155	82107	22066
65751	91049	94127	47558	99880	79667	86254	72797	67117	44699
72064	62102	39155	79462	82975	02638	00302	79476	72656	84003
01227	35821	80607	61734	02600	45564	72344	71034	48370	96826
44768	56504	13993	59701	88238	92483	09497	66058	36651	37927
69838	91226	85736	72247	64099	86305	49877	76215	66980	30228
01800	39313	57730	84410	47637	81369	51830	43536	58937	91901
11756	45441	59948	57975	92422	70057	50210	30345	55912	31638
39056	86614	53643	62909	27198	04454	33789	86463	66603	48083
88086	93172	68311	39164	42012	10447	45933	28844	36844	57684
12648	27948	76750	19915	66815	34015	43011	27150	94264	89516
16254	87661	66181	68609	58626	58428	75051	27558	49463	66646
69682	19109	94189	94626	09299	10649	55405	54571	57855	54921
61336	86663	13010	40412	50139	30769	13048	61407	41056	60510
65727	66488	12304	70011	93324	58764	87274	43103	96002	06984
55705	34418	99410	32635	42984	40981	91750	27431	05142	77950
95402	51746	98184	38830	97590	00066	82770	42325	28778	83571
79228	94510	57711	64366	89040	43278	69072	22003	89465	61483
48103	56760	82564	33649	35176	32278	51357	05489	47462	55931
70969	27677	99621	63065	73194	70462	19316	77945	45004	39895
69931	20237	75246	59124	12484	22012	79731	82435	56301	99752
37208	22741	41946	74109	03760	24094	40210	76617	52317	50643
60151	92327	85150	27728	64813	47667	66078	03628	95240	03808
46210	47674	53747	95354	67757	75477	26396	09592	96239	50854
55399	48142	12284	95298	56399	61358	87541	12998	79639	63633
23677	64950	97041	43088	80143	34294	91468	01066	90350	78891
41947	70066	90311	17133	11674	00826	75760	37586	33621	14199

Source of data: RAND Corporation. A Million Random Digits with 100,000 Normal Deviates. New York, Free Press, 1955. Copyright 1955 and 1983 by the RAND Corporation. Used by permission.
*Instructions for use of the table: Decide in advance how many table columns will be used and in what direction the numbers will be assigned after the starting point is identified. Then blindly put a pencil on the table, and start with the numbers nearest the pencil point, moving in the predetermined direction (e.g., moving up the columns).

TABLE B. STANDARD NORMAL-TAIL PROBABILITIES (TABLE OF z VALUES)*

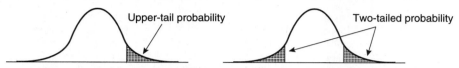

z	Upper-Tail Probability	Two-Tailed Probability	z	Upper-Tail Probability	Two-Tailed Probability
0.00	0.5000	1.0000	0.36	0.3594	0.7188
0.01	0.4960	0.9920	0.37	0.3557	0.7114
0.02	0.4920	0.9840	0.38	0.3520	0.7039
0.0251	0.49	0.98	0.3853	0.35	0.70
0.03	0.4880	0.9761	0.39	0.3483	0.6965
0.04	0.4840	0.9681	0.40	0.3446	0.6892
0.05	0.4801	0.9601	0.41	0.3409	0.6818
0.0502	0.48	0.96	0.4125	0.34	0.68
0.06	0.4761	0.9522	0.42	0.3372	0.6745
0.07	0.4721	0.9442	0.43	0.3336	0.6672
0.0753	0.47	0.94	0.4399	0.33	0.66
0.08	0.4681	0.9362	0.44	0.3300	0.6599
0.09	0.4641	0.9283	0.45	0.3264	0.6527
0.10	0.4602	0.9203	0.46	0.3228	0.6455
0.1004	0.46	0.92	0.4677	0.32	0.64
0.11	0.4562	0.9124	0.47	0.3192	0.6384
0.12	0.4522	0.9045	0.48	0.3156	0.6312
0.1257	0.45	0.9	0.49	0.3121	0.6241
0.13	0.4483	0.8966	0.4959	0.31	0.62
0.14	0.4443	0.8887	0.50	0.3085	0.6171
0.15	0.4404	0.8808	0.51	0.3050	0.6101
0.1510	0.44	0.88	0.52	0.3015	0.6031
0.16	0.4364	0.8729	0.5244	0.3	0.6
0.17	0.4325	0.8650	0.53	0.2981	0.5961
0.1764	0.43	0.86	0.54	0.2946	0.5892
0.18	0.4286	0.8571	0.55	0.2912	0.5823
0.19	0.4247	0.8493	0.5534	0.29	0.58
0.20	0.4207	0.8415	0.56	0.2877	0.5755
0.2019	0.42	0.84	0.57	0.2843	0.5687
0.21	0.4168	0.8337	0.58	0.2810	0.5619
0.22	0.4129	0.8259	0.5828	0.28	0.56
0.2275	0.41	0.82	0.59	0.2776	0.5552
0.23	0.4090	0.8181	0.60	0.2743	0.5485
0.24	0.4052	0.8103	0.61	0.2709	0.5419
0.25	0.4013	0.8026	0.6128	0.27	0.54
0.2533	0.40	0.80	0.62	0.2676	0.5353
0.26	0.3974	0.7949	0.63	0.2643	0.5287
0.27	0.3936	0.7872	0.64	0.2611	0.5222
0.2793	0.39	0.78	0.6433	0.26	0.52
0.28	0.3897	0.7795	0.65	0.2578	0.5157
0.29	0.3859	0.7718	0.66	0.2546	0.5093
0.30	0.3821	0.7642	0.67	0.2514	0.5029
0.3055	0.38	0.76	0.6745	0.25	0.50
0.31	0.3783	0.7566	0.68	0.2483	0.4956
0.32	0.3745	0.7490	0.69	0.2451	0.4902
0.33	0.3707	0.7414	0.70	0.2420	0.4839
0.3319	0.37	0.74	0.7063	0.24	0.48
0.34	0.3669	0.7339	0.71	0.2389	0.4777
0.35	0.3632	0.7263	0.72	0.2358	0.4715
0.3585	0.36	0.72	0.73	0.2327	0.4654

Continued

TABLE B. Standard Normal-Tail Probabilities (Table of z Values)* *(Continued)*

z	Upper-Tail Probability	Two-Tailed Probability	z	Upper-Tail Probability	Two-Tailed Probability
0.7388	0.23	0.46	1.13	0.1292	0.2585
0.74	0.2296	0.4593	1.14	0.1271	0.2543
0.75	0.2266	0.4533	1.15	0.1251	0.2501
0.76	0.2236	0.4473	1.16	0.1230	0.2460
0.77	0.2206	0.4413	1.17	0.1210	0.2420
0.7722	0.22	0.44	1.175	0.12	0.24
0.78	0.2177	0.4354	1.18	0.1190	0.2380
0.79	0.2148	0.4295	1.19	0.1170	0.2340
0.80	0.2119	0.4237	1.20	0.1151	0.2301
0.8064	0.21	0.42	1.21	0.1131	0.2263
0.81	0.2090	0.4179	1.22	0.1112	0.2225
0.82	0.2061	0.4122	1.227	0.11	0.22
0.83	0.2033	0.4065	1.23	0.1093	0.2187
0.84	0.2005	0.4009	1.24	0.1075	0.2150
0.8416	0.20	0.40	1.25	0.1056	0.2113
0.85	0.1977	0.3953	1.26	0.1038	0.2077
0.86	0.1949	0.3898	1.27	0.1020	0.2041
0.87	0.1922	0.3843	1.28	0.1003	0.2005
0.8779	0.19	0.38	1.282	0.10	0.20
0.88	0.1894	0.3789	1.29	0.0985	0.1971
0.89	0.1867	0.3735	1.30	0.0968	0.1936
0.90	0.1841	0.3681	1.31	0.0951	0.1902
0.91	0.1814	0.3628	1.32	0.0934	0.1868
0.9154	0.18	0.36	1.33	0.0918	0.1835
0.92	0.1788	0.3576	1.34	0.0901	0.1802
0.93	0.1762	0.3524	1.341	0.09	0.18
0.94	0.1736	0.3472	1.35	0.0885	0.1770
0.95	0.1711	0.3421	1.36	0.0869	0.1738
0.9542	0.17	0.34	1.37	0.0853	0.1707
0.96	0.1685	0.3371	1.38	0.0838	0.1676
0.97	0.1660	0.3320	1.39	0.0823	0.1645
0.98	0.1635	0.3271	1.40	0.0808	0.1615
0.99	0.1611	0.3222	1.405	0.08	0.16
0.9945	0.16	0.32	1.41	0.0793	0.1585
1.00	0.1587	0.3173	1.42	0.0778	0.1556
1.01	0.1562	0.3125	1.43	0.0764	0.1527
1.02	0.1539	0.3077	1.44	0.0749	0.1499
1.03	0.1515	0.3030	1.45	0.0735	0.1471
1.036	0.15	0.3	1.46	0.0721	0.1443
1.04	0.1492	0.2983	1.47	0.0708	0.1416
1.05	0.1469	0.2937	1.476	0.07	0.14
1.06	0.1446	0.2891	1.48	0.0694	0.1389
1.07	0.1423	0.2846	1.49	0.0681	0.1362
1.08	0.1401	0.2801	1.50	0.0668	0.1336
1.080	0.14	0.28	1.51	0.0655	0.1310
1.09	0.1379	0.2757	1.52	0.0643	0.1285
1.10	0.1357	0.2713	1.53	0.0630	0.1260
1.11	0.1335	0.2670	1.54	0.0618	0.1236
1.12	0.1314	0.2627	1.55	0.0606	0.1211
1.1264	0.13	0.26	1.555	0.06	0.12

Continued

TABLE B. Standard Normal-Tail Probabilities (Table of z Values)* (Continued)

z	Upper-Tail Probability	Two-Tailed Probability	z	Upper-Tail Probability	Two-Tailed Probability
1.56	0.0594	0.1188	2.03	0.0212	0.0424
1.57	0.0582	0.1164	2.04	0.0207	0.0414
1.58	0.0571	0.1141	2.05	0.0202	0.0404
1.59	0.0559	0.1118	2.054	0.02	0.04
1.60	0.0548	0.1096	2.06	0.0197	0.0394
1.61	0.0537	0.1074	2.07	0.0192	0.0385
1.62	0.0526	0.1052	2.08	0.0188	0.0375
1.63	0.0516	0.1031	2.09	0.0183	0.0366
1.64	0.0505	0.1010	2.10	0.0179	0.0357
1.645	0.05	0.10	2.11	0.0174	0.0349
1.65	0.0495	0.0989	2.12	0.0170	0.0340
1.66	0.0485	0.0969	2.13	0.0166	0.0332
1.67	0.0475	0.0949	2.14	0.0162	0.0324
1.68	0.0465	0.0930	2.15	0.0158	0.0316
1.69	0.0455	0.0910	2.16	0.0154	0.0308
1.70	0.0446	0.0891	2.17	0.0150	0.0300
1.71	0.0436	0.0873	2.18	0.0146	0.0293
1.72	0.0427	0.0854	2.19	0.0143	0.0285
1.73	0.0418	0.0836	2.20	0.0139	0.0278
1.74	0.0409	0.0819	2.21	0.0136	0.0271
1.75	0.0401	0.0801	2.22	0.0132	0.0264
1.751	0.04	0.08	2.23	0.0129	0.0257
1.76	0.0392	0.0784	2.24	0.0125	0.0251
1.77	0.0384	0.0767	2.25	0.0122	0.0244
1.78	0.0375	0.0751	2.26	0.0119	0.0238
1.79	0.0367	0.0734	2.27	0.0116	0.0232
1.80	0.0359	0.0719	2.28	0.0113	0.0226
1.81	0.0352	0.0703	2.29	0.0110	0.0220
1.82	0.0344	0.0688	2.30	0.0107	0.0214
1.83	0.0336	0.0672	2.31	0.0104	0.0209
1.84	0.0329	0.0658	2.32	0.0102	0.0203
1.85	0.0322	0.0643	2.326	0.01	0.02
1.86	0.0314	0.0629	2.33	0.0099	0.0198
1.87	0.0307	0.0615	2.34	0.0096	0.0193
1.88	0.0301	0.0601	2.35	0.0094	0.0188
1.881	0.03	0.06	2.36	0.0091	0.0183
1.89	0.0294	0.0588	2.37	0.0089	0.0178
1.90	0.0287	0.0574	2.38	0.0087	0.0173
1.91	0.0281	0.0561	2.39	0.0084	0.0168
1.92	0.0274	0.0549	2.40	0.0082	0.0164
1.93	0.0268	0.0536	2.41	0.0080	0.0160
1.94	0.0262	0.0524	2.42	0.0078	0.0155
1.95	0.0256	0.0512	2.43	0.0075	0.0151
1.960	0.025	0.05	2.44	0.0073	0.0147
1.97	0.0244	0.0488	2.45	0.0071	0.0143
1.98	0.0239	0.0477	2.46	0.0069	0.0139
1.99	0.0233	0.0466	2.47	0.0068	0.0135
2.00	0.0228	0.0455	2.48	0.0066	0.0131
2.01	0.0222	0.0444	2.49	0.0064	0.0128
2.02	0.0217	0.0434	2.50	0.0062	0.0124

Continued

TABLE B. Standard Normal-Tail Probabilities (Table of z Values)* *(Continued)*

z	Upper-Tail Probability	Two-Tailed Probability	z	Upper-Tail Probability	Two-Tailed Probability
2.51	0.0060	0.0121	2.90	0.0019	0.0037
2.52	0.0059	0.0117	2.95	0.0016	0.0032
2.53	0.0057	0.0114	3.00	0.0013	0.0027
2.54	0.0055	0.0111	3.05	0.0011	0.0023
2.55	0.0054	0.0108	3.090	0.001	0.002
2.56	0.0052	0.0105	3.10	0.0010	0.0019
2.57	0.0051	0.0102	3.15	0.0008	0.0016
2.576	0.005	0.01	3.20	0.0007	0.0014
2.58	0.0049	0.0099	3.25	0.0006	0.0012
2.59	0.0048	0.0096	3.291	0.0005	0.001
2.60	0.0047	0.0093	3.30	0.0005	0.0010
2.61	0.0045	0.0091	3.35	0.0004	0.0008
2.62	0.0044	0.0088	3.40	0.0003	0.0007
2.63	0.0043	0.0085	3.45	0.0003	0.0006
2.64	0.0041	0.0083	3.50	0.0002	0.0005
2.65	0.0040	0.0080	3.55	0.0002	0.0004
2.70	0.0035	0.0069	3.60	0.0002	0.0003
2.75	0.0030	0.0060	3.65	0.0001	0.0003
2.80	0.0026	0.0051	3.70	0.0001	0.0002
2.85	0.0022	0.0044	3.75	0.0001	0.0002
			3.80	0.0001	0.0001

Source of data: National Bureau of Standards. Applied Mathematics Series—23. US Government Printing Office, Washington, D. C., 1953. Abstracted by Shott, S. Statistics for Health Professionals. Philadelphia, W. B. Saunders Company, 1990. Used by permission.

*Instructions for use of the table to determine the p value that corresponds to a calculated z value:** In the left-hand column (headed z), look up the value of z found from calculations. Look at the first column to the right (for a one-tailed p value) or the second column to the right (for a two-tailed p value) that corresponds to the value of z obtained. For example, a z value of 1.74 corresponds to a two-tailed p value of 0.0819. **Instructions for use of the table to determine the z value that corresponds to a chosen p value:** To find the appropriate z value for use in confidence limits or sample size determinations, define the one-tailed or two-tailed p value desired, look that up in the second or third column, respectively, and determine the z value on the left that corresponds. For example, for a two-tailed alpha at 0.05, the corresponding z is 1.960; and for a one-tailed beta of 0.20, the corresponding z is 0.8416.

TABLE C. UPPER PERCENTAGE POINTS FOR t DISTRIBUTIONS*

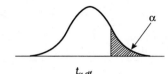

df	Upper-Tail Probability						
	0.40	0.30	0.20	0.15	0.10	0.05	0.025
1	0.325	0.727	1.376	1.963	3.078	6.314	12.706
2	0.289	0.617	1.061	1.386	1.886	2.920	4.303
3	0.277	0.584	0.978	1.250	1.638	2.353	3.182
4	0.271	0.569	0.941	1.190	1.533	2.132	2.776
5	0.267	0.559	0.920	1.156	1.476	2.015	2.571
6	0.265	0.553	0.906	1.134	1.440	1.943	2.447
7	0.263	0.549	0.896	1.119	1.415	1.895	2.365
8	0.262	0.546	0.889	1.108	1.397	1.860	2.306
9	0.261	0.543	0.883	1.100	1.383	1.833	2.262
10	0.260	0.542	0.879	1.093	1.372	1.812	2.228
11	0.260	0.540	0.876	1.088	1.363	1.796	2.201
12	0.259	0.539	0.873	1.083	1.356	1.782	2.179
13	0.259	0.537	0.870	1.079	1.350	1.771	2.160
14	0.258	0.537	0.868	1.076	1.345	1.761	2.145
15	0.258	0.536	0.866	1.074	1.341	1.753	2.131
16	0.258	0.535	0.865	1.071	1.337	1.746	2.120
17	0.257	0.534	0.863	1.069	1.333	1.740	2.110
18	0.257	0.534	0.862	1.067	1.330	1.734	2.101
19	0.257	0.533	0.861	1.066	1.328	1.729	2.093
20	0.257	0.533	0.860	1.064	1.325	1.725	2.086
21	0.257	0.532	0.859	1.063	1.323	1.721	2.080
22	0.256	0.532	0.858	1.061	1.321	1.717	2.074
23	0.256	0.532	0.858	1.060	1.319	1.714	2.069
24	0.256	0.531	0.857	1.059	1.318	1.711	2.064
25	0.256	0.531	0.856	1.058	1.316	1.708	2.060
26	0.256	0.531	0.856	1.058	1.315	1.706	2.056
27	0.256	0.531	0.855	1.057	1.314	1.703	2.052
28	0.256	0.530	0.855	1.056	1.313	1.701	2.048
29	0.256	0.530	0.854	1.055	1.311	1.699	2.045
30	0.256	0.530	0.854	1.055	1.310	1.697	2.042
40	0.255	0.529	0.851	1.050	1.303	1.684	2.021
60	0.254	0.527	0.848	1.045	1.296	1.671	2.000
120	0.254	0.526	0.845	1.041	1.289	1.658	1.980
∞	0.253	0.524	0.842	1.036	1.282	1.645	1.960

Continued

TABLE C. Upper Percentage Points for t Distributions* (Continued)

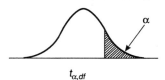

$t_{\alpha,df}$

df	Upper-Tail Probability						
	0.02	0.015	0.01	0.0075	0.005	0.0025	0.0005
1	15.895	21.205	31.821	42.434	63.657	127.322	636.590
2	4.849	5.643	6.965	8.073	9.925	14.089	31.598
3	3.482	3.896	4.541	5.047	5.841	7.453	12.924
4	2.999	3.298	3.747	4.088	4.604	5.598	8.610
5	2.757	3.003	3.365	3.634	4.032	4.773	6.869
6	2.612	2.829	3.143	3.372	3.707	4.317	5.959
7	2.517	2.715	2.998	3.203	3.499	4.029	5.408
8	2.449	2.634	2.896	3.085	3.355	3.833	5.041
9	2.398	2.574	2.821	2.998	3.250	3.690	4.781
10	2.359	2.527	2.764	2.932	3.169	3.581	4.587
11	2.328	2.491	2.718	2.879	3.106	3.497	4.437
12	2.303	2.461	2.681	2.836	3.055	3.428	4.318
13	2.282	2.436	2.650	2.801	3.012	3.372	4.221
14	2.264	2.415	2.624	2.771	2.977	3.326	4.140
15	2.249	2.397	2.602	2.746	2.947	3.286	4.073
16	2.235	2.382	2.583	2.724	2.921	3.252	4.015
17	2.224	2.368	2.567	2.706	2.898	3.222	3.965
18	2.214	2.356	2.552	2.689	2.878	3.197	3.922
19	2.205	2.346	2.539	2.674	2.861	3.174	3.883
20	2.197	2.336	2.528	2.661	2.845	3.153	3.849
21	2.189	2.328	2.518	2.649	2.831	3.135	3.819
22	2.183	2.320	2.508	2.639	2.819	3.119	3.792
23	2.177	2.313	2.500	2.629	2.807	3.104	3.768
24	2.172	2.307	2.492	2.620	2.797	3.091	3.745
25	2.167	2.301	2.485	2.612	2.787	3.078	3.725
26	2.162	2.296	2.479	2.605	2.779	3.067	3.707
27	2.158	2.291	2.473	2.598	2.771	3.057	3.690
28	2.154	2.286	2.467	2.592	2.763	3.047	3.674
29	2.150	2.282	2.462	2.586	2.756	3.038	3.659
30	2.147	2.278	2.457	2.581	2.750	3.030	3.646
40	2.123	2.250	2.423	2.542	2.704	2.971	3.551
60	2.099	2.223	2.390	2.504	2.660	2.915	3.460
120	2.076	2.196	2.358	2.468	2.617	2.860	3.373
∞	2.054	2.170	2.326	2.432	2.576	2.807	3.291

Source of data: Shott, S. Statistics for Health Professionals. Philadelphia, W. B. Saunders Company, 1990. Used by permission.

*Instructions for use of the table: To determine the p value that corresponds to a calculated t value, first find the line that corresponds to the number of degrees of freedom (df) on the left. Then in the center of the table find the value that most closely corresponds to the value of t found from calculations. **(1) For a paired (one-tailed) t-test:** Look at the top row to find the corresponding probability. For example, a t value of 2.147 on 30 df corresponds to a p value of 0.02. If the observed value of t falls between values given, state the two p values between which the results of the t-test fall. For example, if a t of 2.160 is found on 30 df, the probability is expressed as follows: $0.015 < p < 0.02$. **(2) For the Student's (two-tailed) t-test:** The procedure is the same as for a one-tailed test, except that the p value obtained must then be *doubled* to include the other tail probability. For example, if the Student's t-test gives a p value of 2.147 on 30 df, the p value of that column (0.02) must be doubled to give the correct p value of 0.04.

TABLE D. UPPER PERCENTAGE POINTS FOR CHI-SQUARE DISTRIBUTIONS*

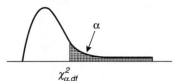

$\chi^2_{\alpha, df}$

df	Probability								
	0.9995	0.995	0.99	0.975	0.95	0.90	0.80	0.70	0.60
1	0.000000393	0.0000393	0.000157	0.000982	0.00393	0.0158	0.0642	0.148	0.275
2	0.00100	0.0100	0.0201	0.0506	0.103	0.211	0.446	0.713	1.022
3	0.0153	0.0717	0.115	0.216	0.352	0.584	1.005	1.424	1.869
4	0.0639	0.207	0.297	0.484	0.711	1.064	1.649	2.195	2.753
5	0.158	0.412	0.554	0.831	1.145	1.610	2.343	3.000	3.655
6	0.299	0.676	0.872	1.237	1.635	2.204	3.070	3.828	4.570
7	0.485	0.989	1.239	1.690	2.167	2.833	3.822	4.671	5.493
8	0.710	1.344	1.646	2.180	2.733	3.490	4.594	5.527	6.423
9	0.972	1.735	2.088	2.700	3.325	4.168	5.380	6.393	7.357
10	1.265	2.156	2.558	3.247	3.940	4.865	6.179	7.267	8.295
11	1.587	2.603	3.053	3.816	4.575	5.578	6.989	8.148	9.237
12	1.934	3.074	3.571	4.404	5.226	6.304	7.807	9.034	10.182
13	2.305	3.565	4.107	5.009	5.892	7.042	8.634	9.926	11.129
14	2.697	4.075	4.660	5.629	6.571	7.790	9.467	10.821	12.078
15	3.108	4.601	5.229	6.262	7.261	8.547	10.307	11.721	13.030
16	3.536	5.142	5.812	6.908	7.962	9.312	11.152	12.624	13.983
17	3.980	5.697	6.408	7.564	8.672	10.085	12.002	13.531	14.937
18	4.439	6.265	7.015	8.231	9.390	10.865	12.857	14.440	15.893
19	4.912	6.844	7.633	8.907	10.117	11.651	13.716	15.352	16.850
20	5.398	7.434	8.260	9.591	10.851	12.443	14.578	16.266	17.809
21	5.896	8.034	8.897	10.283	11.591	13.240	15.445	17.182	18.768
22	6.404	8.643	9.542	10.982	12.338	14.041	16.314	18.101	19.729
23	6.924	9.260	10.196	11.689	13.091	14.848	17.187	19.021	20.690
24	7.453	9.886	10.856	12.401	13.848	15.659	18.062	19.943	21.652
25	7.991	10.520	11.524	13.120	14.611	16.473	18.940	20.867	22.616
26	8.538	11.160	12.198	13.844	15.379	17.292	19.820	21.792	23.579
27	9.093	11.808	12.879	14.573	16.151	18.114	20.703	22.719	24.544
28	9.656	12.461	13.565	15.308	16.928	18.939	21.588	23.647	25.509
29	10.227	13.121	14.256	16.047	17.708	19.768	22.475	24.577	26.475
30	10.804	13.787	14.953	16.791	18.493	20.599	23.364	25.508	27.442
35	13.787	17.192	18.509	20.569	22.465	24.797	27.836	30.178	32.282
40	16.906	20.707	22.164	24.433	26.509	29.051	32.345	34.872	37.134
45	20.137	24.311	25.901	28.366	30.612	33.350	36.884	39.585	41.995
50	23.461	27.991	29.707	32.357	34.764	37.689	41.449	44.313	46.864
60	30.340	35.534	37.485	40.482	43.188	46.459	50.641	53.809	56.620
70	37.467	43.275	45.442	48.758	51.739	55.329	59.898	63.346	66.396
80	44.791	51.172	53.540	57.153	60.391	64.278	69.207	72.915	76.188
90	52.276	59.196	61.754	65.647	69.126	73.291	78.558	82.511	85.993
100	59.896	67.328	70.065	74.222	77.929	82.358	87.945	92.129	95.808
120	75.467	83.852	86.923	91.573	95.705	100.624	106.806	111.419	115.465
140	91.391	100.655	104.034	109.137	113.659	119.029	125.758	130.766	135.149
160	107.597	117.679	121.346	126.870	131.756	137.546	144.783	150.158	154.856
180	124.033	134.884	138.820	144.741	149.969	156.153	163.868	169.588	174.580
200	140.660	152.241	156.432	162.728	168.279	174.835	183.003	189.049	194.319

Continued

TABLE D. Upper Percentage Points for Chi-Square Distributions* *(Continued)*

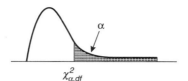

df	Probability									
	0.50	0.40	0.30	0.20	0.10	0.05	0.025	0.01	0.005	0.0005
1	0.455	0.708	1.074	1.642	2.706	3.841	5.024	6.635	7.879	12.116
2	1.386	1.833	2.408	3.219	4.605	5.991	7.378	9.210	10.597	15.202
3	2.366	2.946	3.665	4.642	6.251	7.815	9.348	11.345	12.838	17.730
4	3.357	4.045	4.878	5.989	7.779	9.488	11.143	13.277	14.860	19.997
5	4.351	5.132	6.064	7.289	9.236	11.070	12.833	15.086	16.750	22.105
6	5.348	6.211	7.231	8.558	10.645	12.592	14.449	16.812	18.548	24.103
7	6.346	7.283	8.383	9.803	12.017	14.067	16.013	18.475	20.278	26.018
8	7.344	8.351	9.524	11.030	13.362	15.507	17.535	20.090	21.955	27.868
9	8.343	9.414	10.656	12.242	14.684	16.919	19.023	21.666	23.589	29.666
10	9.342	10.473	11.781	13.442	15.987	18.307	20.483	23.209	25.188	31.420
11	10.341	11.530	12.899	14.631	17.275	19.675	21.920	24.725	26.757	33.137
12	11.340	12.584	14.011	15.812	18.549	21.026	23.337	26.217	28.300	34.821
13	12.340	13.636	15.119	16.985	19.812	22.362	24.736	27.688	29.819	36.478
14	13.339	14.685	16.222	18.151	21.064	23.685	26.119	29.141	31.319	38.109
15	14.339	15.733	17.322	19.311	22.307	24.996	27.488	30.578	32.801	39.719
16	15.338	16.780	18.418	20.465	23.542	26.296	28.845	32.000	34.267	41.308
17	16.338	17.824	19.511	21.615	24.769	27.587	30.191	33.409	35.718	42.879
18	17.338	18.868	20.601	22.760	25.989	28.869	31.526	34.805	37.156	44.434
19	18.338	19.910	21.689	23.900	27.204	30.144	32.852	36.191	38.582	45.973
20	19.337	20.951	22.775	25.038	28.412	31.410	34.170	37.566	39.997	47.498
21	20.337	21.991	23.858	26.171	29.615	32.671	35.479	38.932	41.401	49.011
22	21.337	23.031	24.939	27.301	30.813	33.924	36.781	40.289	42.796	50.511
23	22.337	24.069	26.018	28.429	32.007	35.172	38.076	41.638	44.181	52.000
24	23.337	25.106	27.096	29.553	33.196	36.415	39.364	42.980	45.559	53.479
25	24.337	26.143	28.172	30.675	34.382	37.652	40.646	44.314	46.928	54.947
26	25.336	27.179	29.246	31.795	35.563	38.885	41.923	45.642	48.290	56.407
27	26.336	28.214	30.319	32.912	36.741	40.113	43.195	46.963	49.645	57.858
28	27.336	29.249	31.391	34.027	37.916	41.337	44.461	48.278	50.993	59.300
29	28.336	30.283	32.461	35.139	39.087	42.557	45.722	49.588	52.336	60.735
30	29.336	31.316	33.530	36.250	40.256	43.773	46.979	50.892	53.672	62.162
35	34.336	36.475	38.859	41.778	46.059	49.802	53.203	57.342	60.275	69.199
40	39.335	41.622	44.165	47.269	51.805	55.758	59.342	63.691	66.766	76.095
45	44.335	46.761	49.452	52.729	57.505	61.656	65.410	69.957	73.166	82.876
50	49.335	51.892	54.723	58.164	63.167	67.505	71.420	76.154	79.490	89.561
60	59.335	62.135	65.227	68.972	74.397	79.082	83.298	88.379	91.952	102.695
70	69.334	72.358	75.689	79.715	85.527	90.531	95.023	100.425	104.215	115.578
80	79.334	82.566	86.120	90.405	96.578	101.879	106.629	112.329	116.321	128.261
90	89.334	92.761	96.524	101.054	107.565	113.145	118.136	124.116	128.299	140.782
100	99.334	102.946	106.906	111.667	118.498	124.342	129.561	135.807	140.169	153.167
120	119.334	123.289	127.616	132.806	140.233	146.567	152.211	158.950	163.648	177.603
140	139.334	143.604	148.269	153.854	161.827	168.613	174.648	181.840	186.847	201.683
160	159.334	163.898	168.876	174.828	183.311	190.516	196.915	204.530	209.824	225.481
180	179.334	184.173	189.446	195.743	204.704	212.304	219.044	227.056	232.620	249.048
200	199.334	204.434	209.985	216.609	226.021	233.994	241.058	249.445	255.264	272.423

Source of data: © 1982 by Ciba-Geigy Corporation. Reprinted with permission from Lentner, C., ed. Geigy Scientific Tables, 8th ed. Vol. 2. Ciba-Geigy, Basle, 1982. All rights reserved. Abstracted by Shott, S. Statistics for Health Professionals. Philadelphia, W. B. Saunders Company, 1990.

*__Instructions for use of the table:__ Determine the degrees of freedom *(df)* appropriate to the chi-square test just calculated, and go to the line that most closely corresponds, using the left-hand column (headed *df*). On that line, move to the right in the body of the table and find the chi-square value that corresponds to what was calculated. The corresponding *p* value is found at the top of that column. For example, on 6 *df*, a calculated chi-square of 12.592 corresponds to a *p* value of 0.05. If the calculated chi-square value falls between two columns in the table, state the two *p* values between which the results of the chi-square test fall. For example, on 6 *df*, the probability of a chi-square of 13.500 is expressed as follows: $0.025 < p < 0.05$.